Advances in Electronic Ceramics

Advances in Electronic Ceramics

Guest Editors

Dawei Wang
Fayaz Hussain

MDPI

Basel • Beijing • Wuhan • Barcelona • Belgrade • Novi Sad • Cluj • Manchester

Guest Editors

Dawei Wang
School of Instrumentation Science and Engineering
Harbin Institute of Technology
Harbin
China

Fayaz Hussain
Department of Materials Engineering
NED University of Engineering&Technology
Karachi
Pakistan

Editorial Office
MDPI AG
Grosspeteranlage 5
4052 Basel, Switzerland

This is a reprint of the Special Issue, published open access by the journal *Ceramics* (ISSN 2571-6131), freely accessible at: www.mdpi.com/journal/ceramics/special_issues/BM5K96SGZQ.

For citation purposes, cite each article independently as indicated on the article page online and as indicated below:

Lastname, A.A.; Lastname, B.B. Article Title. *Journal Name* **Year**, *Volume Number*, Page Range.

ISBN 978-3-7258-3922-3 (Hbk)
ISBN 978-3-7258-3921-6 (PDF)
https://doi.org/10.3390/books978-3-7258-3921-6

Contents

About the Editors

Dawei Wang

Dr. Dawei Wang is a Professor at the Harbin Institute of Technology and was previously a Research Associate at the Department of Materials Science and Engineering of the University of Sheffield (2014-2020) and a joint PhD student/visiting scholar at the Materials Research Institute of Pennsylvania State University (2010.9-2011.9/2016.4-6). He received his PhD degree in Materials Processing Engineering from the Beijing Institute of Technology in 2012. He is an Associate Editor for the *Journal of American Ceramic Society, Frontiers in Materials,* and *Ceramics* and an Editorial board member for *Materials Today Communications, Journal of Advanced Ceramics, Journal of Advanced Dielectrics, Crystals,* and *Microstructures.* His research focuses on advanced electronic ceramics for energy storage, conversion, harvesting, and the translation of new materials to prototype devices/components for electronic systems. He has published 200 refereed papers, with a total citation of 11000 and a Google Scholar h-index of 60. Also, he holds 20 issued patents and has given 50+ invited talks at international conferences.

Fayaz Hussain

Fayaz Hussain joined the Department of Materials Engineering in 2007, first as a Lecturer and then as an Assistant Professor in 2010. He was promoted as an Associate Professor in 2020 and then again promoted as a Professor and Chairman of the Department of Materials Engineering in 2022. Prior to this, he worked three years in the metal industry. He is also an Editorial board member of the journals *Frontiers in Materials, Electroactive Materials,* and *Advanced Dielectrics.* He completed his PhD from the University of Sheffield, England, UK, in 2016-2017, working on "KNN based lead-oxide free piezoelectric ceramics". This ABO3 system has been studied from the perspective of optimizing its performance for multilayer actuators, potentially for energy-harvesting applications under the supervision of Professor Ian Reaney at the University of Sheffield. To fabricate the multilayers, a novel Wet Multilayer Method (WMM) was also developed to overcome the issues of delamination during the firing of multilayer actuators. He authored/co-authored publications in well-reputed journals, around 63 papers, including key articles on piezoelectric, capacitor, and microwave dielectric ceramics in bulk and multilayers with 876 citations, with an h-index of 15 and an i10-index of 22 during the last five years. Current research interests and development include the following: the synthesis of piezoelectric ceramics and their multilayers, multiferroics, thermoelectric ceramics, and microwave dielectrics through cold sintering/microwave sintering/ 3D printing/conventional sintering. Characterisation methods include the following: LCR, impedance spectroscopy, a d33 meter for piezoelectric coefficients, a vibrating sample magnetometer for magnetic properties, XRD analysis, SEM/EDX, ferroelectric testing, etc.

ceramics

MDPI

Article

Effect of Thickness on Ferroelectric Properties of $Bi_{3.25}La_{0.75}Ti_3O_{12}$ Thin Films

Wenfeng Yue [1], Yali Cai [2], Quansheng Guo [2], Dawei Wang [1,*] and Tingting Jia [2,*]

1 School of Instrumentation Science and Engineering, Harbin Institute of Technology, Harbin 150001, China
2 School of Materials Science and Engineering, Hubei University, Wuhan 430062, China
* Correspondence: wangdawei102@gmail.com (D.W.); jia.tingting@hubu.edu.cn (T.J.)

Abstract: The pursuit of low-power/low-voltage operation in devices has prompted a keen interest in the mesoscale effects within ferroelectric thin films. The downsizing of ferroelectrics can significantly influence performance; for instance, the remanent polarization and coercive field are susceptible to alterations based on thickness. In this study, randomly oriented $Bi_{3.25}La_{0.75}Ti_3O_{12}$ thin films were fabricated on $Pt/Ti/SiO_2/Si$ substrates using the sol–gel method, and SEM observations revealed rod-like grains in all thin films. The investigation delved into the correlation between dielectric and ferroelectric properties with thin film thickness. The thin film exhibited an increased remanent polarization and a reduced coercive electric field. Additionally, the ferroelectric domain structure was scrutinized through PFM, and the resistor properties of the BLT4 thin film were studied, which shows the potential of BLT thin films in non-volatile memory and memristor.

Keywords: ferroelectric; $Bi_{3.25}La_{0.75}Ti_3O_{12}$; thickness; thin films

Citation: Yue, W.; Cai, Y.; Guo, Q.; Wang, D.; Jia, T. Effect of Thickness on Ferroelectric Properties of $Bi_{3.25}La_{0.75}Ti_3O_{12}$ Thin Films. *Ceramics* **2024**, *7*, 29–38. https://doi.org/10.3390/ceramics7010003

Academic Editor: Gilbert Fantozzi

Received: 10 November 2023
Revised: 30 December 2023
Accepted: 4 January 2024
Published: 6 January 2024

1. Introduction

Ferroelectric thin films exhibit diverse properties, such as ferroelectricity, piezoelectricity, dielectricity, pyroelectricity, and photoelectric effects, rendering them suitable for various applications including ferroelectric memory, piezoelectric drivers, infrared detectors, and optoelectronic devices [1–5]. Traditional ferroelectric materials, predominantly lead-based compounds, are exemplified by the widely used lead zirconate titanate (PZT) in commercial applications due to their robust ferroelectric and piezoelectric characteristics [6,7]. However, the environmental and health concerns arising from the volatilization of lead in lead-based materials necessitate the exploration of lead-free alternatives. Bismuth-based titanate and bismuth ferrite emerge as promising candidates for lead-free piezoelectric materials. Notably, bismuth titanate ($Bi_4Ti_3O_{12}$, BIT) exhibits favorable properties, including a small coercive field, a low leakage current density, excellent fatigue resistance, and applicability in non-volatile memory [8–10].

BIT, classified as a typical three-layer bismuth layered ferroelectric (BLSF) material, consists of a bismuth oxide layer $(Bi_2O_3)^{2+}$ and a perovskite-like layer $(A_{n-1}B_nO_{3n+1})^{2-}$ along the c-axis direction [11,12]. Characterized by strong anisotropy, BIT demonstrates spontaneous polarization of approximately 50 $\mu C/cm^2$ along the a-axis and 4 $\mu C/cm^2$ along the c-axis. Additionally, it features a low film deposition temperature and a high Curie temperature (Tc = 675 °C) [13,14]. Despite its favorable properties, BIT faces challenges such as instability in the oxidation state of Ti ions and volatility in Bi ions during the sintering process. These issues lead to defects, resulting in a high leakage current and domain pinning, thereby affecting its practical applications [9,15]. To enhance its ferroelectric properties, researchers have explored the substitution of Bi^{3+} ions in the perovskite-like layer $(Bi_2Ti_3O_{12})^{2+}$ with trivalent rare-earth ions such as Pr^{3+}, Nd^{3+}, Sm^{3+}, Eu^{3+}, and La^{3+} [1,16–19]. In 1999, Park et al. reported a ferroelectric thin film of La-doped BIT, $Bi_{3.25}La_{0.75}Ti_3O_{12}$ (BLT) [11]. Wu et al. demonstrated that the application of tensile or compressive stress on the surface of BLT thin films in the high-field region (>75 kV/cm) effectively mitigates the leakage current [20]. Furthermore, substituting high-valent cations for Ti^{4+} at the B site proves instrumental in eliminating oxygen vacancy defects, thereby enhancing remanent polarization and reducing the leakage current. BIT thin films co-doped with Sm and Ta, denoted as BSTTO, exhibited markedly improved ferroelectric properties, boasting a higher remanent polarization ($2P_r$ = 46.2 μC/cm^2) compared to BIT thin films ($2P_r$ = 26 μC/cm^2) [21]. Various techniques, including pulsed laser deposition (PLD) [22], magnetron sputtering (MS) [23], metal–organic chemical vapor deposition (MOCVD) [24], and the sol–gel method [25], have been employed for the preparation of BLT ferroelectric thin films. Among these, the sol–gel process stands out for its cost-effectiveness, facile stoichiometry control, and uniform deposition over large areas, making it versatile for widespread applications. Notably, factors such as the precursor solution, annealing conditions, film thickness, and substrate properties exert substantial influence on film orientation [25–27]. Optimal film layer thicknesses of 30, 50, and 100 nm corresponded to preferentially oriented thin films with (001), (100), and (117) orientations, respectively [28]. By adjusting the grain size of BLT through annealing temperature modulation, Yang et al. achieved a higher energy storage density, rendering BLT ferroelectric thin films suitable for high-energy-density storage devices, thereby expanding their application scope [29]. In addition, information storage technology stands as one of the most rapidly advancing fields within the realm of integrated circuits. Among its components, memory stands as the pivotal core in information storage technology. However, conventional storage devices have reached a point where they can no longer meet the burgeoning demands of information storage technology. Consequently, the pursuit of developing a new generation of memristors has captured academic interest in recent years. The development of ferroelectric materials with resistive switching characteristics holds significant importance for memristors.

In this study, lead-free BLT thin films were synthesized using the sol–gel method, and the adjustment of spin coating times yielded BLT thin films with exceptional ferroelectric properties. The dielectric and ferroelectric characteristics of thin films subjected to varying spin coating times were systematically compared across different electric field strengths and frequencies. The microscopic ferroelectricity of the BLT film and the microscopic evolution of the domain wall of the BLT film were studied through PFM observation. The resistor characteristics of the BLT4 thin film were studied, demonstrating its potential application in memristor.

2. Materials and Methods

The $Bi_{3.25}La_{0.75}Ti_3O_{12}$ thin films were fabricated using the sol–gel method on $Pt/Ti/SiO_2/Si$ substrates. The precursor solution was composed of bismuth(III) acetate (Aladdin, Shanghai, China), lanthanum(III) acetate (Aladdin, Shanghai, China), and tetrabutyl titanate (Aladdin, Shanghai, China) as primary raw materials, with propionic acid (Aladdin, Shanghai, China) as the solvent and ethanolamine (Aladdin, Shanghai, China) as the stabilizer. The quantities of bismuth acetate, lanthanum acetate, and tetrabutyl titanate were determined according to stoichiometric proportions, with an additional 5% of bismuth acetate to compensate for Bi volatilization during annealing. The synthesis process entailed the gradual addition of bismuth acetate and lanthanum acetate to propionic acid, followed by stirring at room temperature until complete dissolution. Subsequently, tetrabutyl titanate was introduced, and a specific amount of ethanolamine was added to stabilize the colloid. The mixture was stirred at room temperature for one day to yield a yellow stock solution, and the concentration of precursor solution was 0.10 mol/L. This solution was spun at 6000 rpm for 20 s and then pyrolyzed on a hot stage (C-MAG HS 7, IKA, Staufen, Germany) at 400 °C for 5 min. This procedure was iterated four, five, six, and seven times to produce the samples denoted as BLT4, BLT5, BLT6, and BLT7, respectively. Finally, the samples underwent annealing in an infrared annealing furnace (IRLA-1200, Wuhan JouleYacht Technology Co., LTD, Wuhan, China) at 700 °C in an air atmosphere for 30 min.

The thin films' phase analysis and crystallographic orientation were determined using X-ray diffraction (XRD, D8 Advance X, Bruker, Germany) with CuKα radiation at room temperature. Surface topography and cross-section were examined using a scanning electron microscope (SEM, Apreo 2, Thermo Scientific, Waltham, MA, USA). The piezoelectric response and domain structure of the BLT thin films were characterized using piezoelectric force microscopy (AFM/PFM, Cypher S, Oxford Instruments, Abingdon, Oxfordshire, UK). Before electrical characterization, Au top electrodes with a 0.5 mm diameter were deposited onto the thin films through magnetron sputtering (ETD-800, Vision Precision Instruments, Beijing, China), forming Au/BLT/Pt capacitors. The hysteresis loop, leakage current characteristic, and current–voltage loop were characterized using a ferroelectric analyzer (Precision Multiferroic II, Radiant, Albuquerque, NM, USA), and the dielectric constant and dielectric loss were determined using an impedance analyzer (WK6500B, Wayne Kerr, Woburn, Massachusetts, UK). The resistor characteristics (I-V curve) of the device were tested using a semiconductor analyzer (4200A-SCS, Keithley, Seattle, WA, USA).

3. Results and Discussion

Figure 1 depicts the X-ray diffraction (XRD) patterns of BLT thin films deposited on (111)Pt/Ti/SiO_2/Si substrates. The thin films exhibit complete crystallization and demonstrate polycrystalline structures (ICDD/PDF No. 00-035-2795), with no observable secondary phases in the XRD scans. Based on the intensity of the (117) XRD reflection, the following formula [28] can be used to estimate the degree of preferred orientation of BLT films:

$$\alpha 117 = \frac{I(117)}{I(008) + I(117) + I(200)} \quad (1)$$

where $I(006)$, $I(117)$, and $I(200)$ stand for the integrated peak intensities for (008), (117), and (200) diffraction lines, respectively. α117 would be about 66% in a fully random film. This indicates that all of the thin films possessed random orientation. Increasing the number of spin coating cycles leads to a significant enhancement in the intensities of (111), (220), (1115), and (317) peaks relative to the background noise level. This is to be expected as a consequence of the film thickness increase with the number of spun layers. The slight decrease in α117 with the number of spin cycles is probably due to this increase in the intensities of the weak (008) and (200/020) lines relative to the background noise in the XRD scans, rather than any change in preferred orientation.

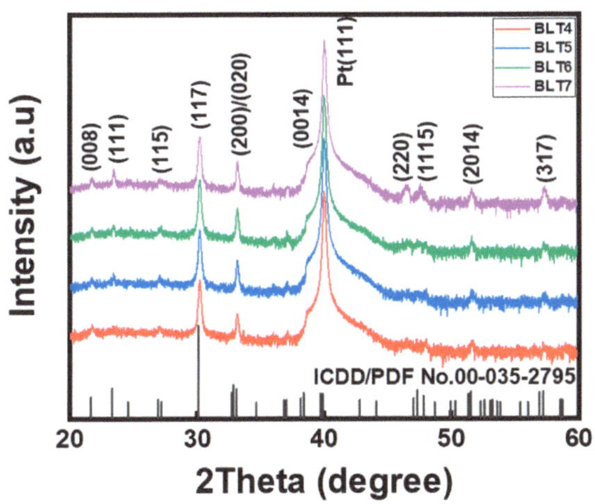

Figure 1. XRD patterns of BLT4, BLT5, BLT6, and BLT7 thin films.

In Figure 2, surface images of the BLT thin films on Pt/Ti/SiO_2/Si substrates with varying spin coating cycles are presented. The micrographs reveal dense and smooth thin films without discernible cracks or bubbles, which indicates that all of the thin films have good quality. With an increasing number of spin coatings, the BLT grains exhibit elongation, which is confirmed by SEM micrographs. The thin films primarily consist of rod-like grains oriented with different lengths. It is not clear whether these grains are rods or plates, which are viewed edge-on. The formation of rod-like grains can be attributed to the anisotropic growth of the BIT-based material, driven by lower c-plane interface energy compared to the a/b surface, resulting in faster growth in the plane perpendicular to the crystallographic c-axis [28]. In the process of film preparation, the shapes of grains depend on nucleation and grain growth. According to Schwartz's theory, in films with interface nucleation, the grains are columnar, while in films with interface and bulk nucleation, the grains are fine [30]. Therefore, it is speculated that when the film thickness is low, interface nucleation is predominant, and as the thickness gradually increases, it transitions to both interface and bulk nucleation. Furthermore, the grain density is influenced by nucleation centers and the surface migration of adsorbed atoms. With an increase in the number of spin-coating layers, i.e., an increase in thickness, the surface diffusion of adsorbed atoms is restricted, and the density of nucleation centers increases, resulting in a more uniform distribution of particle aggregation, and as the width of the rod increases, it is easier for them to make contact with each other, which also leads to the smaller width of the grain during the growth process [31]. The cross-sectional images in Figure 2e–h depict thin films with varying layer counts. The thicknesses for four, five, six, and seven layers are 140 nm, 180 nm, 210 nm, and 240 nm, respectively, with each layer measuring approximately 35 nm.

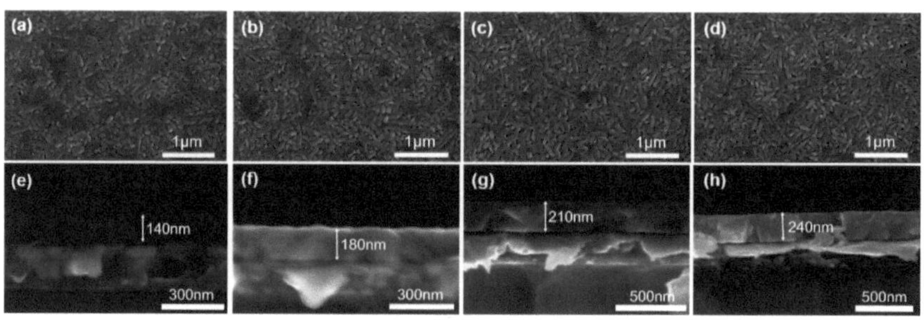

Figure 2. SEM micrographs and cross-sections of $Bi_{3.25}La_{0.75}Ti_4O_{12}$ thin films; (**a**,**e**) BLT4, (**b**,**f**) BLT5, (**c**,**g**) BLT6, (**d**,**h**) BLT7.

The polarization–voltage hysteresis loop was measured at 1 kHz for each thin film sample, and the remanent polarization (P_r) was extracted from the hysteresis loop. The hysteresis loops for BLT thin films with different thicknesses are presented in Figure 3a, and the corresponding thickness-dependent P_r values are illustrated. It was observed that as the number of spin coatings increased, P_r gradually decreased. Furthermore, all samples exhibited well-defined saturation hysteresis loops, which indicated that all of the films had good ferroelectric properties. However, significant asymmetries were observed in the hysteresis loop of the BLT4 thin film, likely stemming from differences in the work function, defect charge, and interface control between the top and bottom electrodes. In Figure 3b, dielectric constant and dielectric loss are shown as functions of frequency for all thin films at room temperature. At 100 Hz, the dielectric constant increased with the film thickness, rising from 258 at 140 nm to 276 at 240 nm. The dielectric constant increases with the film thickness, influenced by both the extrinsic and intrinsic properties of the ferroelectric film. This relationship can be further elucidated. Meanwhile, the dielectric constants of all thin films decrease as the test frequency rises due to the suppression of the space charge effect. In the frequency range from 10^2 to 10^5 Hz, the dielectric loss remains nearly constant (<0.10). The abrupt changes in the dielectric constant and dielectric loss beyond 10^5 Hz result from the L-C resonance induced by the stray inductance at the probe–film contact [32].

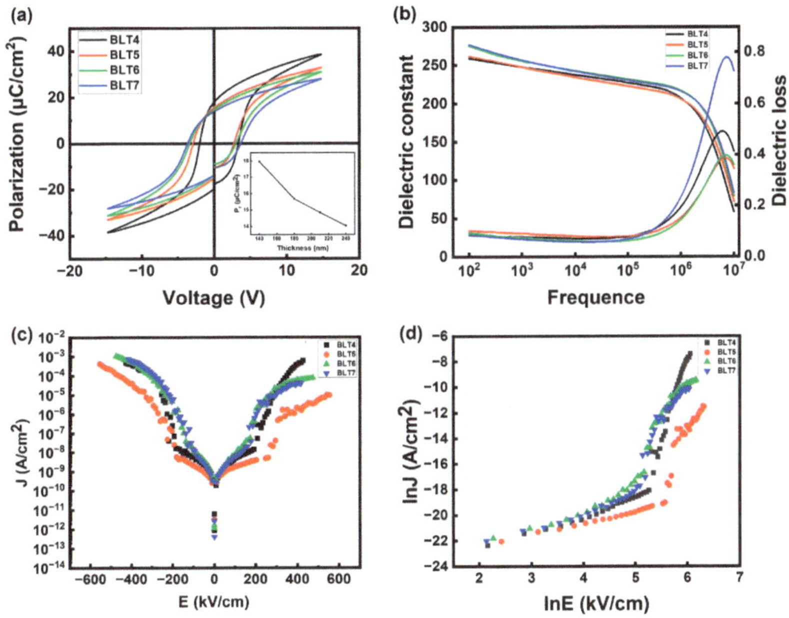

Figure 3. (**a**) Polarization voltage loops, (**b**) dielectric constant and dielectric loss as functions of frequency, (**c**) leakage current curves, and (**d**) corresponding ln E-ln J curves of BLT thin films with different film thicknesses.

It is widely recognized that BIT thin films exhibit high leakage currents, which are primarily attributed to defects like oxygen vacancies. The incorporation of La into BIT (BLT) can ameliorate the leakage current characteristics. In Figure 3c, the leakage current density versus the applied electric field (J-E) for BLT thin films with varying spin-coating times is depicted. Irrespective of the spin-coating times, all BLT thin films exhibit reduced leakage current densities, measuring less than 4.5×10^{-8} A/cm^2 at an electric field of 150 kV/cm. The J-E curves of all Au/BLT/Pt samples display asymmetry around x = 0, which is attributable to differing Schottky heights of the top and bottom electrodes in the ferroelectric heterostructures [10]. The leakage current curve manifests two distinct regions. In the low-voltage region, the leakage current densities of all BLT thin films demonstrate linear growth, while in the high-voltage region, they increase exponentially. The boundary between these two regions for BLT thin films is observed at approximately 180 kV/cm. In Figure 3d, the ln J-ln E curves for all BLT thin films illustrate a similar trend in which the current density increases with the electric field, indicating that the leakage current characteristics are independent of the film thickness. Detailed discussions on the leakage current characteristics in the low electric field and high electric field regions will be provided for the BLT4 film.

Figure 4 presents the electrical characteristics of BLT4 thin films. In Figure 4a, the hysteresis loop of the BLT4 film at 1 kHz is shown as a function of the electric field. Notably, an increase in the electric field leads to a rapid rise in both the remanent polarization and coercive field. Once the saturation voltage is reached, the coercive field remains constant, while the remanent polarization exhibits a linear increase with the electric field. In Figure 4b, we depict the frequency-dependent hysteresis loops of BLT4, revealing a decrease in maximal polarization and an increase in the corresponding remanent polarization as the frequency varies. Importantly, the hysteresis loops remain relatively stable, indicating good frequency stability. Figure 4c displays the leakage current behavior of BLT4, and the inset exhibits the ln J-ln E curve. The leakage conductance mechanism of ln J-ln E was analyzed by the slope α of ln J-ln E. In general, when α~1, the conduction mechanism is an ohmic conduction mechanism, and when α~2, the conduction mechanism is a space charge limited current mechanism (SCLC). The ln J-ln E curve of BLT4 under a positive electric field can be divided into two segments; when the electric field is low, α~1.4, this indicates that the conduction mechanism of the BLT4 film is the combined action of an ohmic conduction mechanism and an SCLC mechanism. When the electric field continues to increase, reaching α~12.4, this may be dominated by the Schottky emission mechanism and the Poole–Frenkel emission mechanism [33]. Lastly, in Figure 4d, we present the hysteresis loop of BLT4 at 8 V and its corresponding current and voltage profiles. The current-voltage curve exhibits a significant current peak near the coercive field, which is associated with ferroelectric domain inversion induced by the applied electric field. The reorientation of most domains under the applied electric field results in higher current densities, underscoring the stability of the thin film domains.

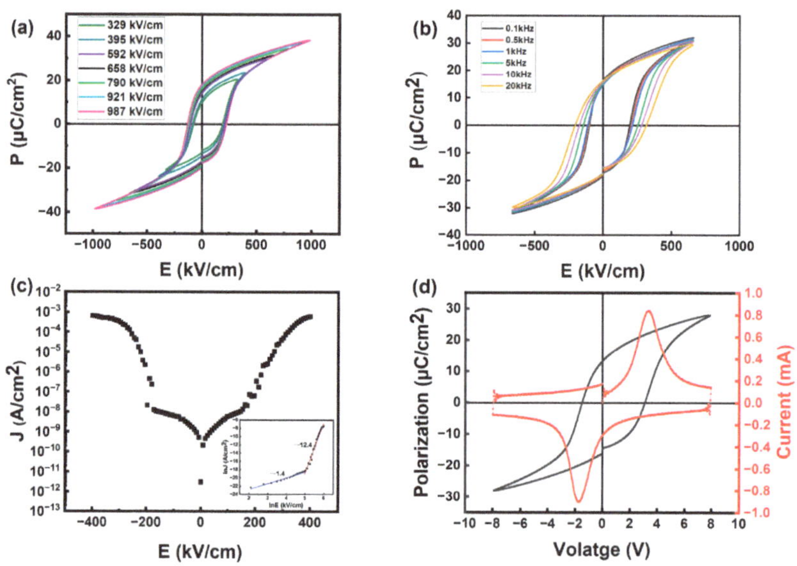

Figure 4. (**a**) The hysteresis loop of BLT4 of different electric fields at 1 kHz; (**b**) the change in BLT4 hysteresis loop with frequency; (**c**) the leakage current curve of BLT4; (**d**) the P-V hysteresis loop of BLT4 at 8 V and the corresponding I-V curve.

To confirm the presence of microscopic ferroelectricity, we utilized piezoelectric response force microscopy to examine the as-prepared BLT4 thin films in randomly selected areas. Figure 5a,b show the original amplitude and phase image of the BLT4 thin film, respectively. Figure 5c,d show the amplitude and phase image of the BLT4 thin film obtained after polarizing an area of approximately 3×3 mm^2 at -9 V and subsequently polarizing the central region with +9 V bias. This process revealed three distinct polarization states: an outer region that remained unbiased during growth, a region biased negatively at -9 V, and a centrally biased region at +9 V. These observations indicate complete polarization, switching between "up" and "down" states. Figure 5e,f show an additional PFM scan of the sample taken 30 min after the writing domain. The results indicate that the sample exhibits robust domain stability, underscoring its potential value in non-volatile ferroelectric memory applications.

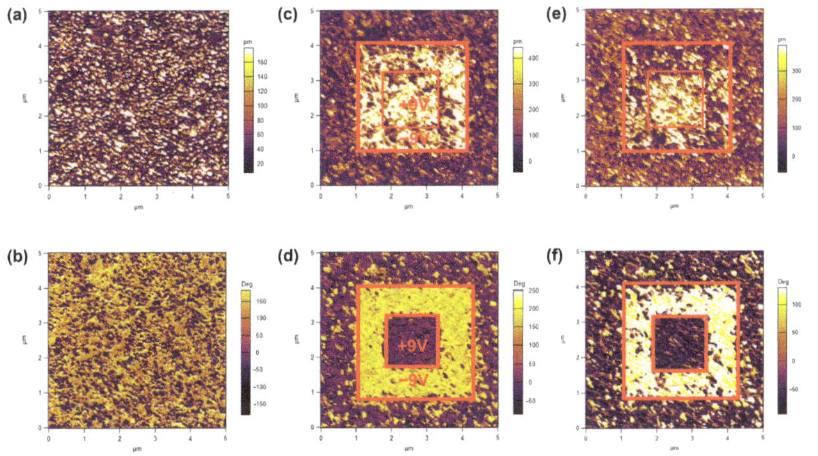

Figure 5. Out-of-plane piezoelectric force microscopy phase diagram of BLT4 film: original state (**a**) amplitude image, and (**b**) phase image; after writing domains (**c**) amplitude image, and (**d**) phase image; after 30 min (**e**) amplitude image, and (**f**) phase image.

Similarly, we conducted piezoelectric response force microscopy characterization on the as-prepared BLT5 thin films in randomly selected regions, as illustrated in Figure 6. Following domain writing, we observed that some shapes in the amplitude map matched the intended designs, yet the phase map did not exhibit clear shape features. This discrepancy arises from the fact that when writing domains, ferroelectric domains are only partially induced, which is a notable contrast to BLT4 thin films.

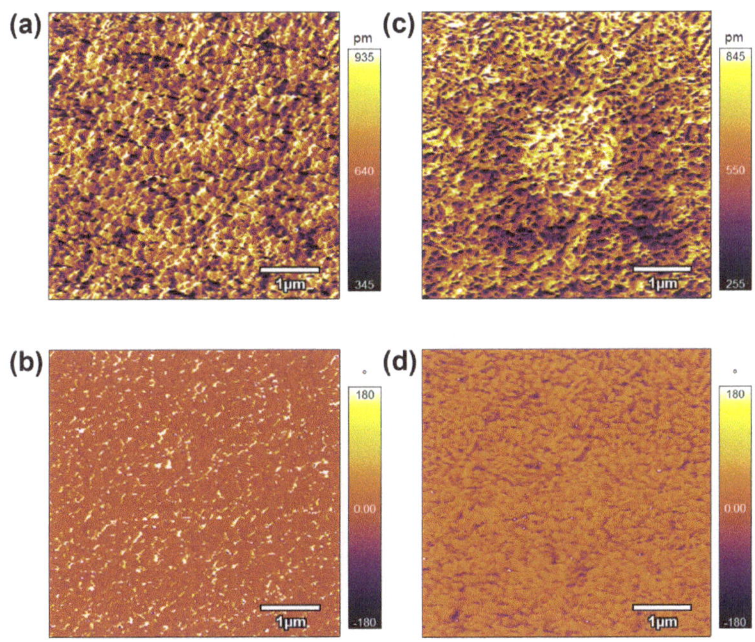

Figure 6. Out-of-plane piezoelectric force microscopy phase diagram of BLT5 film: original state (**a**) amplitude image, and (**b**) phase image; after writing domains (**c**) amplitude image, and (**d**) phase image.

To investigate the resistance characteristics of Au/BLT/Pt devices, I-V curves were measured at room temperature according to the schematic shown in Figure 7a. Figure 7b shows a semilog plot of the I-V measurement, where the arrow indicates the voltage sweep. The voltage scanning sequence is $1 \rightarrow 2 \rightarrow 3 \rightarrow 4 \rightarrow 5$, and the corresponding voltage is $0\ V \rightarrow 6\ V \rightarrow 0\ V \rightarrow -6\ V \rightarrow 0\ V$. It can be found that the device has resistance switching characteristics. With the increase in the forward sweep voltage, an SET occurs at about 3.2 V, and the device current increases significantly. The device switches from the high-resistance state (HRS) to the low-resistance state (LRS) [34,35]. As the negative sweep voltage increases, and as the sweep voltage returns from −6 V to 0 V, the device switches from the LRS to the HRS. This shows that Au/BLT/Pt memory devices have resistor characteristics and have good application prospects in memristor.

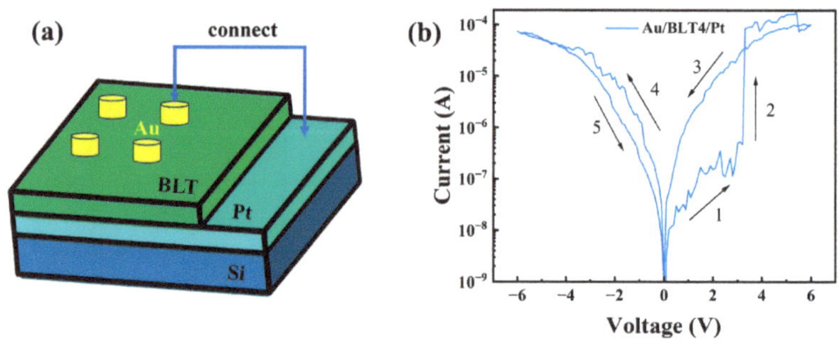

Figure 7. (**a**) Schematic drawing of the Au/BLT/Pt device; (**b**) semilogarithmic I-V curve.

4. Conclusions

In summary, $Bi_{3.25}La_{0.75}Ti_3O_{12}$ thin films were synthesized on $Pt/Ti/SiO_2/Si$ substrates using the sol–gel method, employing various spin-coating times. All thin films show random orientations and have the appearance of rod-like grains. The findings indicate a proportional increase in the film thickness, approximately 35 nm per layer, with each additional spin-coating application. Different spin-coating durations influence the ferroelectric properties and dielectric properties of BLT thin films. The remnant polarization experienced a decrease, and the dielectric constant increased with the increased film thickness. The BLT4 thin film has a good domain flip and keeps its characteristics and resistance characteristics, which show that it has potential applications in non-volatile memory and memristors.

Author Contributions: Conceptualization, W.Y. and T.J.; methodology, W.Y., Y.C. and D.W.; software, W.Y.; validation, Y.C.; formal analysis, W.Y.; investigation, W.Y.; resources, T.J.; data curation, W.Y.; writing—original draft preparation, W.Y.; writing—review and editing, T.J. and Q.G.; visualization, Y.C.; supervision, T.J.; project administration, T.J.; funding acquisition, T.J. All authors have read and agreed to the published version of the manuscript.

Funding: This research was funded by the National Natural Science Foundation of China, grant numbers 51702351 and 51777209; the Basic and Applied Basic Research Foundation of Guangdong Province, grant number 2020B1515120019; and the Shenzhen Science and Technology Innovation Committee, grant numbers JCYJ20170413152832151 and KQTD20170810160424889.

Institutional Review Board Statement: Not applicable.

Informed Consent Statement: Not applicable.

Data Availability Statement: The data that support the finding of this study are available from the corresponding author upon reasonable request.

Conflicts of Interest: The authors declare no conflicts of interest.

References

1. Ahn, Y.; Son, J.Y. Mixed grains and orientation-dependent piezoelectricity of polycrystalline Nd-substituted $Bi_4Ti_3O_{12}$ thin films. *Ceram. Int.* **2016**, *42*, 13061–13064. [CrossRef]
2. Roy, S.; Majumder, S. Recent advances in multiferroic thin films and composites. *J. Alloy Compd.* **2012**, *538*, 153–159. [CrossRef]
3. Scott, J.F.; de Araujo, C.A.P. Ferroelectric Memories. *Science* **1989**, *246*, 1400–1405. [CrossRef] [PubMed]
4. Cheng, H.; Ouyang, J.; Zhang, Y.-X.; Ascienzo, D.; Li, Y.; Zhao, Y.-Y.; Ren, Y. Demonstration of ultra-high recyclable energy densities in domain-engineered ferroelectric films. *Nat. Commun.* **2017**, *8*, 1999. [CrossRef] [PubMed]
5. Grinberg, I.; West, D.V.; Torres, M.; Gou, G.; Stein, D.M.; Wu, L.; Chen, G.; Gallo, E.M.; Akbashev, A.R.; Davies, P.K.; et al. Perovskite oxides for visible-light-absorbing ferroelectric and photovoltaic materials. *Nature* **2013**, *503*, 509–512. [CrossRef] [PubMed]
6. Peng, G.-G.; Zheng, D.-Y.; Cheng, C.; Zhang, J.; Zhang, H. Effect of rare-earth addition on morphotropic phase boundary and relaxation behavior of the PNN-PZT ceramics. *J. Alloy Compd.* **2017**, *693*, 1250–1256. [CrossRef]
7. Kumari, S.; Ortega, N.; Pradhan, D.K.; Kumar, A.; Scott, J.F.; Katiyar, R.S. Effect of thickness on dielectric, ferroelectric, and optical properties of Ni substituted $Pb(Zr_{0.2}Ti_{0.8})O_3$ thin films. *J. Appl. Phys.* **2015**, *118*, 184103. [CrossRef]
8. Cho, S.W.; Lee, J.I.; Jeong, Y.H. Microstructure, ferroelectric and piezoelectric properties of $Bi_4Ti_3O_{12}$ platelet incorporated $0.36BiScO_3$-$0.64PbTiO_3$ thick films for high temperature piezoelectric device applications. *Ceram. Int.* **2021**, *47*, 23880–23887. [CrossRef]
9. Du, X.; Huang, W.; He, S.; Kumar, T.S.; Hao, A.; Qin, N.; Bao, D. Dielectric, ferroelectric, and photoluminescent properties of Sm-doped $Bi_4Ti_3O_{12}$ thin films synthesized by sol-gel method. *Ceram. Int.* **2018**, *44*, 19402–19407. [CrossRef]
10. Ma, S.; Cheng, X.; Ma, Z.; Ali, T.; Xu, Z.; Chu, R. Effect of thickness and crystalline morphology on electrical properties of rf-magnetron sputtering deposited $Bi_4Ti_3O_{12}$ thin films. *Ceram. Int.* **2018**, *44*, 20465–20471. [CrossRef]
11. Park, B.H.; Kang, B.S.; Bu, S.D.; Noh, T.W.; Lee, J.; Jo, W. Lanthanum-substituted bismuth titanate for use in non-volatile memories. *Nature* **1999**, *401*, 682–684. [CrossRef]
12. Subbarao, E. A family of ferroelectric bismuth compounds. *J. Phys. Chem. Solids* **1962**, *23*, 665–676. [CrossRef]
13. Long, C.; Chang, Q.; Fan, H. Differences in nature of electrical conductions among $Bi_4Ti_3O_{12}$-based ferroelectric polycrystalline ceramics. *Sci. Rep.* **2017**, *7*, 4193. [CrossRef] [PubMed]
14. Chen, Y.; Xie, S.; Wang, H.; Chen, Q.; Wang, Q.; Zhu, J.; Guan, Z. Dielectric abnormality and ferroelectric asymmetry in W/Cr co-doped $Bi_4Ti_3O_{12}$ ceramics based on the effect of defect dipoles. *J. Alloy Compd.* **2017**, *696*, 746–753. [CrossRef]
15. Roselin, A.A.; Karkuzhali, R.; Anandhan, N.; Gopu, G. Bismuth titanate ($Bi_4Ti_3O_{12}$, BTO) sol–gel spin coated thin film for heavy metal ion detection. *J. Mater. Sci. Mater. Electron.* **2021**, *32*, 24801–24811. [CrossRef]
16. Du, X.; Huang, W.; Thatikonda, S.K.; Qin, N.; Bao, D. Improved ferroelectric and dielectric properties of Sm, La co-doped $Bi_4Ti_3O_{12}$ multifunctional thin films with orange-red emission. *J. Mater. Sci. Mater. Electron.* **2019**, *30*, 13158–13166. [CrossRef]
17. Zhang, S.-T.; Chen, Z.; Zhang, C.; Yuan, G.-L. Temperature-dependent ferroelectric and dielectric properties of $Bi_{3.25}La_{0.75}Ti_3O_{12}$ thin films. *Appl. Surf. Sci.* **2010**, *256*, 2468–2473. [CrossRef]
18. Zhu, Z.; Chen, Y.-B.; Zheng, X.-J. Nanoscale domain switching mechanism of $Bi_{3.15}Eu_{0.85}Ti_3O_{12}$ thin film under the different mechanical forces. *Chin. Phys. B* **2015**, *24*, 107702. [CrossRef]
19. Zhou, H.; Wu, G.; Qin, N.; Bao, D. Dual enhancement of photoluminescence and ferroelectric polarization in Pr^{3+}/La^{3+}-codoped bismuth titanate thin films. *J. Am. Ceram. Soc.* **2010**, *93*, 2109–2112. [CrossRef]
20. Wu, D.; Li, A.; Ming, N. Leakage current characteristics of $Pt/Bi_{3.25}La_{0.75}Ti_3O_{12}/Pt$ thin-film capacitors. *J. Appl. Phys.* **2005**, *97*, 1915533. [CrossRef]
21. Kao, M.-C.; Chen, H.-Z.; Young, S.-L. The microstructure and ferroelectric properties of Sm and Ta-doped bismuth titanate ferroelectric thin films. *Thin Solid Films* **2013**, *529*, 143–146. [CrossRef]
22. Su, L.; Lu, X.; Chen, L.; Wang, Y.; Yuan, G.; Liu, J.-M. Flexible, Fatigue-Free, and Large-Scale $Bi_{3.25}La_{0.75}Ti_3O_{12}$ Ferroelectric Memories. *ACS Appl. Mater. Interfaces* **2018**, *10*, 21428–21433. [CrossRef] [PubMed]
23. Ma, S.; Cheng, X.; Ma, Z.; Xu, Z.; Chu, R. Characterization of highly (117)-oriented $Bi_{3.25}La_{0.75}Ti_3O_{12}$ thin films prepared by rf-magnetron sputtering technique. *Solid State Commun.* **2018**, *278*, 31–35. [CrossRef]
24. Xue, K.-H.; de Araujo, C.A.P.; Celinska, J. A comparative study on $Bi_4Ti_3O_{12}$ and $Bi_{3.25}La_{0.75}Ti_3O_{12}$ ferroelectric thin films derived by metal organic decomposition. *J. Appl. Phys.* **2010**, *107*, 3428968. [CrossRef]
25. Wu, A.; Soares, M.R.; Salvado, I.M.M.; Vilarinho, P.M. Sol–gel synthesis and electrical characterization of $Bi_{3.25}La_{0.75}Ti_3O_{12}$ thin films. *Mater. Res. Bull.* **2012**, *47*, 3819–3824. [CrossRef]
26. Liu, B.; Zhang, W.; Wang, Y.; Sun, H.; Li, F.; Yan, Z.; Du, J.; Zhao, Q. Impact of Pt bottom electrode on the properties of ferroelectric $Bi_{3.25}La_{0.75}Ti_3O_{12}$ capacitors. *Mater. Lett.* **2007**, *61*, 1933–1936. [CrossRef]
27. Sun, S.; Yuan, J.; Guo, W.; Duan, X.; Jia, D.; Lin, H. Thickness effects on the sinterability, microstructure, and nanohardness of SiC-based ceramics consolidated by spark plasma sintering. *J. Am. Ceram. Soc.* **2023**, *107*, 777–784. [CrossRef]
28. Zhang, W.L.; Tang, M.H.; Xiong, Y.; Wang, K.; Wang, Z.P.; Xiao, Y.G.; Yan, S.A.; Li, Z.; He, J. Influence of the annealing temperature of the $Bi_4Ti_3O_{12}$ seeding layer on the structural and electrical properties of $Bi_{3.15}Nd_{0.85}Ti_{2.99}Mn_{0.01}O_{12}$ thin films. *RSC Adv.* **2016**, *6*, 88668–88673. [CrossRef]
29. Yang, B.B.; Guo, M.Y.; Song, D.P.; Tang, X.W.; Wei, R.H.; Hu, L.; Yang, J.; Song, W.H.; Dai, J.M.; Lou, X.J.; et al. $Bi_{3.25}La_{0.75}Ti_3O_{12}$ thin film capacitors for energy storage applications. *Appl. Phys. Lett.* **2017**, *111*, 183903. [CrossRef]

30. Schwartz, R.W.; Voigt, J.A.; Tuttle, B.A.; Payne, D.A.; Reichert, T.L.; DaSalla, R.S. Comments on the effects of solution precursor characteristics and thermal processing conditions on the crystallization behavior of sol-gel derived lead zirconate titanate thin films. *J. Mater. Res.* **1997**, *12*, 444–456. [CrossRef]
31. Iljinas, A.; Stankus, V. Influence of deposition temperature on structural and ferroelectric properties of $Bi_4Ti_3O_{12}$ thin films. *Appl. Surf. Sci.* **2016**, *381*, 2–5. [CrossRef]
32. Fan, Q.; Liu, M.; Ma, C.; Wang, L.; Ren, S.; Lu, L.; Lou, X.; Jia, C.-L. Significantly enhanced energy storage density with superior thermal stability by optimizing $Ba(Zr_{0.15}Ti_{0.85})O_3/Ba(Zr_{0.35}Ti_{0.65})O_3$ multilayer structure. *Nano Energy* **2018**, *51*, 539–545. [CrossRef]
33. Zhang, W.; Gao, Y.; Kang, L.; Yuan, M.; Yang, Q.; Cheng, H.; Pan, W.; Ouyang, J. Space-charge dominated epitaxial $BaTiO_3$ heterostructures. *Acta Mater.* **2015**, *85*, 207–215. [CrossRef]
34. Zhou, H.-C.; Jiang, Y.-P.; Tang, X.-G.; Liu, Q.-X.; Li, W.-H.; Tang, Z.-H. Excellent Bipolar Resistive Switching Characteristics of $Bi_4Ti_3O_{12}$ Thin Films Prepared via Sol-Gel Process. *Nanomaterials* **2021**, *11*, 2705. [CrossRef]
35. Bu, Y.; Xu, T.; Geng, S.; Fan, S.; Li, Q.; Su, J. Ferroelectrics-Electret Synergetic Organic Artificial Synapses with Single-Polarity Driven Dynamic Reconfigurable Modulation. *Adv. Funct. Mater.* **2023**, *33*, 2213741. [CrossRef]

ceramics

MDPI

Article

Effect of Superstoichiometric Bismuth Addition on the Structure and Dielectric Characteristics of the Solid Solutions $(1-x)BiFeO_3$-$xBaTiO_3$

Nikita A. Boldyrev [1,*], Egor S. Esin [1], Lidia A. Shilkina [1], Svetlana I. Dudkina [1], Alexander V. Nagaenko [2] and Larisa A. Reznichenko [1]

[1] Research Institute of Physics, Southern Federal University, Rostov-On-Don 344090, Russia
[2] Scientific Design and Technological Bureau "Piezopribor", Southern Federal University, Rostov-On-Don 344090, Russia
* Correspondence: nboldyrev@sfedu.ru

Abstract: Ceramic samples of solid solutions of the binary system $(1-x)BiFeO_3$-$xBaTiO_3$ + 2 wt.% Bi_2O_3 ($0.29 \leq x \leq 0.33$, $\Delta x = 0.01$) were prepared using the conventional solid-phase reaction method with and without mechanical activation. Using X-ray studies, it was found that the objects have a pseudocubic crystal structure, and limited solubility occurs in solid solutions of the studied composition, as evidenced by the presence of regions with an increased Bi or Ba content and similar cell parameters. A diffuse phase transition occurred from the FE to PE state in the temperature ranges of (650–850) K. Relaxor-like behavior and the smearing of the phase transition in the studied ceramics can be associated with the presence of non-interacting regions with an increased content of Bi or Ba, different modulation, and crystal lattice symmetry. The grain morphology and dielectric characteristics of the selected solid solutions were investigated. The highest piezoelectric coefficient, d_{33} = 120 pC/N, was obtained in the mechanically activated ceramics $0.71BiFeO_3$-$0.29BaTiO_3$ + 2 wt.% Bi_2O_3.

Keywords: dielectric properties; perovskites; multiferroics; piezoelectric properties; solid solutions

Academic Editors: Dawei Wang and Fayaz Hussain

Received: 14 November 2024
Revised: 25 December 2024
Accepted: 9 January 2025
Published: 13 January 2025

Citation: Boldyrev, N.A.; Esin, E.S.; Shilkina, L.A.; Dudkina, S.I.; Nagaenko, A.V.; Reznichenko, L.A. Effect of Superstoichiometric Bismuth Addition on the Structure and Dielectric Characteristics of the Solid Solutions $(1-x)BiFeO_3$-$xBaTiO_3$. *Ceramics* **2025**, *8*, 7. https://doi.org/10.3390/ceramics8010007

1. Introduction

Multiferroic materials, which have coexisting electric, magnetic, and elastic orderings, are currently among the most intensively studied objects in materials science [1] due to their wide range of potential applications, including the production of magnetic field sensors, memory elements, and spintronics devices [2–4]. Bismuth ferrite ($BiFeO_3$, BF), a representative of this class of materials, has Curie and Néel temperatures of 1123 K and 643 K, respectively, and is being considered for use in magnetoelectric structures. However, its use is limited due to several factors, such as the difficulty in obtaining it in a single phase, the presence of Fe^{2+}/Fe^{3+} ions and oxygen vacancies that cause leakage currents, and the high electric coercive field required for domain switching [5,6]. However, modification by rare-earth elements or the creation of solid solutions based on BF allows for stabilizing the structure and improving the multiferroic properties of the obtained materials [7–9]. Among $BiFeO_3$-based solid solutions, the lead-free $BiFeO_3$-$BaTiO_3$ (BF-BT) system has been proposed as a potential replacement for lead zirconate titanate (PZT) ceramics, which have been widely used in piezoelectric devices due to their excellent dielectric and piezoelectric properties near the morphotropic phase boundary (MPB) between the rhombohedral (Rh) and tetragonal (T) phases [10,11]. The Curie temperature (T_C) of commercial PZT ceramics ranges from 450 K to 770 K, depending on the composition. However, with the development of modern industry, the need for devices with a wider operating range is growing. Solid solutions of BF-BT with low BT content exhibit T_C values exceeding 650 K, indicating their potential for high-temperature applications [12]. The properties of this solid solution can be further enhanced through mechanical activation during the production process and by incorporating additional elements (Table 1).

Table 1. The values of the piezoelectric modulus d_{33} and the Curie temperature for modified binary ceramics BF-BT from the works [13–18] and this work.

Modificator	Composition	d_{33}, pC/N	T_C, K
Zn^{2+}	$0.7Bi_{0.4}Zn_{0.6}FeO_3$-$0.3BaTiO_3$ [13]	160	723
Mn^{4+}	$0.71BiFeO_3$-$0.29BaTiO_3$ + 1.2 mol%MnO_2 [14]	169	779
Bi^{3+}	$0.7Bi_{1.02}FeO_3$-$0.3BaTiO_3$ [15]	214	801
Ga^{3+}	$0.7Bi(Fe_{0.98}Ga_{0.02})O_3$-$0.29BaTiO_3$ [16]	157	740
Zr^{4+}	$0.75BiFeO_3$-$0.25Ba(Zr_{0.1}Ti_{0.9})O_3$ [17]	138	729
Sc^{3+}	$0.7Bi(Fe_{0.98}Sc_{0.02})O_3$-$0.3BaTiO_3$ [18]	165	778
Bi^{3+}	$0.71BiFeO_3$-$0.29BaTiO_3$ [this work]	120	650–750

As is well-known, Bi is a volatile element at elevated temperatures due to its high vapor pressure, which can result in deviations from the optimal composition [19,20]. This can obviously impact the electrical properties of piezoelectric materials based on bismuth. It has been demonstrated that mechanical activation during the manufacturing process or the addition of excess bismuth to the system can effectively address this issue [21,22]. On the other hand, Bi_2O_3 is frequently used as a sintering aid for piezoelectric materials to lower their sintering temperature and modify their dielectric and piezoelectric properties. In this context, it is essential to comprehend how mechanical activation and the addition of excess bismuth affect the crystal structure, as this determines the macroscopic characteristics of the ceramic material being studied.

In this paper, $(1-x)BiFeO_3$-$xBaTiO_3$ (BF-xBT) ceramic samples with additional Bi content were prepared with a conventional solid-state reaction method using mechanical activation and without it, and the structure, dielectric, ferroelectric, and piezoelectric properties of the obtained ceramics were studied.

2. Materials and Methods

Ceramic samples of the solid solutions of the binary system $(1-x)BiFeO_3$-$xBaTiO_3$ ($0.29 \leq x \leq 0.33$, $\Delta x = 0.01$) were fabricated using conventional ceramic technology with double solid-phase synthesis at temperatures T_1 = 1123 K and T_2 = 1143 K and holding times $\tau_1 = \tau_2$ = 10 h with following sintering at T_{sin1} = 1343 K for samples prepared with mechanical activation (BF-xBTm) and T_{sin2} = 1323 K for other samples (BF-xBT) within 2 h. The experimental sample density ρ_{exp} was determined with hydrostatic weighing; octane was used as a liquid medium. The density was calculated with the formula $\rho_{exp} = (\rho_{oct} \times m_1)/(m_2 - m_3 + m_4)$, where ρ_{oct} is the density of the octane, m_1 is the mass of the dry workpiece, m_2 is the mass of the workpiece saturated with octane, m_3 is the mass of the saturated workpiece suspended in octane with suspension, and m_4 is the suspension mass for the workpiece. Experimental and relative densities are shown in Table 2. The initial reagents were Bi_2O_3, Fe_2O_3, TiO_2, $BaCO_3$ with a base substance content of at least 99.95%. The mechanical activation of synthesized powders was carried out at the stage of manufacturing press powders prepared for sintering. For mechanical activation, a high-energy grinding ball mill AGO-2 manufactured by NOVITS (Novosibirsk, Russia) was used. The prepared powder was loaded into drums with an internal diameter of 63 mm together with ZrO_2 balls with a diameter of 8 mm with a total weight of 200 g. The drum with the mixture was placed in the AGO-2 planetary mill. Grinding was carried out in an alcoholic environment for 15 min; the rotation speed of the drum was 1800 rpm.

Table 2. Experimental and relative densities of the studied ceramics.

Sample	ρ_{exp}, g/cm^3	ρ_{rel}, %
BF-0.29BT	7.03	91.06
BF-0.3BT	7.00	90.67
BF-0.31BT	7.03	91.66
BF-0.32BT	7.05	92.40
BF-0.33BT	7.04	92.63
BF-0.29BTm	7.28	94.33
BF-0.3BTm	7.28	94.32
BF-0.31BTm	7.24	94.45
BF-0.32BTm	7.17	94.17
BF-0.33BTm	7.21	94.90

Ceramics were sintered in the form of cylinders, which were cut into disks Ø = 10 mm and h = 1 mm. Electrodes were applied to the flat surfaces of the disks by the stepwise burning of a silver-containing paste: 573 K for 20 min, 773 K for 30 min, and 1073 K for 20 min.

The X-ray study was carried out using the method of powder diffraction on the DRON-3 diffractometer (IC "Bourevestnik", Saint Petersburg, Russia) (Bragg-Brentano focusing) using CoKα radiation.

The microstructure of the samples was studied using the scanning electron microscope JSM-6390L.

Temperature dependences of the relative complex permittivity $\varepsilon^*/\varepsilon_0 = \varepsilon'/\varepsilon_0 - i\varepsilon''/\varepsilon_0$ ($\varepsilon'/\varepsilon_0$ and $\varepsilon''/\varepsilon_0$ are the real and imaginary parts of $\varepsilon^*/\varepsilon_0$, respectively; ε_0 – dielectric permittivity of vacuum) in the temperature range (300–873) K and the frequency range f = (50–2 × 10^6) Hz were obtained using a measuring bench based on the Agilent 4980A LCR meter (Keysight Technologies, Inc., Santa Rosa, CA, USA).

Samples were polarized at T = 400 K in the polyethylene siloxane fluid under applied fields of 3–6 kV. Piezoelectric coefficients of samples were measured using a quasistatic YE2730A d33 METER, (APC International Ltd., West Kingston, RI, USA) at f = 110 Hz.

3. Results and Discussion

3.1. XRD Analysis

A powder X-ray diffraction study of the binary system $(1-x)BiFeO_3$-$xBaTiO_3$ ($0.00 \leq x \leq 0.5$, $\Delta x = 0.1$) [23] showed that the solid solutions demonstrate rhombohedral symmetry in the range of $0 \leq x < 0.4$, pseudocubic (Psc) in the range of $0.3 \leq x \leq 0.5$, and at x = 0.3, both phases coexist in the same solid solution. In this study, we investigated in more detail the concentration region $0.29 \leq x \leq 0.33$ containing MPB for compositions of the binary system $(1-x)BiFeO_3$-$xBaTiO_3$ ($0.29 \leq x \leq 0.33$, $\Delta x = 0.01$) prepared with and without mechanical activation and with the superstoichiometric addition of Bi_2O_3.

Before analyzing the results of X-ray studies of solid solutions, it is important to compare the crystallochemical characteristics of the ions involved. These include the X-ray atomic scattering factor (f), at $\theta = 0$, for positive ions; ionic radii (R); cation electronegativity (EN); and their correspondence to the isomorphism criteria [24]. According to the isomorphism rules, the difference in ionic radii should not exceed 15% when compared to the smaller value. Similarly, the difference in electronegativities should be less than 0.4. This ensures that the ions can form stable compounds within the solid solution

From the data presented in Table 3, it is evident that both conditions are fulfilled for B-cations. For A-cations, the ionic radii also meet the above condition, but the ΔEN exceeds the permissible value by a factor of 2.5. This indicates that BF and BT may not form a continuous series of solid solutions. Instead, only limited solubility is possible, and the solid solution may disintegrate into several solid solutions with increased concentrations of Bi or Ba and similar cell parameters.

Table 3. Atomic scattering factors, f, ionic radii, R, electronegativity, $\Delta R/R_{min}$, EN, ΔEN.

Cation	f [25]	R, Å [26]	$\Delta R/R_{min} \times 100$, %	EN [27]	ΔEN
Ba^{2+}	54	1.54 for CN 12	15	0.9	1.0
Bi^{3+}	80	1.34 for CN 12		1.9	
Ti^{4+}	18	0.64 for CN 6	4	1.5	0.3
Fe^{3+}	23	0.67 for CN6		1.8	

A similar effect has been observed in the $Pb_{1-x}BaTiO_3$ system, which has been described in detail in [28]. It would be interesting to investigate whether it is possible to improve the homogeneity of the solid solution by incorporating the mechanical activation of the synthesized product into the sintering process. In this study, we compare X-ray data with electrophysical properties of solid solutions that were prepared without mechanical activation, and those that were mechanically activated before sintering.

The heterogeneity of the solid solution should be manifested in the X-ray diffraction pattern as a slight splitting of the diffraction peaks and the emergence of diffuse maxima near them. These diffuse maxima indicate the presence of atomic segregation, as described by [29]. If an ordered arrangement of regions with a higher concentration of Bi and Ba occurs, the diffuse maxima transform into satellites of the main reflections. The intensity of these satellites is proportional to the square of the difference in atomic scattering factors $(\Delta f)^2$ of the atoms that form the planar defect. In our case, this planar defect represents the boundary between regions with different chemical compositions.

Figure 1 shows X-ray diffraction patterns for studied ceramics in the angle range $20 \leq 2\theta \leq 85$ (°). In Figure 1a weak peaks of the impurity phase $Ba_5Fe_2O_8$ can be seen. After mechanical activation, the pikes of the impurity phase disappeared. No distortion of diffraction reflections corresponding to any one symmetry was detected, so all ceramics have Psc symmetry.

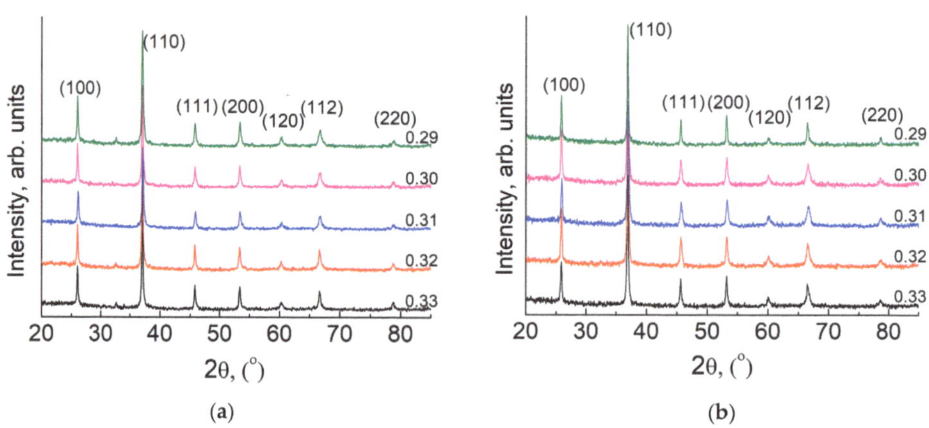

Figure 1. X-ray diffraction patterns for BF-xBT (**a**) and BF-xBTm (**b**) ceramics. Admixture—$Ba_5Fe_2O_8$ [JCPDS set 29, card 172].

Figures 2–4 show the diffraction peaks of solid solutions with different compositions with x = 0.29, 0.31, and 0.33, which were prepared both with and without mechanical activation of the synthesized powder.

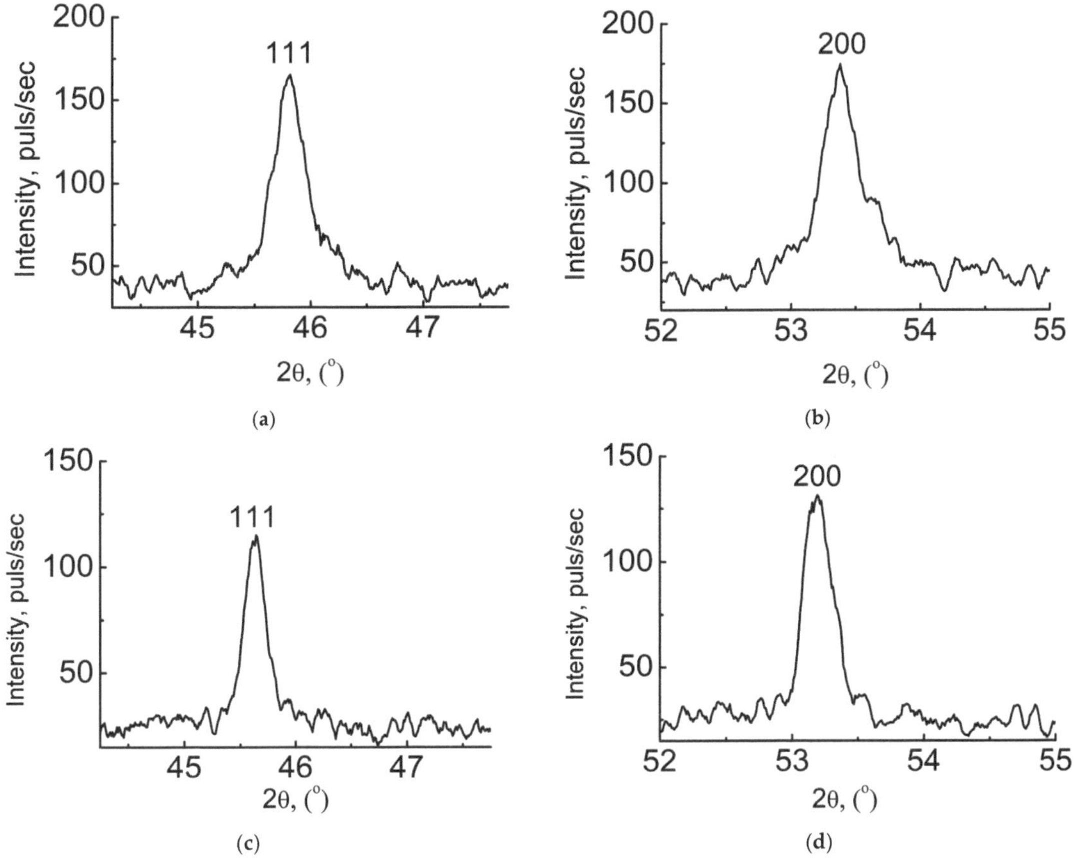

Figure 2. Diffraction peaks (111) and (200) on a larger scale for BF-0.29BT (**a**,**b**) and BF-0.29BTm (**c**,**d**).

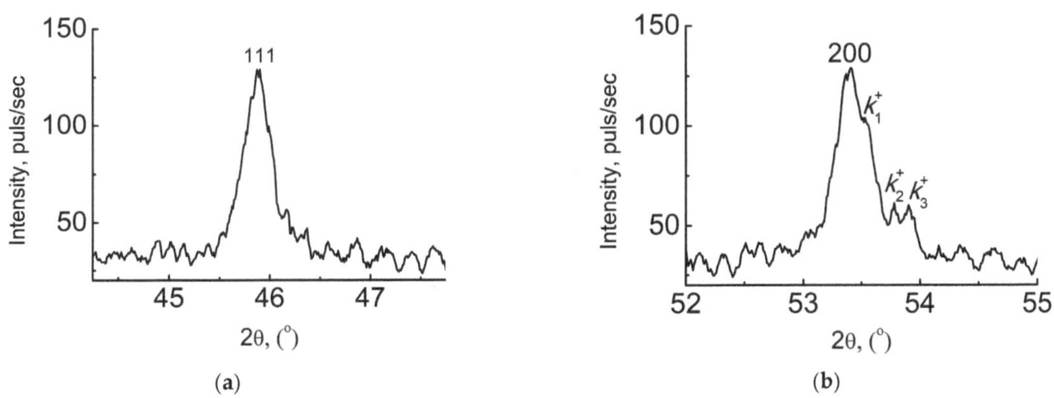

Figure 3. *Cont.*

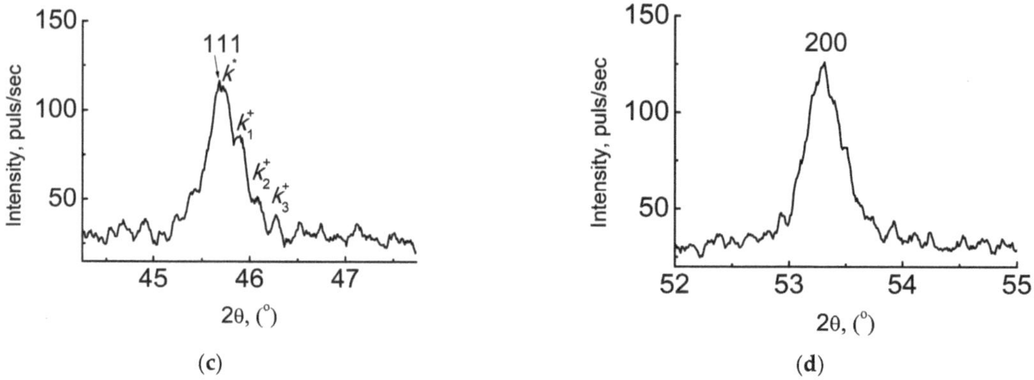

Figure 3. Diffraction peaks (111) and (200) on a larger scale for BF-0.31BT (**a**,**b**) and BF-0.31BTm (**c**,**d**).

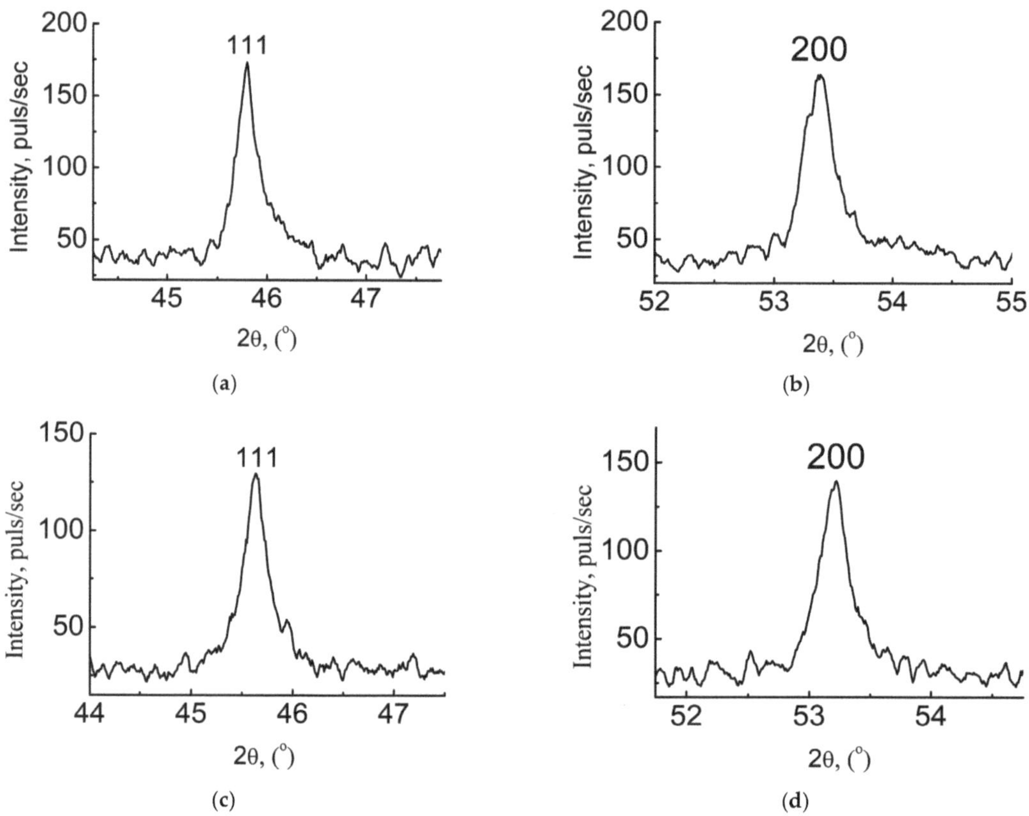

Figure 4. Diffraction peaks (111) and (200) on a larger scale for BF-0.33BT (**a**,**b**) and BF-0.33BTm (**c**,**d**).

Figure 2a,b show the diffraction pattern for the sample prepared without mechanical activation. There is a diffuse scattering at the base of the 111 and 200 diffraction peaks, indicating the presence of atomic segregation. This means that there are regions in the crystal structure that have a different chemical composition from the main matrix. The diffuse maxima are blurred, suggesting some order in the arrangement of these segregated regions. In contrast, Figure 2c,d show the diffraction pattern of the mechanically activated sample.

Here, the signs of atomic segregation have disappeared, but a splitting has appeared at the top of the diffraction peaks. This suggests that after the mechanical activation, there was a more uniform distribution of BT within the BF matrix, leading to the formation of two solid solutions with similar cell parameters.

Figure 3 shows fragments of the X-ray diffraction patterns for samples with x = 0.31. Satellite maxima, denoted by the letter "k^{+}_{1-3}", are visible near the main diffraction peaks and on them (200 in Figure 3b and 111 in Figure 3c). The subscript indicates the position of the satellite relative to the main peak ("+" or "−" from the side of a large angle θ), and the superscript indicates its serial number.

The following modulation parameters were calculated: the wave number k, the modulation wavelength Λ, and the number of perovskite cells N in the modulation wavelength. The relationships between the wave numbers were also determined, and the character of the modulation was identified. The modulation wavelength, Λ, was calculated using the Formula (1):

$$\Lambda = \left| \frac{1}{d_{hkl}} - \frac{1}{d_s} \right|^{-1}, \quad (1)$$

where d_{hkl} and d_s are the interplanar spacings of the main peak and satellite, respectively [29]. The calculation results are presented in Table 4.

Table 4. Wave number, k, modulation wavelength Λ, number of cells in a wavelength, N, and the relationship between the wave numbers of the solid solutions with x = 0.31.

Diffraction Reflection and Composition	Wave Number	k	Λ, Å	N	Relation Between the Wave Numbers k
200 BF-0.31BT	k_1^+	0.00104	963	241	$k_3^+ = k_1^+ + k_2^+$
	k_2^+	0.00314	318	79.6	
	k_3^+	0.00424	235	59	
111 BF-0.31BTm	k_1^+	0.0018	556	80.5	$k_1^+ = 1/2\ k_2^+ = 1/3\ k_3^+$
	k_2^+	0.00359	278	40	
	k_3^+	0.00539	185.6	27	
	k^*	0.00036	2781	403	$k^* = 1/5\ k_1^+ = 1/10\ k_2^+ = 1/15\ k_3^+$

In BF-0.31BT the modulation wave propagated along < 100 > direction. The satellites of the diffraction peak at 200 (Table 4) were the first-, third-, and fourth-order satellites, as the wave numbers associated with them were related by the ratio $k_1^+ = \frac{1}{3}k_2^+ = \frac{1}{4}k_3^+$. However, the lack of a second-order satellite and the strong intensity of the fourth-order satellite suggest that the wave number, k_3^+, is the total wave number. These satellites occur when two modulation waves are superimposed in a polydomain crystal in one crystallographic direction, as described by [30]. In our experiment, crushed ceramics can be viewed as polydomain crystals, with regions of increased Bi and Ba concentration acting as domains.

For the BF-0.31BTm (Figure 3c,d), only traces of modulation were observed on the (200) diffraction peak, but modulation was also present along the < 111 > direction. The (111) diffraction peak satellites were related by a ratio $k_1^+ = \frac{1}{2}k_2^+ = \frac{1}{3}k_3^+$, which corresponds to the harmonic modulation characteristic of the concentration wave. Mechanical activation not only changed the modulation direction but also reduced the modulation wavelength. Since the atomic displacements that determine the symmetry of the crystal are modulated, we can conclude that at x = 0.31, there are regions with very small tetragonal distortion (modulation in the < 001 > direction) and regions with very small rhombohedral distortion, (modulation in the < 111 > direction). In BF-0.31BT, a phase Rh→T transition has already occurred in the bulk of the rhombohedral distorted regions, and mechanoactivation shifts BF-0.31BTm towards the Re phase, thus shifting this transition to a region with a higher $BaTiO_3$ content. Figure 3 shows that all the diffraction peaks have split tops, and the widths of the split peaks are approximately equal, except for the (111) peak in Figure 3c, which has a narrower and more intense peak marked by an arrow. Assuming that this wide peak is a satellite of the peak indicated by the arrow, a modulation wavelength can be calculated. An interesting result was found, as shown at the bottom of Table 4. This ratio of wave numbers is not accidental. It shows that mechanical activation of BF-0.31BTm allows us to obtain a long-range ordered structure consisting of regions with increased concentrations of Bi and Ba. Each of these regions contains approximately 400 perovskite cells. In fact, the solid solution decomposes with the spinodal mechanism, resulting in the formation of two solid solutions with similar cell parameters.

The diffraction peaks (111) and (200) of ceramics with x = 0.33 are shown in Figure 4. The X-ray diffraction pattern of BF-0.33BT clearly shows the asymmetry of the diffraction peaks. The tails on the side with larger θ angles indicate the presence of clusters in the structure with a cell parameter smaller than that of the matrix. BF-0.33BTm is more homogeneous. However, the diffuse maxima at the bases of the peaks suggest segregation.

Figure 5 shows the dependence of the cell parameter a(x) for solid solutions prepared using two methods and the linear dependence according to Vegard's law. For the initial compounds, the parameters were obtained from the JCPDS database (BF-Set 14, card 181; BT-Set 3, card 725). For x = 0.1, 0.2, 0.4, and 0.5, the cell parameters were used from a previous study of the BF-BT system [23]. Since the cell parameter for x = 0.3, which was prepared without adding 2% Bi_2O_3, differs by 0.0001 Å from the values in this study, this difference is acceptable.

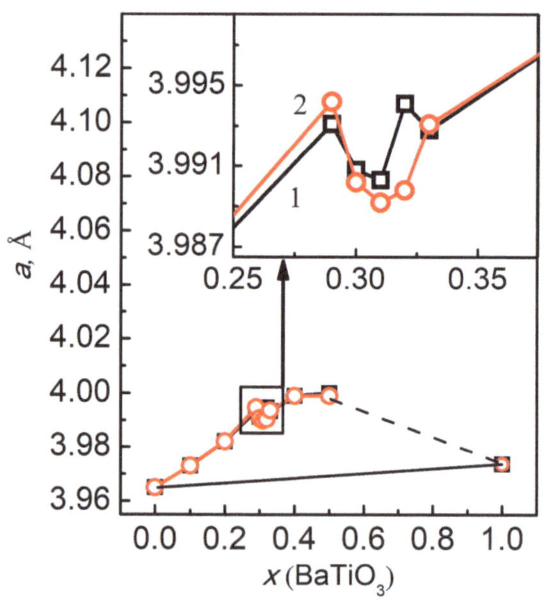

Figure 5. Dependency a(x) for $(1-x)BiFeO_3$-$xBaTiO_3$ + 2 wt.% Bi_2O_3 at $0 \leq x \leq 0.5$ (for $BaTiO_3$ $V^{1/3}$ was used), direct line—Vegard's rule; on the inset—a(x) cell parameter break area. 1—samples without mechanical activation, 2—samples with mechanical activation.

Inset on Figure 5 shows that a break in the a(x) dependence occurs in the range 0.30 < x < 0.32. The width of this break for samples manufactured without mechanical activation is $\Delta x = 0.03$ (0.29–0.32) with the jump $\Delta a = 0.003$ Å. For mechanically activated ceramics, $\Delta x = 0.04$ (0.29–0.33) and $\Delta a = 0.005$ Å. In both cases, the parameter a has a minimum value at x = 0.31.

In Figure 5, it is also clear that the curve a(x) deviates positively from Vegard's rule. This type of dependence of a(x) is observed in non-metallic solid solutions with immiscibility regions within the homogeneous region of a given system [31]. Thus, the appearance of the X-ray diffraction patterns and the dependence of the cell parameter on x suggest that a continuous series of solid solutions is not formed in the BF-BT system with Bi_2O_3 addition.

3.2. Microstructural Characterization

Figure 6 shows micrographs of chips of the studied ceramics. The microstructure of the BF-(1−x)BT samples (Figure 6a,c,e,g,i) is primarily composed of two types of grains: small grains with a diameter of less than 1.5 μm, which form compact aggregates, and larger grains with a size of more than 3 μm in the shape of irregular polyhedra. Despite a slight reduction in the average grain size, an increase in the BT concentration has no significant impact on the microstructure of ceramics produced without mechanical activation. The grain structure of BF-0.29BTm ceramics (Figure 6b) consists of crystallites with an average size of ~5 μm. The chip passes along the grain boundaries. An increase in the BT concentration leads to the appearance of segregations of small grains (Figure 6d,f) as in BF-(1−x)BT ceramics and a decrease in the average grain size to ~1.5 μm in BF-0.32BTm and BF-0.33BTm (Figure 6h,j). The presence of a homogeneous mass of very small grains ~1 μm arranged in an ordered manner on separate fragments of the chip is characteristic of solid solutions after spinodal decomposition. This indirectly confirms the conclusions of the X-ray analysis.

3.3. Dielectric and Piezoelectric Characteristics

Figure 7 shows the temperature dependences of the real part of the complex permittivity and the dielectric loss tangent of the studied ceramics. In all samples at T < 600 K, an increase in the permittivity is observed with increasing temperature, ending in a diffuse frequency-dependent maximum corresponding to the phase transition from the paraelectric phase to the ferroelectric one. With an increase in the frequency of the measuring field, the maximum shifts to the high-temperature region. For all samples except BF-0.29BTm, the presence of BT also affects the position of the maximum, shifting it to the low-temperature region with an increase in the BT concentration from ~740 K (BF-0.29BT) and ~820 K (BF-0.30BTm) to 680 K (BF-0.33BT) and 685 K (BF-0.33BTm) (Figure 7a,c). Increasing the BT concentration shifts the T_C in the BF-BT solid solutions to the low-temperature region since the T_C of BT is much lower than the T_C of BF, which can be seen in Figure 7a. In Figure 7c, the $\varepsilon'/\varepsilon_0$ curves are divided into the two groups. Curves of the samples with x = 0.30, 0.31, 0.32 correspond to the solid solutions from the morphotropic phase boundary. Other curves correspond to the solid solution located outside the morphotropic phase transition region. In each of the groups, T_C decreases with increasing BT concentration. Almost all samples demonstrate extremely high conductivity, which is expressed in a sharp increase in the dielectric loss tangent after 400 K (Figure 7b,d). BF-0.29BTm was the only sample that demonstrated acceptable tangent values. The reason for the observed effects may be the already mentioned oxidation of Fe-ions and volatilization of Bi_2O_3, but we consider that such a large increase in conductivity in the samples of one research section is due precisely to Maxwell–Wagner relaxation arising from the accumulation of free charges at the interfaces of Bi- or Ba-rich regions. In the BF-0.29BTm sample, a more uniform dissolution of $BaTiO_3$ into $BiFeO_3$ occurred after mechanical activation, thus avoiding these problems.

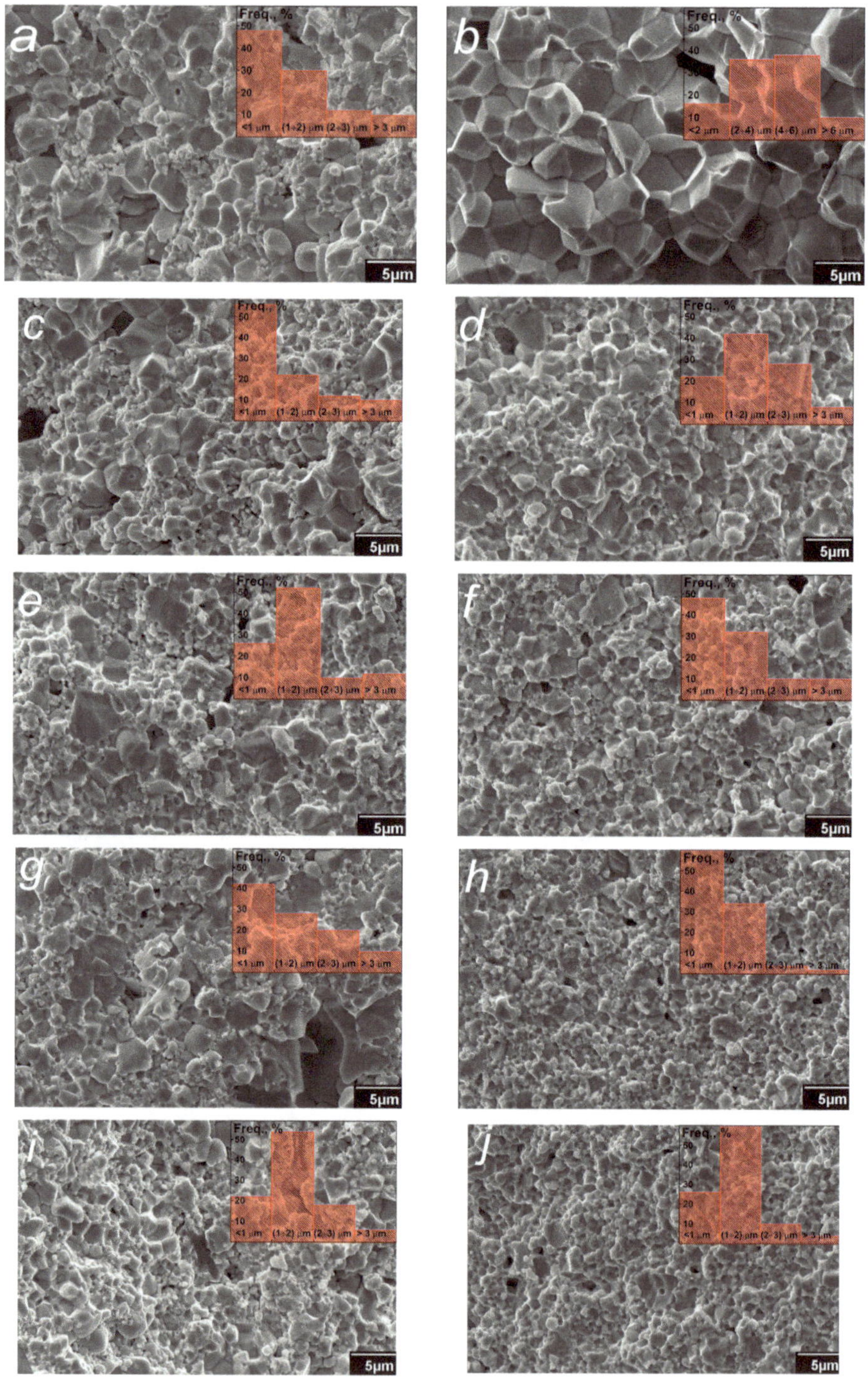

Figure 6. Microstructure of the studied ceramics: (**a**) BF-0.29BT, (**b**) BF-0.29BTm, (**c**) BF-0.30BT, (**d**) BF-0.30BTm, (**e**) BF-0.31BT, (**f**) BF-0.31BTm, (**g**) BF-0.32BT, (**h**) BF-0.32BTm, (**i**) BF-0.33BT, (**j**) BF-0.33BTm. Histograms with grain size distribution are shown in the insets.

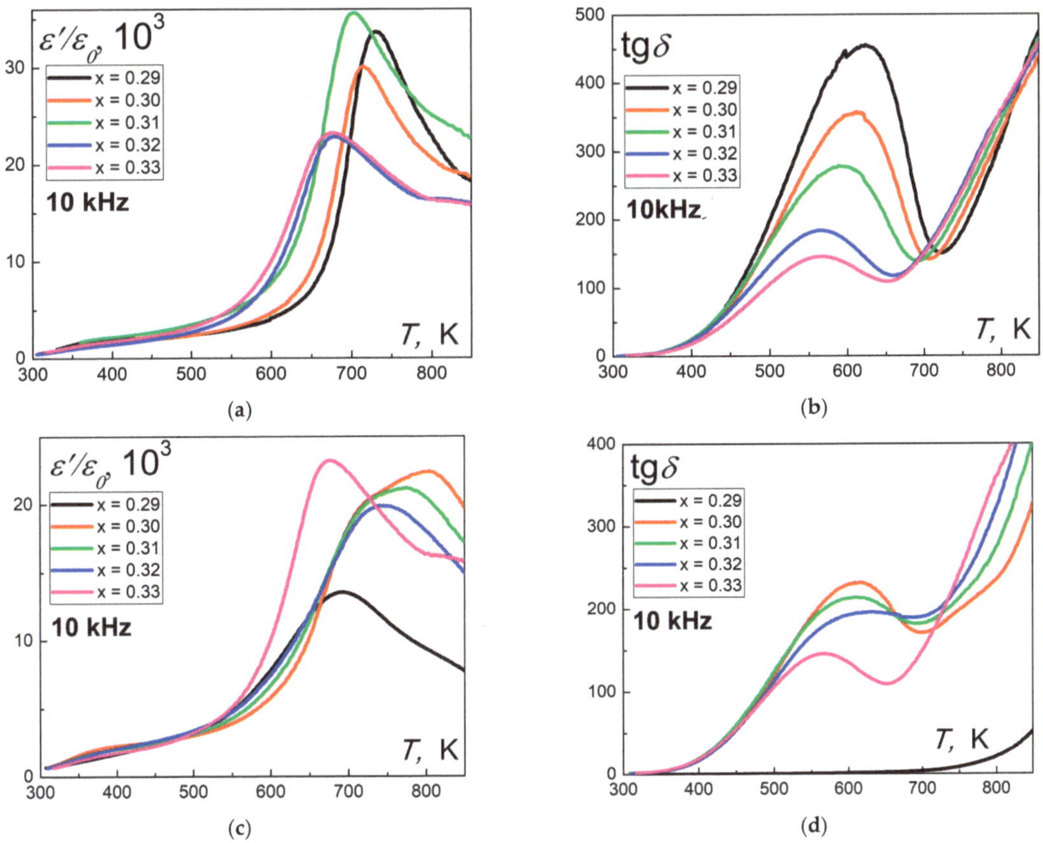

Figure 7. Temperature dependences of dielectric constant and dielectric loss tangent in the temperature range (300–850) K on the frequency 10 kHz ((**a**,**b**)–BF-(1−*x*)BT, (**c**,**d**)–BF-(1−*x*)BTm).

Figure 8 shows the temperature dependences of the real part of the complex permittivity and the dielectric loss tangent in a wide range of temperatures and research frequencies. As can be seen from the figure, in the range of (650–750) K, a diffuse phase transition occurs in the sample, the temperature of which, T_m, depends on the frequency. Similar behavior is observed in the tgδ(T), but the growth of the electrical conductivity of the material at T~500 K leads to a sharp increase of the tgδ values. As a result, in the vicinity of the FE→PE phase transition, we observe insignificant anomalies at high f values. When approximating the T_m(f) dependence in the whole frequency range, we used the Vogel–Fulcher relation (2):

$$f = f_0 exp \frac{E_a}{k(T_m - T_f)} \tag{2}$$

where f_0 is the frequency of attempts to overcome a potential barrier E_a, k is the Boltzmann constant, and T_f is the Vogel–Fulcher temperature, interpreted as the temperature of "static freezing" of electric dipoles or transition to the state of a dipole glass. The best result was achieved with the parameters $f_0 \sim 4 \cdot 10^{12}$ Hz, $E_a \approx 0.29$ eV, and the Vogel–Fulcher temperature $T_f \approx 595$ K. The obtained results indicate that the solid solution BF-0.29BTm demonstrates relaxor behavior that may be associated with the presence of regions rich in Bi^{+2} or Ba^{+2} in the ceramic crystal structure.

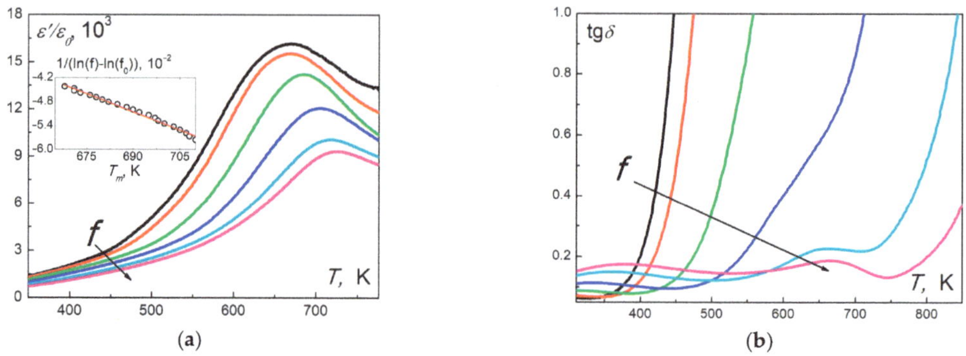

Figure 8. Temperature dependences of dielectric constant (**a**) and dielectric loss tangent (**b**) in the temperature range (300–850) K and the frequency range (150–1.5 × 10^6) Hz. The dependence $(\ln(f_0)-\ln(f))^{-1}$ on T_m, in the inset, illustrates the Vogel–Fulcher law.

Despite the fact that solid solutions from MPB in many cases exhibit extreme electromechanical properties, we did not find stable piezoresponse after poling most of the samples studied. However, it was possible to observe the piezoelectric characteristics that were stable in time in the BF-0.29BTm ceramics. The highest piezoelectric constant of 120 pC/N is much larger than d_{33} = 65 pC/N achieved in BF-BT ceramics fabricated by conventional solid-phase reaction methods [23]. However, this value is lower than presented in Table 1, indicating that complex modification and good solubility of the solid solution components are required to improve the piezoelectric properties of the investigated system.

4. Conclusions

Ceramic samples of solid solutions of the binary system $(1-x)BiFeO_3$-$xBaTiO_3$ + 2 wt.% Bi_2O_3 ($0.29 \leq x \leq 0.33$, $\Delta x = 0.01$) were prepared using the conventional solid-phase reaction method with and without mechanical activation. Using X-ray studies, it was found that in the investigated concentration range the structure of solid solutions is modulated, in the interval $0.30 \leq x \leq 0.32$ there is a morphotropic phase transition at the local level. Clusters with very small tetragonal cell distortion predominate in BF-0.31BT, and clusters with rhombohedral distortion predominate in BF-0.31BTm. It is concluded that the solid solutions are heterogeneous and contain regions rich in Bi or Ba. The reason for this is the violation of the rules of isomorphism during the substitution of Bi↔Ba. Mechanical activation led to a limited improvement in the homogeneity of the obtained ceramics only for the BF-0.29BTm composition. Data from the study of the microstructure and dielectric properties of the ceramics confirmed the incomplete solubility and heterogeneity of the studied compositions. The average grain size in the BF-0.29BTm sample was ~5 μm. In the remaining compositions, the average grain size varied from 1 μm to 3 μm, which is characteristic of solid solutions in the region of morphotropic phase transition or solid solution decomposition. At the same time, these samples had high electrical conductivity, which prevented the production of a stable piezoactive state. A diffuse phase transition occurred from the FE to PE state in the temperature ranges of (650–750 K) for BF-(0.29–0.33)BT samples and (650–850 K) for BF-(0.29–0.33)BTm samples. For the BF-0.29BTm sample, the relaxation process was approximated using the Vogel–Fulcher relation, with parameters E_{act} = 0.29 eV and T_f = 595K. We assume that relaxor-like behavior and the smearing of the phase transition in the studied ceramics can be associated with the presence of non-interacting regions with increased content Bi or Ba, different modulation, crystal lattice symmetry, and chemical composition identified by X-ray analysis. Thus, it is shown that the mismatch of the solid solution composition with the rules of isomorphism does not allow obtaining ceramics with high piezocharacteristics, and even additional mechanical treatment can only improve the technological characteristics of ceramics to a limited extent. The maximum values of the piezoelectric modulus were observed in the BF-0.29BTm ceramics (~120 pC/N). The results show that mechanically activated BF-BT ceramics can be a promising material for future applications, but a comprehensive modification considering the isomorphism rule is required to seriously improve its performance.

Author Contributions: Conceptualization, N.A.B.; Methodology, S.I.D. and A.V.N.; Formal analysis, L.A.S.; Investigation, N.A.B. and E.S.E.; Resources, S.I.D.; Data curation, L.A.S.; Writing—original draft, N.A.B. and L.A.R.; Writing—review & editing, E.S.E.; Project administration, L.A.R.; Funding acquisition, L.A.R. All authors have read and agreed to the published version of the manuscript.

Funding: This research was funded by the Russian Science Foundation. Project No.24-22-00415.

Institutional Review Board Statement: Not applicable.

Informed Consent Statement: Not applicable.

Data Availability Statement: The data presented in this study are available on request from the corresponding author due to the Terms of Agreement with the funding organization.

Acknowledgments: Work was performed using the equipment of the Center for Collective Use "Electromagnetic, Electromechanical and Thermal Properties of Solids", Research Institute of Physics, Southern Federal University.

Conflicts of Interest: The authors declare no conflicts of interest.

References

1. Pyatakov, A.P.; Zvezdin, A.K. Magnetoelectric and multiferroic media. *Phys. Uspekhi* **2012**, *55*, 557–581. [CrossRef]
2. Zhai, J.; Xing, Z.; Dong, S.; Li, J.; Viehland, D. Detection of pico-Tesla magnetic fields using magneto-electric sensors at room temperature. *Appl. Phys. Lett.* **2006**, *88*, 062510. [CrossRef]
3. Tehrani, S.; Slaughter, J.M.; Deherrera, M.; Engel, B.N.; Rizzo, N.D.; John Salter, J.; Durlam, M.; Dave, R.W.; Janesky, J.; Butcher, B.; et al. Magnetoresistive random access memory using magnetic tunnel junctions. *Proc. IEEE* **2003**, *91*, 703–714. [CrossRef]
4. Borders, W.A.; Akima, H.; Fukami, S.; Moriya, S.; Kurihara, S.; Horio, Y.; Sato, S.; Ohno, H. Analogue spin-orbit torque device for artificial-neural-network-based associative memory operation. *Appl. Phys. Express* **2016**, *10*, 013007. [CrossRef]
5. Rojac, T.; Bencan, A.; Malic, B.; Tutuncu, G.; Jones, J.L.; Daniels, J.E.; Damjanovic, D. $BiFeO_3$ ceramics: Processing, Electrical, and Electromechanical Properties. *J. Am. Ceram. Soc.* **2014**, *97*, 1993–2011. [CrossRef]
6. Catalan, G.; Scott, J.F. Physics and applications of bismuth ferrite. *Adv. Mater.* **2009**, *21*, 2463–2485. [CrossRef]
7. Boldyrev, N.A.; Sitalo, E.I.; Shilkina, L.A.; Nazarenko, A.V.; Ushakov, A.D.; Shur, V.Y.; Reznichenko, L.A.; Glazunova, E.V. Structure and Relaxor Behavior of $(0.5-x)BiFeO_3$-$0.5PbFe_{0.5}Nb_{0.5}O_3$-$xPbTiO_3$ Ternary Ceramics. *Ceramics* **2023**, *6*, 1735–1748. [CrossRef]
8. Tian, Y.; Xue, F.; Fu, Q.; Zhou, L.; Wang, C.; Gou, H.; Zhang, M. Structural and physical properties of Ti-doped $BiFeO_3$ nanoceramics. *Ceram. Int.* **2018**, *44*, 4287–4291. [CrossRef]
9. Khasbulatov, S.V.; Pavelko, A.A.; Shilkina, L.A.; Reznichenko, L.A.; Gadjiev, G.G.; Bakmaev, A.G.; Magomedov, M.-R.M.; Omarov, Z.M.; Aleshin, V.A. Phase composition, microstructure, and thermophysical and dielectric properties of multiferroic $Bi_{1-x}Dy_xFeO_3$. *Thermophys. Aeromech.* **2016**, *23*, 445–450. [CrossRef]
10. Ahart, M.; Somayazulu, M.; Cohen, R.E.; Ganesh, P.; Dera, P.; Mao, H.; Hemley, R.J.; Ren, Y.; Liermann, P.; Wu, Z. Origin of morphotropic phase boundaries in ferroelectrics. *Nature* **2008**, *451*, 545–548. [CrossRef]
11. Jaffe, B.; Cook, W.R., Jr.; Jaffe, H. *Piezoelectric Ceramics*; Academic Press: New York, NY, USA, 1971.
12. Kumar, M.M.; Srinivas, A.; Suryanarayana, S.V. Structure property relations in $BiFeO_3/BaTiO_3$ solid solutions. *J. Appl. Phys.* **2000**, *87*, 855–862. [CrossRef]
13. Tong, K.; Zhou, C.; Wang, J.; Li, Q.; Yang, L.; Xu, J.; Zeng, W.; Chen, G.; Yuan, C.; Rao, G. Enhanced piezoelectricity and high-temperature sensitivity of Zn-modified BF-BT ceramics by in situ and ex situ measuring. *Ceram. Int.* **2017**, *43*, 3734–3740. [CrossRef]
14. Li, Q.; Wei, J.; Cheng, J.; Chen, J. High temperature dielectric, ferroelectric and piezoelectric properties of Mn-modified $BiFeO_3$-$BaTiO_3$ lead-free ceramics. *J. Mater. Sci.* **2017**, *52*, 229–237. [CrossRef]

15. Xun, B.-W.; Wang, N.; Zhang, B.-P.; Chen, X.-Y.; Zheng, Y.-Q.; Jin, W.-S.; Mao, R.; Liang, K. Enhanced piezoelectric properties of 0.7$BiFeO_3$-0.3$BaTiO_3$ lead-free piezoceramics with high Curie temperature by optimizing Bi self-compensation. *Ceram. Int.* **2019**, *45*, 24382–24391. [CrossRef]
16. Zhou, Q.; Zhou, C.; Yang, H.; Yuan, C.; Chen, G.; Cao, L.; Fan, Q. Piezoelectric and ferroelectric properties of Ga modified $BiFeO_3$–$BaTiO_3$ lead-free ceramics with high Curie temperature. *J. Mater. Sci. Mater. Electron.* **2014**, *25*, 196–201. [CrossRef]
17. Chen, J.; Cheng, J. High Electric-Induced Strain and Temperature-Dependent Piezoelectric Properties of 0.75BF–0.25BZT Lead-Free Ceramics. *J. Am. Ceram. Soc.* **2016**, *99*, 536–542. [CrossRef]
18. Xun, B.; Tang, Y.-C.; Chen, J.Y.; Zhang, B.P. Enhanced resistance in $Bi(Fe_{1-x}Sc_x)O_3$-0.3BaTiO3 lead-free piezoelectric ceramics: Facile analysis and reduction of oxygen vacancy. *J. Eur. Ceram. Soc.* **2019**, *39*, 4085–4095. [CrossRef]
19. Sehirlioglu, A.; Sayir, A.; Dynys, F. High temperature properties of $BiScO_3$-$PbTiO_3$ piezoelectric ceramics. *Appl. Phys.* **2009**, *106*, 014102. [CrossRef]
20. Zhou, C.; Yang, H.; Zhou, Q.; Chen, G.; Li, W.; Wang, H. Effects of Bi excess on the structure and electrical properties of high-temperature $BiFeO_3$–$BaTiO_3$ piezoelectric ceramics. *J. Mater. Sci. Mater. Electron.* **2013**, *24*, 1685–1689. [CrossRef]
21. Zhang, Y.-R.; Li, J.-F.; Zhang, B.-P.; Peng, C.-E. Piezoelectric and ferroelectric properties of Bi-compensated $(Bi_{1/2}Na_{1/2})TiO_3$–$(Bi_{1/2}K_{1/2})TiO_3$ lead-free piezoelectric ceramics. *J. Appl. Phys.* **2008**, *103*, 074109. [CrossRef]
22. Wang, X.X.; Tang, X.G.; Kwok, K.W.; Chan, H.L.W.; Choy, C.L. Effect of excess Bi_2O_3 on the electrical properties and microstructure of $(Bi_{1/2}Na_{1/2})TiO_3$ ceramics. *Appl. Phys. A Mater. Sci. Process.* **2005**, *80*, 1071–1075. [CrossRef]
23. Boldyrev, N.A.; Pavlenko, A.V.; Shilkina, L.A.; Nazarenko, A.V.; Reznichenko, L.A.; Miller, A.I. Structure, microstructure, and dielectric responses of $(1-x)BiFeO_3$-$xBaTiO_3$ solid solutions. *Bull. Russ. Acad. Sci. Phys.* **2016**, *80*, 1303–1305. [CrossRef]
24. Urusov, V.S. *Theory of Isomorphic Miscibility*; M.: Science: Moscow, Russia, 1977.
25. Mirkin, L.I. *Handbook of X-Ray Structural Analysis of Polycrystals*; State Publishing House of Physics and Mathematics Literature: Moscow, Russia, 1961.
26. Bokiy, G.B. *Introduction to Crystal Chemistry*; M.: MSU Publishing House: Moscow, Russia, 1954.
27. Pauling, L. *General Chemistry*; W.H. Freeman and company: San Francisco, CA, USA, 1970.
28. Shilkina, L.A.; Talanov, M.V.; Shevtsova, S.I.; Grin', P.G.; Kozakov, A.T.; Dudkina, S.I.; Nikol'skii, A.V.; Reznichenko, L.A. Isomorphism problems in lead-barium titanate. *J. Alloys Compd.* **2020**, *829*, 154589. [CrossRef]
29. Guinier, A. *X-Ray Diffraction of Crystals*; Publishing House of Physics and Mathematics Literature: Moscow, Russia, 1961.
30. Ustinov, A.I.; Olikhovskaya, L.A.; Shmyt'ko, I.M. X-ray diffraction in polydomain crystals modulated by transverse waves of atomic displacements. 2. Two-wave modulation of crystals. *Crystallogr. Rep.* **2000**, *45*, 374–379. [CrossRef]
31. West, A.R. *Solid State Chemistry and Its Applications*; John Wiley & Sones: Chichester, UK; New York, NY, USA; Brisbane, Australia; Toronto, ON, Canada; Singapore, 1991.

ceramics

MDPI

Article

Enhancement of Microwave Dielectric Properties in Mixed-Phase Ceramics Through CuB_2O_4 Doping: Achieving Ultra-Low Loss and High Dielectric Constant

Yuan-Bin Chen *, Siyi Xiong and Jie Peng *

School of Electronics and Electrical Engineering, Zhaoqing University, Zhaoqing 526061, China; xsy2471436141@163.com

* Correspondence: n2890103@outlook.com (Y.-B.C.); pengjie@zqu.edu.cn (J.P.)

Abstract: The microwave dielectric properties of $(1-x)Ca_{0.6}(La_{0.9}Y_{0.1})_{0.2667}TiO_3$-$x(Nd_{1/2}La_{1/2})(Mg_{(1+\delta)1/2}Ti_{1/2})O_3$ ((1−x)CYTO-xNLMTO) ceramics were investigated in this study. It was discovered that the addition of 1 wt% CuB_2O_4 effectively enhanced the densification and improved the microwave dielectric properties of (1−x)CYTO-xNLMTO, where $\delta = 0.02$. The new ceramic systems of (1−x)CYTO-xNLMTO could achieve ultra-low loss and a high dielectric constant. The novel ceramic systems comprising (1−x)CYTO-xNLMTO exhibited remarkably low loss and a significantly high dielectric constant.

Keywords: ultra-low dielectric loss; $(1-x)Ca_{0.6}(La_{0.9}Y_{0.1})_{0.2667}TiO_3$-$x(Nd_{1/2}La_{1/2})(Mg_{(1+\delta)1/2}Ti_{1/2})O_3$; CuB_2O_4 doping; high dielectric constant

Citation: Chen, Y.-B.; Xiong, S.; Peng, J. Enhancement of Microwave Dielectric Properties in Mixed-Phase Ceramics Through CuB_2O_4 Doping: Achieving Ultra-Low Loss and High Dielectric Constant. *Ceramics* **2024**, *7*, 1895–1904. https://doi.org/10.3390/ceramics7040119

Academic Editors: Dawei Wang and Fayaz Hussain

Received: 14 September 2024
Revised: 8 November 2024
Accepted: 4 December 2024
Published: 11 December 2024

1. Introduction

With the rapidly increasing demands for 5G communication and satellite broadcasting, there is an urgent need to advance dielectric resonator materials in microwave frequency applications. The dielectric ceramics utilized in these applications must exhibit exceptional dielectric properties, including high permittivity (ε_r) [1–3], low dielectric loss (high Q × f) [4–8], and a temperature coefficient for the resonant frequency (τ_f) that approaches zero.

Designing high-quality devices is of paramount importance, with a focus on achieving optimal efficiency and stability while incorporating compact dimensions. Therefore, it is imperative to utilize dielectric materials with exceptional qualities for microwave resonators [9–11]. Specifically, the incorporation of high-k (ε_r) dielectric materials plays a crucial role in fabricating RF passive devices such as antenna filters and oscillator resonators, enabling reductions in resonator size. Furthermore, it is essential for these microwave materials to exhibit low-loss characteristics, leading to extensive research efforts aimed at developing dielectric materials with minimal microwave loss. The presence of this characteristic is essential for achieving high-frequency selectivity and stability in both microwave transmitters and receiver components. Ensuring near-zero temperature coefficients (τ_f) becomes imperative for maintaining stability across various operating temperatures in microwave components [12,13].

One promising approach involves utilizing two distinct ceramics with different τ_f values to form a solid solution or mixed phases, thereby effectively achieving a zero temperature coefficient of resonant frequency. This method addresses the demands of microwave circuit designs where precise control over each dielectric property is indispensable [14,15]. The dielectric ceramics utilized in microwave applications should consistently exhibit exceptional dielectric performance, encompassing a high permittivity (ε_r), low dielectric loss, and a near-zero temperature coefficient (τ_f) [16–18]. Perovskites, characterized by the general ABO_3 structure, have been extensively investigated for their application in dielectric ceramics. They play a pivotal role in the fabrication of electronic components due to their high permittivity and reliability in practical scenarios [11,12].

Typically, the investigation of phase formation factors primarily involves considerations such as sintering temperature and thermal calculations [19,20], both of which have a significant impact on the manufacturing process cost [21]. To address this issue, extensive research efforts have been focused on minimizing sintering temperatures through various methods. Noteworthy approaches include the utilization of sol–gel processes, reduction in particle size, and liquid-phase sintering. The sol–gel process has emerged as a promising technique for reducing sintering temperatures by converting a solution or colloidal system into a gel, followed by drying and thermal treatment. This versatile route is known for producing high-purity materials with reduced processing temperatures. Another strategy involves decreasing the particle size of starting materials to enhance reactivity during sintering, thereby promoting densification at lower temperatures [22].

This approach not only contributes to energy efficiency but also mitigates the overall cost of the manufacturing process. Introducing a liquid phase by incorporating glass or other low-melting-point materials has proven effective in lowering the sintering temperature. Liquid-phase sintering facilitates the rearrangement of particles and enhances the diffusion process, leading to improved densification at reduced temperatures [23–26]. This method offers a practical solution for cost-effective manufacturing while maintaining the desired material properties.

After conducting a comprehensive literature review, it becomes evident that the sol–gel process is characterized by inherent complexity, while the alternative approach of reducing particle size in initial powders tends to increase fabrication costs [19,27–29]. Considering both cost efficiency and expedited fabrication as dual considerations, we have chosen to conduct experimental investigations utilizing liquid-phase sintering supplemented with the inclusion of other low-melting-point materials.

The compound $(Nd_{1/2}La_{1/2})(Mg_{1/2}Ti_{1/2})O_3$, referred to as NLMTO, possesses remarkable dielectric properties, characterized by a high dielectric constant ($\varepsilon_r \sim 27.6$), an impressive quality factor ($Q \times f$ value ~ 4550 GHz), and a negative temperature coefficient of resonant frequency (τ_f) of −48 ppm/°C. To counterbalance the positive τ_f value, the introduction of $Ca_{0.6}(La_{0.9}Y_{0.1})_{0.2667}TiO_3$, known as CYTO, with $\varepsilon_r = 111$, $Q \times f = 23{,}100$ GHz, and $\tau_f = +374.6$ ppm/°C [30,31], into the composite system results in the formation of a solid solution, denoted as (1−x)CYTO-xNLMTO. In this investigation, CuB_2O_4 was selected as a sintering aid to effectively lower the sintering temperature of the (1−x)CYTO-xNLMTO ceramics. Moreover, the combination of NLMTO and CYTO yielded a novel dielectric system characterized by a high dielectric constant, a remarkable $Q \times f$ value, and a temperature coefficient of resonant frequency close to zero [32–34].

2. Experimental Procedure

The compositions CYTO and NLMTO were synthesized through mixed oxide solid-state reactions employing high-purity chemical powders, namely $CaCO_3$, TiO_2, La_2O_3, Nd_2O_3, Y_2O_3, TiO_2, and MgO. To address the hygroscopic nature of magnesium oxide, it underwent a drying process at 600 °C for 2 h. Stoichiometric ratios were maintained during the mixing of compositions, and the resulting mixture was ground in distilled water for 10 h using a baling mill with agate balls. Additionally, 1 wt% of the sintering aid $Cu\text{-}B_2O_3$ (CuB_2O_4) was introduced to (1−x) CYTO-x NLMTO ceramics. Sintering of the ceramics was carried out at temperatures ranging from 1200 to 1350 °C for 4 h. Specifically, CYTO underwent calcination at 1000 °C for 2 h, while NLMTO was calcined at 1100 °C for 2 h. Polyvinyl alcohol (PVA 500; Showa, Tokyo, Japan), serving as a binder, was incorporated into the calcined powder. After achieving complete granulation uniformity, the mixture underwent screening with a 100-mesh screen and was pressed with a pressure of 200 MPa to form cylinders, each with a height of 0.5 cm and a diameter of 1.1 cm. Sintering temperatures (STs) for the cylinders were set at various temperatures for 4 h in an air environment, maintaining a consistent temperature rise and fall rate of 10 degrees per minute across all specimens. Crystallization-phase observations of the pre-phased powder and the mixed compositions were conducted using a Siemens D5000 X-Ray Powder diffractometer (XRD, Munich, Germany) with Cu-Kα radiation (at 40 kV and 40 mA). The XRD analysis of the sample was executed at a scan speed of 2 degrees/minute and a step size of 0.06 degrees. Utilizing Scanning Electron Microscopy (SEM, SEM4000Pro, ZEISS, Oberkochen, Germany) technology, the grain size, shape, and distribution of the sample were meticulously observed. Meanwhile, through the application of Energy Dispersive Spectroscopy (EDS) analysis, the types and concentrations of elements contained within the sample were determined. Bulk densities of the sintered pellets were determined using the Archimedes method. Measurements of ε_r and Q × f at radio frequencies were carried out using the Hakki–Coleman [35] dielectric resonator method, employing a network analyzer (Anritsu MS4647B, Morgan Hill, CA, USA). The same technique was utilized to measure the temperature coefficient of resonant frequency (τ_f). The test set was subjected to a thermostat-controlled environment in a temperature range from +25 °C to +80 °C. Additionally, the temperature coefficient of resonant frequency (τ_f) was calculated using Equation (1).

$$\tau_f = \frac{f_{t2} - f_{t1}}{(t_2 - t_1) f_{t1}} \times 10^6 \ (\text{ppm}/^\circ\text{C}) \tag{1}$$

In the above equation, f_{t2} and f_{t1} represent the resonant frequencies at 80 °C and 25 °C, respectively.

3. Results and Discussion

The XRD analyses were performed after incorporating 1 wt% of multiple sintering aids, specifically CuO-B_2O_3 (CuB_2O_4), into the (1−x)CYTO-xNLMTO ceramics. These ceramics were sintered at temperatures ranging from 1200 to 1325 °C for a duration of 4 h. Due to the close resemblance in lattice parameters between the two ceramics, a solid solution was formed. Figure 1 shows peaks that confirm the presence of NLMTO and CYTO as crystalline phases in the composition of 0.15CYTO-0.85NLMTO. All identified peaks were successfully indexed based on the perovskite structure, with no evidence of a second phase observed across all compositions examined in this experiment.

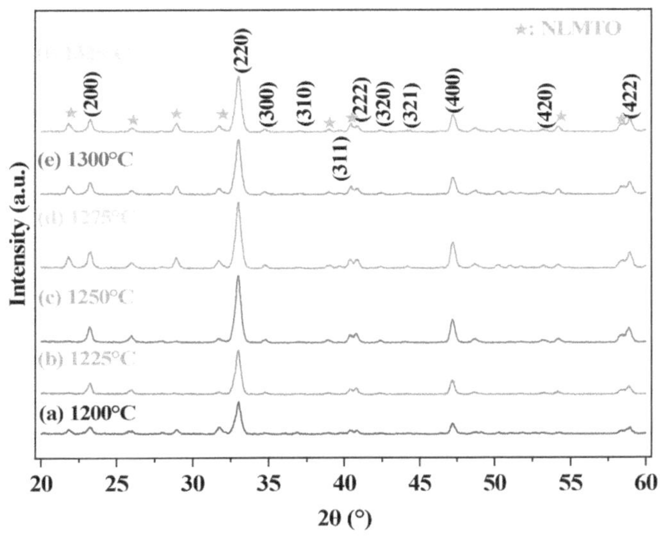

Figure 1. X-ray diffraction patterns of the $0.15Ca_{0.6}(La_{0.9}Y_{0.1})_{0.2667}TiO_3$-$0.85(Nd_{1/2}La_{1/2})(Mg_{(1+\delta)1/2}Ti_{1/2})O_3$ system with 1 wt% CuB_2O_4 additive sintered at various temperatures: (**a**) 1200 °C; (**b**) 1225 °C; (**c**) 1250 °C; (**d**) 1275 °C; (**e**) 1300 °C; (**f**) 1325 °C for 4 h.

The SEM results for the 0.15CYTO-0.85NLMTO ceramics with single additives in varying amounts, sintered at temperatures ranging from 1200 to 1325 °C for a duration of four hours, are presented in Figure 2. The density of the ceramics exhibits an increasing trend with higher sintering temperatures, reaching its peak value at 1250 °C due to enhanced particle mobility and optimal packing efficiency. However, further elevation of temperature can lead to detrimental effects such as excessive crystal growth, grain coarsening, and defect formation. The findings obtained from EDS, as depicted in Figure 2c with labeled spots A–D, have been meticulously documented in Table 1. These spots (A–D) correspond to grains exhibiting distinct classifications: spot A represents large cubic structures; spot B displays polygonal morphology; while spots C and D exhibit smaller rounded shapes. Importantly, the atomic percentages determined through analysis of spots A–D closely align with the expected ratio of 0.15CYTO-0.85NLMTO composition. These outcomes are consistent with those obtained from XRD analysis and affirm the absence of any discernible formation of secondary phases.

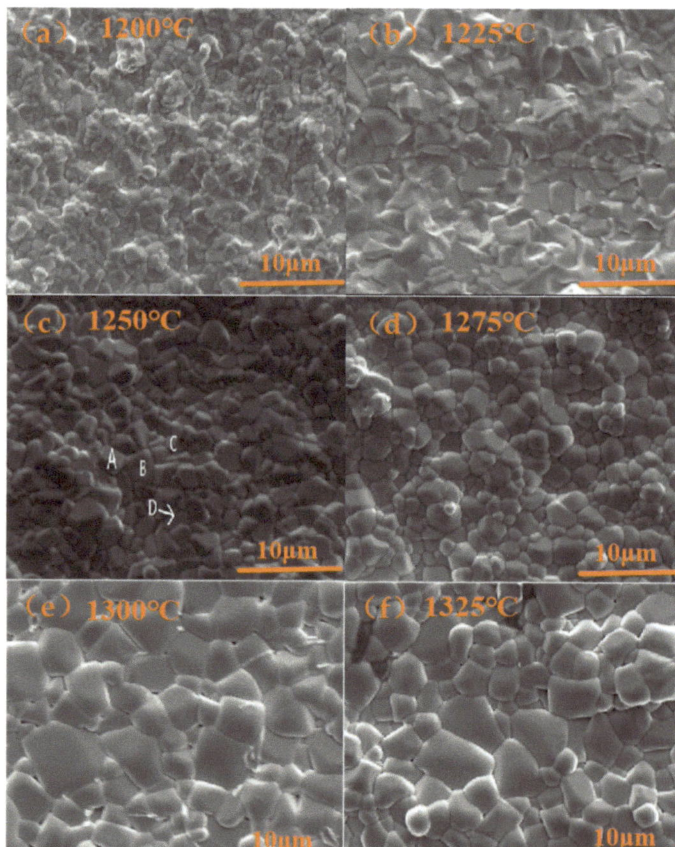

Figure 2. SEM photographs of $0.15Ca_{0.6}(La_{0.9}Y_{0.1})_{0.2667}TiO_3$-$0.85(Nd_{1/2}La_{1/2})(Mg_{(1+\delta)1/2}Ti_{1/2})O_3$ ceramics with 1 wt% CuB_2O_4 additive sintered at (**a**) 1200 °C; (**b**) 1225 °C; (**c**) 1250 °C; (**d**) 1275 °C; (**e**) 1300 °C; (**f**) 1325 °C.

Figure 3 displays the variation in relative density of the (1−x)CYTO-x NLMTO ceramics as a function of sintering temperature and the parameter x. Ceramics with different x values exhibit a similar trend in response to temperature changes. The relative density of the samples peaks at 1250 °C, highlighting a critical temperature where optimal densification is achieved across all compositions. When x = 0.85, the observed relative density reaches an impressive 95%, with an apparent density of 5.85 g/cm^3, indicating an exceptionally robust crystal structure. Such a high relative density suggests effective packing and arrangement of particles, which positively contributes to the integrity and quality of the crystal lattice. Additives of 1 wt% CuB_2O_4 were introduced into the sintering process. It serves a multi-faceted role in influencing the material's properties during the sintering phase. By lowering the material's melting point, it instigates the creation of a liquid phase amid sintering, which leads to the enhancement of particle mobility and the formation of larger crystal grains upon solidification [26]. Furthermore, CuB_2O_4 actively promotes the bonding of adjacent particles throughout the sintering process, which results in the development of a more interconnected and denser structure. Additionally, CuB_2O_4 facilitates the diffusion of atoms within the powder particles, improving atomic mobility. This augmented diffusion process further stimulates the rearrangement of atoms during sintering, ultimately fostering the growth of larger crystal grains. Moreover, the CuB_2O_4 additives may play a crucial role in reducing the surface energy of the particles. This reduction in surface energy facilitates the convergence of particles, making it easier for them to come together and form larger grains. The lower surface energy actively encourages particles to merge and coalesce during the sintering process.

Table 1. EDS analysis results for $(1-x)Ca_{0.6}(La_{0.9}Y_{0.1})_{0.2667}TiO_3$-$x(Nd_{1/2}La_{1/2})(Mg_{(1+\delta)1/2}Ti_{1/2})O_3$ ceramics with x = 0.85 sintered at 1250 °C.

Spot	Element Atomic (%)						
	Ca	La	Y	Ti	O	Nd	Mg
A	10.2	5.5	0.45	18.5	60	1.5	3.85
B	9.8	4.9	0.61	17.1	61.74	1.91	3.94
C	9.9	5.2	0.49	16.2	62.93	1.45	3.83
D	10.1	6	0.48	15.3	62.63	1.39	4.1

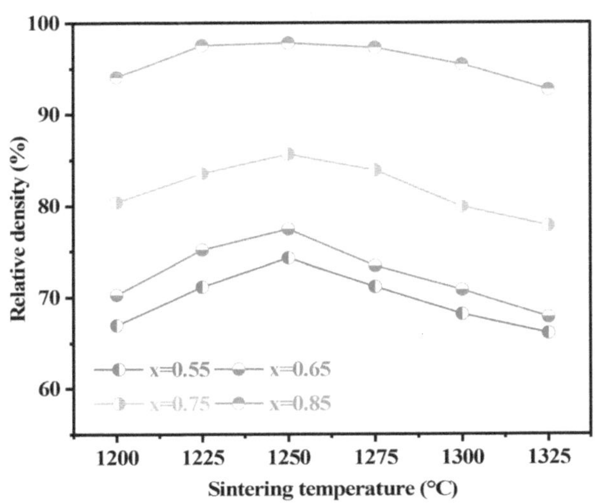

Figure 3. Relative density of $(1-x)Ca_{0.6}(La_{0.9}Y_{0.1})_{0.2667}TiO_3$-$x(Nd_{1/2}La_{1/2})(Mg_{(1+\delta)1/2}Ti_{1/2})O_3$ ceramics with 1 wt% CuB_2O_4 additive system sintered at various temperatures for 4 h.

The dielectric constants of the (1−x)CYTO-xNLMTO ceramics, with 1 wt% CuB_2O_4 additives, sintered at various temperatures for 4 h and considering different x values, are presented in Figure 4. ε_r and densities consistently exhibit parallel trends across the range of sintering temperatures. Ceramics with higher density show elevated ε_r due to reduced pore presence, thereby mitigating decay effects. Specifically, the ε_r value for the 0.15CYTO-0.85NLMTO ceramics reaches saturation around 35 when sintered at a temperature of 1250 °C for 4 h. The dielectric constants of CYTO and NLMTO were measured as 111 and 27.6, respectively [32,33]. Notably, there is a decreasing trend in ε_r with increasing x value attributed to lower dielectric constants observed in NLMTO ceramics; specifically, the dielectric constants decrease from 48.52 ± 0.13 to 35.01 ± 0.14 as x values increase from 0.55 to 0.85, respectively. This observed relationship aligns with the trends seen in densities and sintering temperatures, where a higher density corresponds to lower porosity levels. The interplay between ε_r values and sintering temperatures underscores the intricate connection between material composition, porosity, and dielectric properties exhibited by these ceramics.

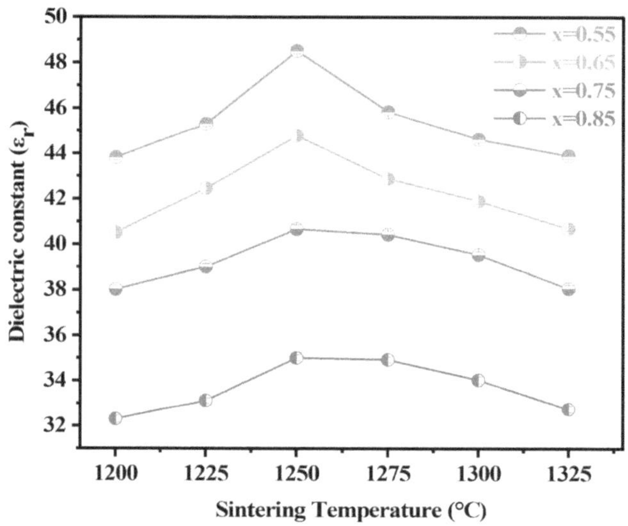

Figure 4. Dielectric constant of the (1-x)$Ca_{0.6}(La_{0.9}Y_{0.1})_{0.2667}TiO_3$-x$(Nd_{1/2}La_{1/2})(Mg_{(1+\delta)1/2}Ti_{1/2})O_3$ ceramics with 1 wt% CuB_2O_4 additive sintered at various temperatures for 4 h.

The correlation between sintering temperature and the value of Q × f for different x values is depicted in Figure 5. As expected, the Q × f value exhibits an increase with the augmentation of NLMTO content, owing to its significantly higher quality factor compared to CYTO. Several factors can influence microwave dielectric loss in dielectric resonators, including lattice vibrational modes, pores, and secondary phases. Generally, a larger grain size and smaller grain boundaries indicate reduced lattice imperfections leading to decreased dielectric loss. Evidence supports the claim that phase transformation primarily governs the dielectric loss in the (1−x)CYTO-xNLMTO ceramics system.

Figure 6 illustrates the temperature coefficients of τ_f for the (1−x)CYTO-xNLMTO ceramics. Specifically, a τ_f value of 12.06 ± 0.6 ppm/°C was achieved for 0.15CYTO-0.85NLMTO ceramics that contained 1 wt% CuB2O4 additive and were sintered at 1250 °C for 4 h. It is widely acknowledged that τ_f is influenced by various factors, including composition, additives, and the presence of secondary phases within the material. Generally, a higher CYTO content tends to result in a more positive τ_f value.

Figure 7 showcases the simulation results for the band-pass filter utilizing a substrate made of 0.15CYTO-0.85NLMTO ceramic material. The experimental implementation of the band-pass filter using the same ceramic substrate is presented in Figure 8. Detailed simulation results can be found in Table 2, whereas Table 3 outlines the performance characteristics of the filter employing the 0.15CYTO-0.85NLMTO ceramic substrate when compared to FR4 and Al_2O_3 substrates. The differences between Figure 7 (Simulation results) and Figure 8 (Measurement results) stem from the fact that the actual fabricated substrate has gaps on its surface, and the printed conductive metal material surface is not absolutely flat. Additionally, the simulation parameters do not account for the loss of surface metal, nor do they include the loss parameters of SMA connectors. Therefore, there will be slight differences between the simulation results and actual measurements. Notably, this design approach leverages an electromagnetic (EM) simulator like IE3D to accurately determine the physical dimensions of the filter.

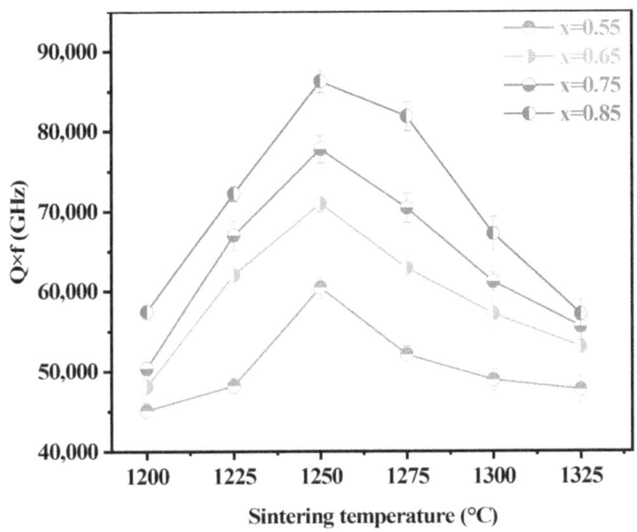

Figure 5. Q × f value of $(1-x)Ca_{0.6}(La_{0.9}Y_{0.1})_{0.2667}TiO_3$-$x(Nd_{1/2}La_{1/2})(Mg_{(1+\delta)1/2}Ti_{1/2})O_3$ ceramics with 1 wt% CuB_2O_4 additive system sintered at various temperatures for 4 h.

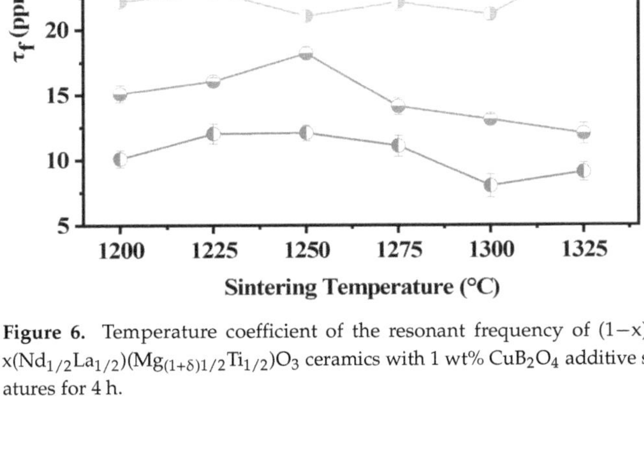

Figure 6. Temperature coefficient of the resonant frequency of $(1-x)Ca_{0.6}(La_{0.9}Y_{0.1})_{0.2667}TiO_3$-$x(Nd_{1/2}La_{1/2})(Mg_{(1+\delta)1/2}Ti_{1/2})O_3$ ceramics with 1 wt% CuB_2O_4 additive sintered at various temperatures for 4 h.

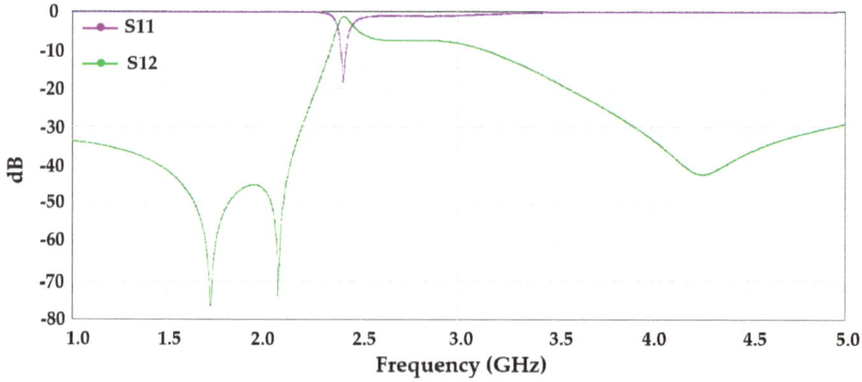

Figure 7. The simulation band-pass filter with $0.15Ca_{0.6}(La_{0.9}Y_{0.1})_{0.2667}TiO_3$–$0.85(Nd_{1/2}La_{1/2})(Mg_{(1+\delta)1/2}Ti_{1/2})O_3$ ceramic substrate (S11: Reflection Loss, S21:Insertion Loss).

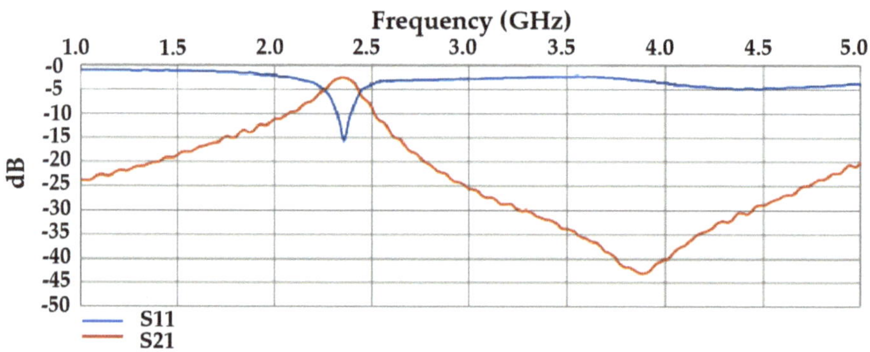

Figure 8. The measured band-pass filter with $0.15Ca_{0.6}(La_{0.9}Y_{0.1})_{0.2667}TiO_3$–$0.85(Nd_{1/2}La_{1/2})(Mg_{(1+\delta)1/2}Ti_{1/2})O_3$ ceramic substrate (S11: Reflection Loss, S21:Insertion Loss).

Table 2. Simulation results of the band-pass filters with various dielectrics.

Substrate	tanδ	ε_r	Size (mm^2)	Insertion Loss (dB)	Return Loss (dB)	f_o (GHz)
FR4	0.015	4.5	24 × 64	2.5	23.9	2.4
Al_2O_3	0.003	9.8	15 × 42	1.5	21	2.4
0.15CLYT-0.85LNMT	0.0001	35	7.51 × 9.5	0.7	21.1	2.4

Table 3. The performance filter with $(1-x)Ca_{0.6}(La_{0.9}Y_{0.1})_{0.2667}TiO_3$-$x(Nd_{1/2}La_{1/2})(Mg_{(1+\delta)1/2}Ti_{1/2})O_3$ ceramic substrate.

		f_0 (GHz)	S_{11} (dB)	S_{21} (dB)	Bandwidth (%)	Size (mm^2)
0.15CLYT-0.85LNMT	simulation	2.4	−19.44	−0.37	10.1%	7.51 × 9.5
	measured	2.4	−15.24	−1.69	10.3%	

4. Conclusions

The dielectric characteristics of (1−x)CLYO-xNLMTO ceramics incorporating CuB_2O_4 sintering aids were systematically investigated in this study. The resulting ceramics exhibited distinct perovskite structures. A non-linear variation in Q × f was observed, indicating an increasing trend for compositions with $x \geq 0.5$. Notably, the 0.15CLYO-0.85NLMTO composition, sintered at 1250 °C for 4 h, demonstrated exceptional dielectric properties including a dielectric constant of 35 and a Q × f value reaching 81,900 GHz. Moreover, it displayed a favorable τ_f at 15 ppm/°C. These findings highlight the promising potential of these ceramics for applications requiring high-performance dielectric materials.

Author Contributions: Methodology, J.P.; formal analysis, Y.-B.C.; investigation, S.X.; data curation, Y.-B.C.; project administration, J.P.; funding acquisition, S.X. All authors have read and agreed to the published version of the manuscript.

Funding: This work was financially supported by Guangdong Provincial Science and Technology Plan 2024A0505050039.This work was supported by Funding Project for Innovative Research Teams at Zhaoqing University (TD202417). This work was supported by the National Natural Science Foundation of China and Advanced Electronic Information Materials and Devices Research Center.

Institutional Review Board Statement: Not applicable.

Informed Consent Statement: Not applicable.

Data Availability Statement: The original contributions presented in this study are included in the article; further inquiries can be directed to the corresponding author(s).

Conflicts of Interest: The authors declare that they have no known competing financial interests or personal relationships that could have appeared to influence the work reported in this paper.

References

1. Zhang, P.; Hao, M.; Xiao, M. Microwave dielectric properties of $Li_3Mg_2NbO_6$-based ceramics with $(M_xW_{1-x})^{5+}$ (M = Li^+, Mg^{2+}, Al^{3+}, Ti^{4+}) substitutions at Nb^{5+} sites. *J. Alloys Compd.* **2021**, *853*, 157386. [CrossRef]
2. Nakagoshi, Y.; Sato, J.; Morimoto, M.; Suzuki, Y. Near-zero volume-shrinkage in reactive sintering of porous $MgTi_2O_5$ with pseudobrookite-type structure. *Ceram. Int.* **2016**, *42*, 9139–9144. [CrossRef]
3. Lin, S.H.; Chen, Y.B. Structure and characterization of B_2O_3 modiffed $yNd(Mg_{1/2}Ti_{1/2})O_3$-(1-y)$Ca_{0.8}Sr_{0.2}TiO_3$ ceramics with a near-zero temperature coefffcient at microwave frequency. *Ceram. Int.* **2017**, *43*, 2368–2371. [CrossRef]
4. Zaman, A.; Uddin, S.; Mehboob, N.; Ali, A. Structural investigation and improvement of microwave dielectric properties in $Ca(Hf_xTi_{1-x})O_3$ ceramics. *Phys. Scr.* **2021**, *96*, 025701. [CrossRef]
5. Wang, K.; Zhou, H.; Luan, X.; Hu, S.; Zhou, X.; He, S.; Wang, X.; Zhou, S.; Chen, X. $NaTaO_3$ microwave dielectric ceramic a with high relative permittivity and as an excellent compensator for the temperature coefffcient of resonant frequency. *Ceram. Int.* **2021**, *47*, 121–129. [CrossRef]
6. Pei, C.; Tan, J.; Li, Y.; Yao, G.; Jia, Y.; Ren, Z.; Liu, P.; Zhang, H. Effect of Sb-site nonstoichiometry on the structure and microwave dielectric properties of $Li_3Mg_2Sb_{1-x}O_6$ ceramics. *J. Adv. Ceram.* **2020**, *9*, 588–594. [CrossRef]
7. Chen, C.; Peng, Z.; Xie, L.; Bi, K.; Fu, X. Microwave dielectric properties of novel $(1-x)MgTiO_3-xCa_{0.5}Sr_{0.5}TiO_3$ ceramics. *J. Mater. Sci. Mater. Electron.* **2020**, *31*, 13696–13703. [CrossRef]
8. Li, J.; Zhang, C.; Liu, H.; Qiu, T.; Fan, C. Structure, morphology, and microwave dielectric properties of $SmAlO_3$ synthesized by stearic acid route. *J. Adv. Ceram.* **2020**, *9*, 558–566. [CrossRef]
9. Hameed, X.; Liu, Q.; Li, L.; Liu, M.Y.; Chen, X.M. Structure evolution and improved microwave dielectric characteristics in $CaTi_{1-x}(Al_{0.5}Nb_{0.5})xO_3$ ceramics. *J. Alloys Compd.* **2020**, *845*, 155435. [CrossRef]
10. Wu, M.; Zhang, Y.; Xiang, M. Synthesis, characterization and dielectric properties of a novel temperature stable $(1-x)CoTiNb_2O_{8-x}ZnNb_2O_6$ ceramic. *J. Adv. Ceram.* **2019**, *8*, 228–237. [CrossRef]
11. Shi, F.; Fu, G.; Xiao, E.; Li, J. Lattice vibrational characteristics and dielectric properties of pure phase $CaTiO_3$ ceramic. *J. Mater. Sci. Mater. Electron.* **2020**, *31*, 18070–18076. [CrossRef]
12. Yang, S.; Liang, B.; Liu, C.; Liu, J.; Fang, C.; Ai, Y. Microwave sintering and microwave dielectric properties of $(1-x)Ca_{0.61}La_{0.26}TiO_3-xNd(Mg_{0.5}Ti_{0.5})O_3$ ceramics. *Materials* **2021**, *14*, 438. [CrossRef] [PubMed]
13. Takahashi, H.; Baba, Y.; Ezaki, K.; Okamoto, Y.; Shibaba, K.; Kuroki, K.; Nakano, S. Dielectric characteristics of $(A^{1+}{}_{0.5}A^{3+}{}_{0.5})TiO_3$ ceramics at microwave frequencies. *Jpn. J. Appl. Phys.* **1991**, *30*, 2339–2342. [CrossRef]
14. Mercurio, J.P.; Manier, M.; Frit, B. Dielectric properties of ceramics within the $BaOLn_2O_3$-TiO_2 system. *Ferroelectrics* **1992**, *127*, 35–40. [CrossRef]
15. Ezaki, K.; Baba, Y.; Takahashi, H.; Shibata, K.; Nakano, S. Microwave dielectric properties of CaO-Li_2O-Ln_2O_3-TiO_2 ceramics. *Jpn. J. Appl. Phys.* **1993**, *1*, 4319–4322. [CrossRef]
16. Kato, J.; Kagata, H.; Nishimoto, K. Dielectric properties of lead alkaline-earth zirconate at microwave frequencies. *Jpn. J. Appl. Phys.* **1991**, *30*, 2343–2346. [CrossRef]
17. Yan, Y.; Li, Z.; Zhang, M.; Sun, P.; Xu, Y. Preparation and microwave dielectric properties of $Ca_{0.6}La_{0.8/3}(Sn_xTi_{1-x})O_3$ ceramics. *Ceram. Int.* **2017**, *43*, 8534–8537. [CrossRef]

18. Wise, P.L.; Reaney, I.M.; Lee, W.E.; Price, T.J.; Iddles, D.M.; Cannell, D.S. Structure–microwave property relations in $(Sr_xCa_{(1-x)})_{n+1}Ti_nO_{3n+1}$. *J. Eur. Ceram. Soc.* **2001**, *21*, 1723–1726. [CrossRef]
19. Takada, T.; Wang, S.F.; Yoshikawa, S.; Jang, S.-J.; Newnham, R.E. Effect of Glass Additions on BaO-TiO_2-WO_3 Microwave Ceramics. *J. Am. Ceram. Soc.* **1994**, *77*, 1909–1916. [CrossRef]
20. Takada, T.; Wang, S.F.; Yoshikawa, S.; Jang, S.-J.; Newnham, R.E. Effects of glass additions on (Zr, Sn) TiO_4 for microwave applications. *J. Am. Ceram. Soc.* **1994**, *77*, 2485–2488. [CrossRef]
21. Hirno, S.; Hayashi, T.; Hattori, A. Chemical Processing and Microwave Characteristics of $(Zr,Sn)TiO_4$ Microwave Dielectrics. *J. Am. Ceram. Soc.* **1991**, *74*, 1320–1324. [CrossRef]
22. Tolmer, V.; Desgardin, G. Low-temperature sintering and influence of the process on the dielectric properties of Ba $(Zn_{1/3}Ta_{2/3})O_3$. *J. Am. Ceram. Soc.* **1997**, *80*, 1981. [CrossRef]
23. Choy, J.H.; Han, Y.S. Microwave characteristics of BaO–TiO_2 ceramics prepared via a citrate route. *J. Am. Ceram. Soc.* **1995**, *78*, 1167–1172. [CrossRef]
24. Weng, M.H.; Liang, T.J.; Huang, C.L. Lowering of sintering temperature and microwave dielectric properties of $BaTi_4O_9$ ceramics prepared by the polymeric precursor method. *J. Eur. Ceram. Soc.* **2002**, *22*, 1693–1698. [CrossRef]
25. Yang, C.F. The microwave characteristics of glass-$BaTi_4O_9$ ceramics. *Jpn. J. Appl. Phys.* **1999**, *38*, 3576–3579. [CrossRef]
26. Yang, J.; Huang, M.; Wang, F.; Zhu, K.; Yan, K. Low temperature sintering, structure and electrical properties of $BiFeO_3$-$BaTiO_3$ based piezoelectric ceramics using CuO-B_2O_3 composite sintering aids. *Mater. Sci. Eng. B* **2023**, *298*, 116872. [CrossRef]
27. Kim, D.W.; Lee, D.G.; Hong, K.S. Low-temperature firing and microwave dielectric properties of $BaTi_4O_9$ with Zn–B–O glass system. *Mater. Res. Bull.* **2001**, *36*, 585–595. [CrossRef]
28. Lu, S.G.; Kwok, K.W.; Chan, H.L.; Choy, C.L. Structural and electrical properties of $BaTi_4O_9$ microwave ceramics incorporated with glass phase. *Mater. Sci. Eng.* **2003**, *B99*, 491–494. [CrossRef]
29. Cheng, C.M.; Yang, C.F.; Lo, S.H.; Tseng, T.Y. Sintering $BaTi_4O_9/Ba_2Ti_9O_{20}$-based ceramics by glass addition. *J. Eur. Ceram. Soc.* **2000**, *20*, 1061–1067. [CrossRef]
30. Seabra, M.P.; Salak, A.N.; Avdeev, M.; Ferreira, V.M. Structure and dielectric characterization of the $La(Mg_{1/2}Ti_{1/2})O_3$-$Nd(Mg_{1/2}Ti_{1/2})O_3$ system. *J. Phys. Condens. Matter* **2003**, *15*, 4229. [CrossRef]
31. Lin, S.-H.; Lin, Z.-Q.; Chen, C.-W. Microwave dielectric characterization of $Ca_{0.6}(La_{1-x}Y_x)_{0.2667}TiO_3$ perovskite ceramics with high positive temperature coefficient. *Ceram. Int.* **2021**, *47*, 16828–16832. [CrossRef]
32. Jhou, M.Z.; Jean, J.H. Low-fire processing of microwave $BaTi4O_9$ with BaO–ZnO–B_2O_3 glass. *J. Am. Ceram. Soc.* **2006**, *89*, 786–791. [CrossRef]
33. Lim, J.B.; Kim, M.H.; Kim, J.C.; Nahm, S.; Paik, J.H.; Kim, J.H. Effect of $BaCuB_2O_5$ additive on the sintering temperature and microwave dielectric properties of $BaTi_4O_9$ ceramics. *Jpn. J. Appl. Phys.* **2006**, *45*, 242–244. [CrossRef]
34. Choi, Y.J.; Park, J.H.; Ko, W.J.; Park, J.H.; Nahm, S.; Park, J.G. Low temperature sintering of $BaTi_4O_9$-based middle-k dielectric composition for LTCC applications. *J. Electroceramics* **2005**, *14*, 157–162. [CrossRef]
35. Hakki, B.W.; Coleman, P.D. A dielectric resonator method of measuring inductive capacities in the millimeter range. *IEEE Trans. Microw. Theory Tech.* **1960**, *8*, 402–410. [CrossRef]

ceramics

Article

Comparing the Efficacies of Electrospun ZnO and TiO_2 Nanofibrous Interlayers for Electron Transport in Perovskite Solar Cells

Abdullah Zafar [1], Waqar Iqbal [2,3], Shahzaib Khan [4], Aiyeshah Alhodaib [5] and Mahvish Fatima [5,*]

1 School of Materials Science and Textile Engineering, Wuhan Textile University, Wuhan 430000, China
2 School of Agricultural Engineering and Food Science, Shandong University of Technology, Zibo 255000, China
3 Shandong Research Center of Engineering and Technology for Clean Energy, Zibo 255000, China
4 School of Textile Science and Engineering, Wuhan Textile University, Wuhan 430000, China
5 Department of Physics, College of Science, Qassim University, Buraydah 51452, Saudi Arabia
* Correspondence: m.aziz@qu.edu.sa

Abstract: ZnO and TiO_2 are both well-known electron transport materials. Their comparison of performance is considered advantageous and novel. Therefore, a viable electrospinning route was considered for the development of highly polycrystalline TiO_2 and ZnO nanofibers as an electron transport material (ETM) for perovskite solar cells. The materials were well-characterized in terms of different analytical techniques. The X-ray diffraction detected polycrystalline structural properties corresponding to TiO_2 and ZnO. Morphological analysis by scanning electron microscopy revealed that the nanofibers are long, uniform, and polycrystalline, having a diameter in the nanometer range. Optoelectronic properties showed that TiO_2 and ZnO exhibit absorption values in the ultraviolet and visible ranges, and band gap values for TiO_2 and ZnO were 3.3 and 3.2 eV, respectively. TiO_2 bandgap and semiconductor nature were more compatible with Electron Transport Layer (ETL) compared to ZnO. Electrical studies revealed that TiO_2 nanofibers have enhanced values of conductivity and sheet carrier mobility compared to ZnO nanofibers. Therefore, higher photovoltaic conversion efficiency was achieved for TiO_2 nanofibers (10.4%) compared to ZnO (8.5%).

Keywords: electrospinning; photoactive materials; metal oxides; charge transportation; solar cells; X-ray diffraction; optoelectronic properties

Citation: Zafar, A.; Iqbal, W.; Khan, S.; Alhodaib, A.; Fatima, M. Comparing the Efficacies of Electrospun ZnO and TiO_2 Nanofibrous Interlayers for Electron Transport in Perovskite Solar Cells. *Ceramics* **2024**, 7, 1742–1757. https://doi.org/10.3390/ceramics7040111

Academic Editors: Dawei Wang and Fayaz Hussain

Received: 4 October 2024
Revised: 24 October 2024
Accepted: 7 November 2024
Published: 13 November 2024

1. Introduction

Since semiconductor nanomaterials have special features in nanoscale devices, there has been significant interest in researching them. In particular, research has evolved to focus on their use in solar energy-harvesting for photocatalysis, electricity, and environmental remediation [1,2]. TiO_2 and ZnO nanostructures, being significant semiconductors, have been thoroughly investigated in the field of photovoltaics, particularly in relation to perovskite solar cells (PSCs) [3]. They are preferred as electron transport materials (ETM) due to their large specific surface area and 1D shape. TiO_2 and ZnO offer a direct path for electron movement. While some researchers claim that ZnO in a wurtzite or hexagonal structure gives remarkable performance, others suggest that the crystalline characteristics of TiO_2 nanostructures are more compatible with ETM [4]. Both ZnO and TiO_2 nanofibers typically have a surface area of 30–40 $m^2\ g^{-1}$ and a broad distribution range of 80–300 nm. Due to their main features of higher electron mobility, mesoporous TiO_2- and ZnO-based ETMs for PSCs exhibit remarkable photovoltaic performance [5]. Higher electron or carrier mobility ideally supports higher current densities and improved open-circuit voltage. Furthermore, a lower recombination rate is required to prevent PSCs from internal losses.

One-dimensional (1D) TiO_2 nanomaterial production is documented in numerous investigations, with the goal of using their large fiber dimensions and high surface areas.

IMA electrospun TiO_2 nanofibers were reported by Mohammed et al. to achieve an acceptable performance (5.41%) in PSCs [4,6]. The primary parameters for adjusting their optoelectronic characteristics are the electrode type, surface area, and photoabsorption contact layer. Functional 1D TiO_2 nanostructures, such as nanowires and nanofibers, were investigated by XL Wang et al. for use in perovskite solar cells. TiO_2 ETM was used in PSCs to produce improved efficiencies of up to 5.9% by customizing their sizes, functional groups, and band gap [7]. TiO_2 nanofibers produced by sol-gel electrospinning were reported to have better contact qualities and larger surface areas. Conversely, a faster rate of photoactivity is induced by increased current density and enhanced electron mobility in the oriented direction, which also results in one-dimensional (1D) ZnO being recorded in perovskite solar cells [8]. Electrospun ZnO nanofibers were utilized as ETM to achieve a sufficient band gap (3.3 eV) and improved electron mobility. ZnO nanofibers have shown compatibility with perovskite solar cells and photoanode materials to achieve an efficiency of up to 8.7% [9].

Fewer energy trap states exist between the valence and conduction bands in ZnO nanofibers, which lowers losses [10]. ZnO surface modification and nanofiber diameter optimization lead to a current density of 14.51 mA/cm^2 in highly efficient perovskite solar cells (PSCs) [11]. When compared to nanoparticles, electrospun TiO_2 and ZnO nanofibers with their 1D structure provide less quantum confinement and greater photovoltaic performance [12]. A large body of research has been done recently on the production of ZnO and TiO_2 nanofibers via electrospinning under various processing conditions [13]. With an ideal dopant concentration of 3%, indium-doped ZnO demonstrated enhanced structural characteristics, adjusted the electronic band structure, and, as a result, demonstrated an improvement in photocatalytic activity [1]. Using an electrospinning approach, highly porous scaffold nanofibers of zinc tin oxide (Zn_2SnO_4) have been synthesized and successfully applied to methyl ammonium lead halide ($CH_3NH_3PbI_3$) perovskite-sensitized solid-state solar cells. The optimized perovskite solar cell devices that were fabricated demonstrated a power conversion efficiency (PCE) of 7.38% under AM 1.5 G sunlight (100 mW cm^{-2}), with an open-circuit voltage (VOC) of 0.986 V, current density (JSC) of 12.68 mA cm^{-2}, and fill factor (FF) of 0.59. This is in comparison to perovskite solar cells based on Zn_2SnO_4 nanoparticles (η = 2.52%) [14]. By careful addition of Mg nanofibers into the ZnO substrate, charge extraction was promoted without sacrificing the charge transport efficiency, leading to enhanced photovoltaic efficiency and an optimized band gap of 3.2 eV [15]. However, no paper has yet comparred the photovoltaic capabilities of ZnO and TiO_2 nanofibers synthesized under similar processing conditions [16].

In this study, electrospun pristine TiO_2 and ZnO nanofibers were fabricated under the same set of circumstances and conditions, including gel flow, syringe gauge, drying conditions, discharge rate, and direct and negative voltages [17]. The materials for the photoanodes were applied to the FTO under the same coating cycles and spin rates. The ability of these potential photoanodes to transport electrons in perovskite solar cells was examined. Through the use of Raman spectroscopy, X-ray diffraction, and scanning electron microscopy, the morphological and structural characteristics of both nanofibers were examined. To determine the fundamental mechanism of improved performance, TiO_2 and ZnO nanofibers' optical, chemical, and electrical characteristics were examined. The suggested 1D TiO_2 and ZnO nanofibers, which were created exactly the same way, work admirably as ETMs for perovskite solar cells.

2. Materials and Methodology

For the manufacture of titanium and zinc oxide nanofiber sol, Sigma-Aldrich, Austin, TX, USA provided absolute ethanol (C_2H_5OH, Merck, Beijing, China) and acetic acid (CH_3CO_2H, 99.7%, Sigma-Aldrich, Austin, TX, USA). Polyvinyl pyrrolidone (PVP, Mw ~29,000 Sigma-Aldrich, Austin, TX, USA), titanium tetra-isopropoxide (TTIP Assay 97%, Sigma-Aldrich, Austin, TX, USA, and zinc acetate di-hydrate ($Zn(CH_3COO)_2$ H_2O: Lab 97%, Aldrich, USA) were used as precursors for the electrospinning process. TiO_2 and

ZnO nanofiber collection was conducted using fluorine-doped tin oxide (FTO) glass. It was decided to employ electrospinning, since it is a popular, quick, and easily controllable technique for creating nanofibers [17]. MaBr (solution parameters such as viscosity, concentration, and surface tension), processing parameters such as applied voltage, spinning distance, and nozzle radius, and environmental parameters such as temperature, humidity, and atmosphere pressure, are the optimized conditions for producing nanofibers in the fabrication of solar cells [12].

2.1. ZnO and TiO_2 Nanofiber Synthesis

Figure 1 illustrates the process of electrospinning ZnO nanofiber synthesis. One hour was spent stirring 40 mL of ethanol with 5 g of zinc acetate added. At 60 °C, 3 mL of acetic acid and 2.5 g of PVP were added while stirring and allowed to stir further for 5 h. In order to ensure optimal gelation, this was left to age for two days. A 0.45-mm needle tip diameter syringe, diffusion pumps, collector and intake stages, and a DC power source make up the electrospinning setup. The entire electrospinning apparatus for the manufacture of nanofibers is shown in Figure 1. Using a needle (21 gauge) and a pumping velocity of 4000 rpm, these mixes were spun in an electric field of up to 20 kV. We used a direct voltage of 18 kV and a negative voltage of 4 kV at the time of discharge. The solutions moved at a rate of 5 mL/min, and there was a gap of 16 to 20 cm between the needle and the substrates. Perovskite and hole transport material were deposited following the creation of nanofiber Electron Transport Layer (ETL), as detailed in the section on the construction of solar cells.

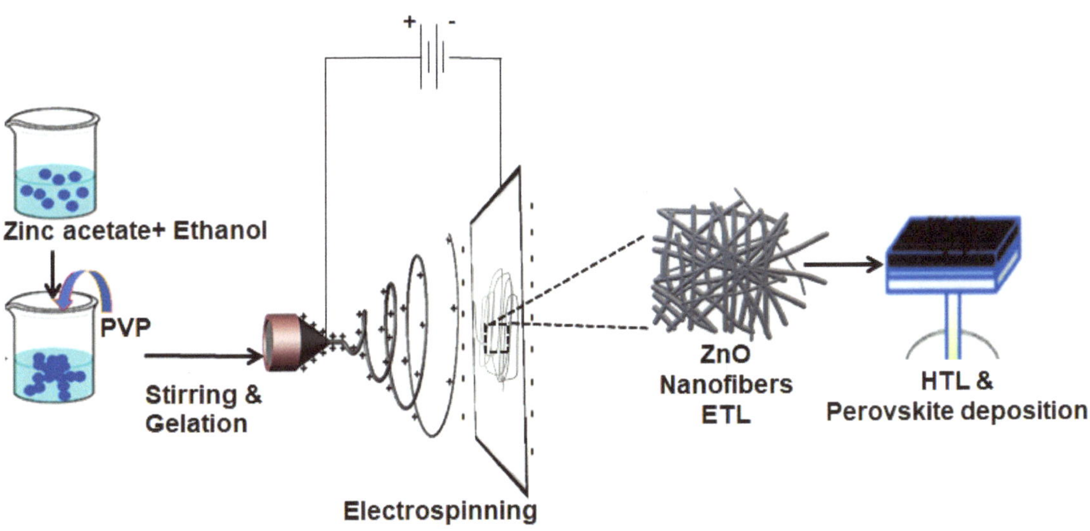

Figure 1. Electrospinning process and PSC fabrication for the ZnO nanofibers.

For the TiO_2 nanofiber sol, 48 mL of ethanol was gradually mixed with 12 mL of TTIP. After two hours of continuous stirring at 400 rpm, 2.5 g of PVP was added in stages and stirred for 5 h at 60 °C. In order to obtain the correct gelation, this was aged for two days, as Figure 2 illustrates.

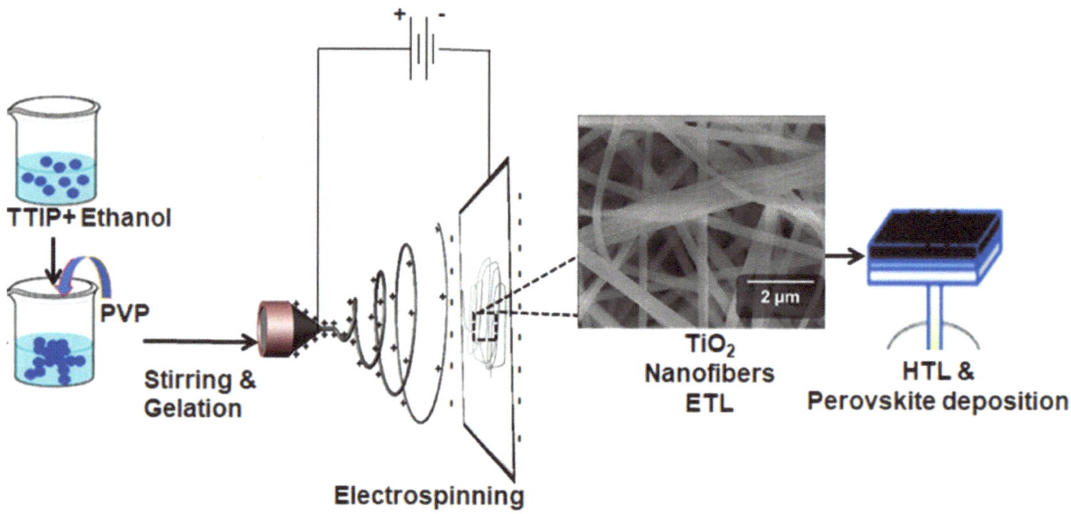

Figure 2. Electrospinning process and PSC fabrication for the TiO_2 nanofibers.

2.2. Solar Cell Fabrication

For solar cell fabrication, firstly, the ETL layer was deposited, followed by a perovskite composition of methyl ammonium bromide (MABr, 0.2 mM). Dimethyl formamide (DMF, 0.8 mL, Sigma Aldrich, Austin, TX, USA) was mixed with dimethyl sulfoxide (DMSO, 0.2 mL, Sigma-Aldrich, Austin, TX, USA) in a 20 µL solution containing $PbBr_2$ and PbI_2 (0.2 mM, Sigma Aldrich, Austin, TX, USA), (1.1 mM, Sigma Aldrich, Austin, TX, USA), with formamidinium iodide (FAI, 1 mM; Sigma Aldrich, Austin, TX, USA) and methyl ammonium bromide (MABr, 0.2 mM, Sigma Aldrich, Austin, TX, USA). In order to deposit this, ZnO and TiO_2 nanofiber ETLs were coated using a spin-coating process for seven seconds at 4500 rpm. Afterward, 90 µL of anhydrous chlorobenzene was added to help the evenly deposited film crystallize. The film was left to dry for fifteen minutes at 100 °C on a hot plate. Next, the HTL layer was coated using a mixture of spiro-OMeTAD (100 mg) and chlorobenzene solution (1.094 mL, Nature chemicals, Islamabad, Pakistan). On the absorber layer, this layer was once again spin-coated for 20 s at 2000 rpm. A glove box filled with N_2 was used for the entire fabrication procedure. Our device's mask aperture measured 15 mm by 15 mm, and its device area measured 25 mm by 25 mm. Finally, a thermal evaporator was utilized to deposit 80 nm gold back electrodes while maintaining a vacuum of 10^{-7} Torr.

2.3. Characterizations

Using an X-ray detector, Sigma Aldrich, Austin, TX, USA with a radiation wavelength of 1.5418 Å and a Bragg–Brentano configuration, the D8 Advance (Bruker Advanced, Berlin, Germany) was used to estimate the structural parameters and assess the impact of the TiO_2 and ZnO structures. The samples were scanned with a step size of 0.05°/5 s, with a range of 2θ = 10° to 80°. To investigate the XRD analysis, DIFFRAC Plus EVA Version 5.0 software was utilized. A 514 nm excitation laser was used to corroborate the phase composition analysis of the nanofibers using RENISHAW Invia 2000 Raman spectroscopy, Bruker Advanced, Berlin, Germany. Using a JEOL JSM6490A scanning electron microscope (SEM) (Astoon, Tokyo, Japan), the morphology of the nanofiber samples was examined. Prior to sample analysis, all samples were coated with gold for one minute in order to obtain clear images free of charge accumulation. The Model UH4150AD UV-Vis-NIR, Bruker Advanced, Berlin, Germany was used to conduct UV-vis spectroscopy (300 nm–800 nm). Attenuated total reflection mode on the Cary 630 (Agilent Technologies, Austin, TX, USA) was utilized to conduct Fourier transform infrared spectrophotometry (FTIR). Electrical tests, conductivity, sheet resistance, and sheet carrier mobility were carried out using a Swin system, Austin, TX, USA with a 5300 G magnetic field at a temperature of 300 K [10]. Under a sunlight simulator (Newport 94043A, Austin, TX, USA, Bruker Advanced, Berlin, Germany), AM 1.5 simulated light (100 mW cm^{-2}) was used to test the properties of solar cells and current density-voltage (J-V). The scan was conducted in reverse, with a speed of 10 mV s^{-1} and a dwell duration of approximately 1 s for each 10 mV step. Using a Xenon lamp, Austin, TX, USA at 1.5 AM sun illumination and a Keithley 2400 source meter, Berlin, Germany calibrated to a silicon reference cell, initial JV parameters were determined. Prior to measurements, a silicon photodiode was used for calibration.

3. Results and Discussion

To investigate the crystal structure, crystallographic phases, and peaks at diffracted sites, XRD analysis was carried out. The fundamental composition of materials and their crystal planes were assessed by this analysis. Following calcination, the XRD spectra of ZnO and TiO_2 nanofibers were measured between 12.5° and 80° (Figure 3a). Zinc oxide nanofibers exhibit the wurtzite phase, while Titania exhibits the anatase phase, upon annealing at 450 °C and the removal of the binder. This provides the crystal's unique crystalline properties and crystallinity. An XRD examination of TiO2 nanofibers shows that peaks at 2θ of 25.5°, 38.52°, 48.49°, 54.09°, 55.62°, and 63.80°, correspond to (101), (004), (200), (105), (211), and (204) planes, respectively. Peaks were identified for a sintered sample at ambient temperature using JCPDS Nos. 21–1272 and 29–1363 [18]. There were no impurity peaks found in the TiO_2 nanofibers, and the peaks of (101) at (2θ = 25.5°) were substantially stronger than the other peaks [19]. The relative intensity and its maxim were aligned with the anatase TiO_2 general perturbation pattern, which shows a preferential growth of the (101) plane. Five primary peaks were seen in the ZnO spectra; these peaks corresponded to crystallographic planes (100), (002), (101), (102), (110), (103), (200), (112), and (201), which generally represent the hexagonal wurtzite structure of ZnO [20]. This observation, which is associated with the structure of polycrystalline ZnO nanofibers, correlates well with ZnO's JCPDS (36–1453) and shows no impurity peaks. (Figure 3a). These nanofibers' distinct, sharp peaks reflected how annealing produces well-oriented nanofibers with specified planes and an appropriate dislocation density. The formation of polycrystalline nanofibers is significantly impacted by the annealing temperature, since this is the stage where the sintering of nanodomains and removal of the binder take place [21].

Raman spectra are used to support the crystallographic analysis obtained via XRD. The Raman spectra of both fibers were obtained in the 250–3250 cm^{-1} range, which is consistent with the XRD results. Photoactive materials can be studied using Raman spectroscopy, which uses the inelastic scattering of light by phonon quanta with the energy of lattice vibrations. The perovskite material absorbs this energy and uses it to create holes and electrons. Raman scattering of the production or annihilation of a phonon affects the mobility and generation of electrons and holes. Due to their higher resonance, the polar bonds of O–O in the Titania lattice, such as the Eg, A1g, and B1g modes, have much greater intensities. The phonon mode of TiO_2, which peaks at 527 cm^{-1}, becomes active due to tensile stresses that impact the solar cell devices' light conversion efficiency. ZnO's Raman spectra showed an A_{1T} phonon mode at 436.5 cm^{-1}, whereas the phonon frequency of T_{2H} indicated a strong mode at 638.5 cm^{-1}. Phonons are scattered and absorbed due to the wurtzite crystal structure's limited phonon frequency. Regarding anatase TiO_2 fibers, three peaks were observed: the Eg, A1g, and B1g modes were located at 648, 405, and 527 cm^{-1}, in that order [22]. These have a structure that is quite similar to anatase Titania. ZnO's Raman spectra showed an A_{1T} phonon mode at 436.5 cm^{-1}, whereas the phonon frequency of T_{2H} indicated a strong mode at 638.5 cm^{-1}. A restricted phonon frequency like this is only seen in the crystal structure of wurtzite. Yang et al. have previously noted the TO phonon frequency redshift in ZnO nanofibers relative to TiO_2 [23]. The second-order phonon mechanism at 1538.5 cm^{-1} is also highlighted, as in Figure 3b.

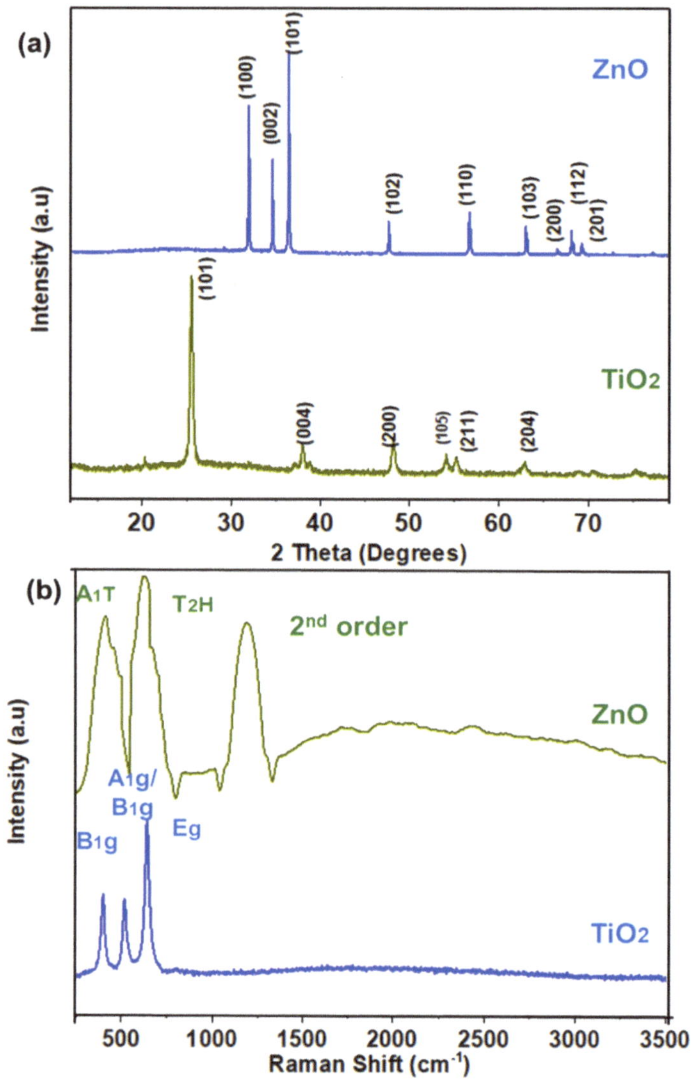

Figure 3. (**a**) XRD spectra for TiO_2 and ZnO nanofibers; (**b**) Raman spectra for TiO_2 and ZnO nanofibers.

Using images from scanning electron microscopy, the surface morphology of the porous network made of nanofibers was examined. Attaining a high interface for electron conduction requires semiporous, fine, elongated, and homogeneous nanofibers. Better light absorption and consequent electron extraction from the perovskite material's conduction band are made possible by these nanofibers' properties. The ideal electrospinning setup settings and a controlled production rate result in the formation of fibers with the ideal shape. The collector at FTO was positioned in accordance with the characteristics of the nanofibers in order to produce a dense and ideal collection of nanofibers. Nanofiber formation is extremely dependent on sol-gel viscosity; at very low concentrations, nanofibers do not form [24]. Taking into consideration the previously mentioned characteristics, we considerably fine-tuned the processing settings to produce fibers suitable for ETL.

Figure 4a–c displays SEM images of the synthesized ZnO and TiO_2 nanofibers following the annealing process. As anticipated, the spontaneous fiber extrusion via the electrospinning jet resulted in the nanofibers being localized at different orientations. There was no aggregation, and every fiber was uniformly produced and extended. Zinc oxide nanofiber films and nanoporous Titania form on the substrate as a result of this. The deposited ZnO and TiO_2 sheet has a large number of nanofibers with diameters of less than 0.5 μm, providing a suitably high interface for charge extraction from the absorber layer.

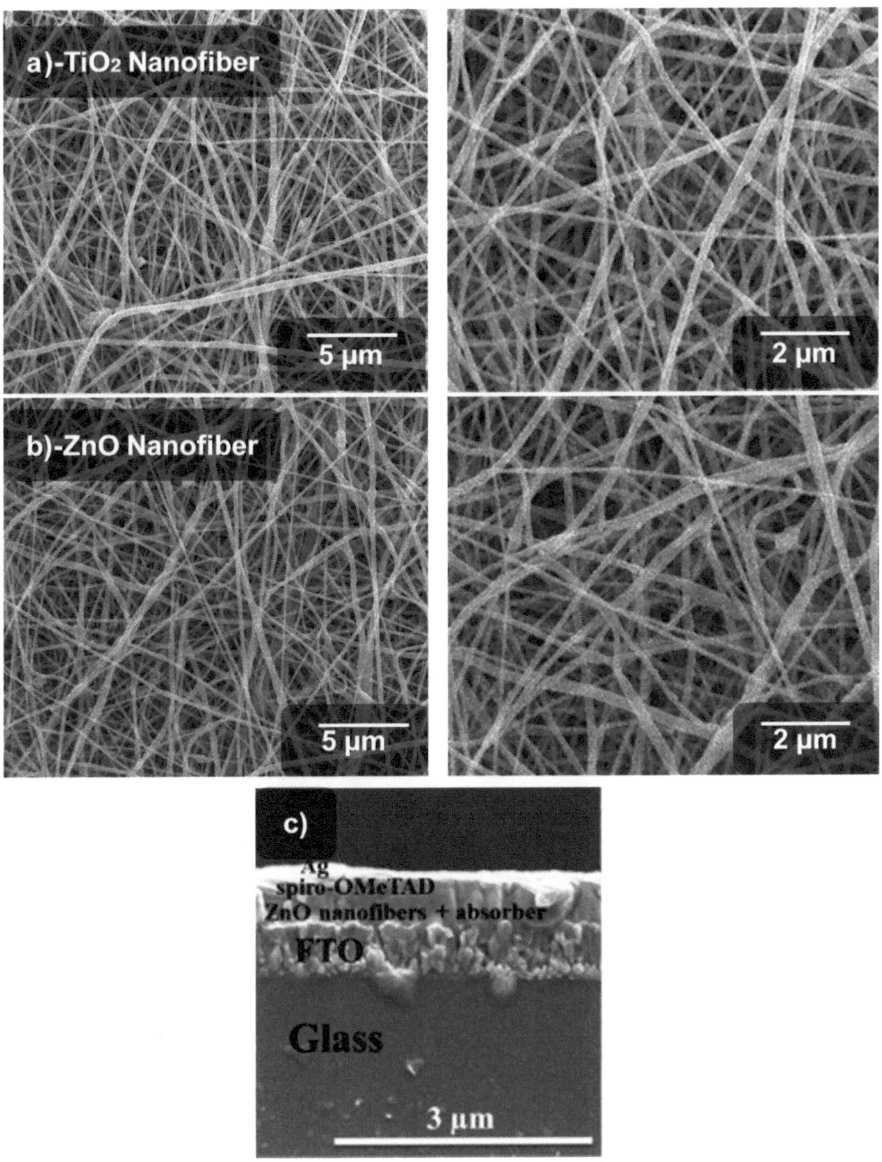

Figure 4. Scanning electron morphology: (**a**) TiO_2 nanofiber, (**b**) ZnO nanofiber, (**c**) cross-section of the fabricated ZnO perovskite.

When compared to ZnO, TiO_2 nanofibers are more homogeneous, elongated, and dense. The maximal conduction of photogenerated electrons from the valence band to the conduction band is achieved by the extremely dense structure of TiO_2 nanofibers. Additionally, the TiO_2 fibrous structure facilitates a more aligned electron path that results in the maximum current densities. Additionally, it causes the electrons from the perovskite layer to diffuse efficiently. The fibers' shape was consistent across the FTO-glass substrate, twisted in certain places, and uniformly porous. The side view of the entire PSC device is displayed in Figure 4c, which prominently displays the glass/FTO/ZnO nanofiber-based ETL/MAPbI3 (absorber layer)/spiro-OMeTAD/Ag, respectively. Figure 4c shows that the perovskite has been properly infused into the ETL nanofibers. Using an optical profilometer, the thickness of the electron transport layer (ETL), including ZnO and TiO_2 nanofibers (440 nm), was determined.

As seen in Figure 5a,b, optical characterization of photoanodes based on TiO_2 and ZnO nanofibers was performed to verify the absorption range, edge, and the band gap analysis for both samples. Photoexcitation is primarily due to semiconductor materials' absorption of ultraviolet (UV); it can also be due to absorption of near-UV visible light. Only a specific percentage of the light—that is, light that is greater than the bandgap—is absorbed. The exciton, or electron-hole pair, that results from photoexcitation initiates photocatalysis. Using a UV-visible spectrophotometer, the radiation absorption of nanofibers was measured in the 220–800 cm^{-1} range. Both ZnO and TiO_2 exhibited high energy spectrum absorption, as seen in the Figure 5, while low energy photons continued to flow through, which is necessary for the perovskite absorber layer [25,26]. A transparent ETL can be formed since the spectra show that a significant portion of absorption occurs in the ultraviolet (UV) and near-UV regions.

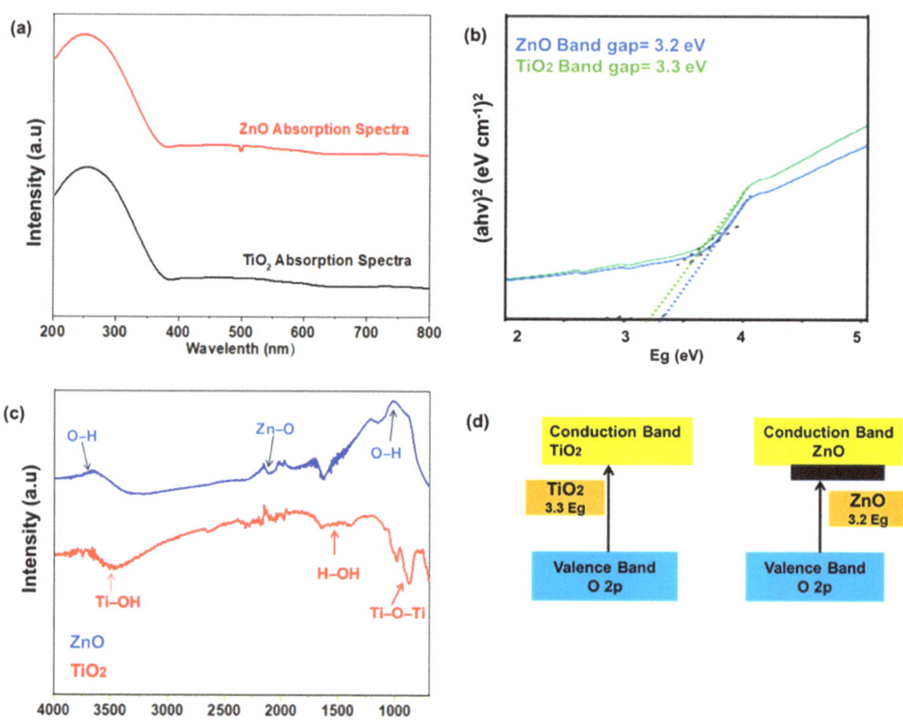

Figure 5. (**a**,**b**) UV vis absorption characterizations for photoanodes, (**c**) FTIR spectra for photoanodes, (**d**) bandgap mechanism for photoanodes based on TiO_2 and ZnO.

The sample bandgap was determined using the Tauc plot, which indicates the lowest energy needed for photoexcitation. Equation (1) was used to convert wavelength into energy and determine the bandgap of each sample:

$$E = hc/\lambda \quad (1)$$

In this case, "E" stands for energy, "h" for Planck's constant, "ε" for wavelength, and "c" for the speed of light. The indirect bandgap of Titania is around 3.3 eV. The exciton band gap (Eg = 3.2 eV) is visible in the ZnO absorption spectra. The photoexcited electrons are transferred from the perovskite to the FTO glass by these appropriate bandgaps, which also supply the necessary Fermi energy level. Improved charge transport reduces energy loss, allowing the solar cell to have higher current densities and a higher open-circuit voltage. In comparison to Titania, ZnO had a marginally smaller band gap; however, Titania gives a more favorable band gap for fermi electron movement from the valence band to the conduction band.

As seen in Figure 5c, Fourier transform infrared spectroscopy analysis was used to find each group of function in the 720–4000 cm^{-1} range. Both nanofibers show characteristics peaks around 3200–3600 cm^{-1} associated that is associated with the OH bond's vibration, while bending of OH appears at 1630 cm^{-1} [27]. On the surface of nanoparticles, dangling bonds give rise to these hydroxyl groups. The extended stay time of excitons and their absorption are caused by these hanging bonds, creating defect states. It is therefore anticipated that the functional surface will have increased photocatalytic activity. The hydrolysis reaction, which remains pendent at the surface, primarily forms Ti–OH bonds. Compared to Titania, ZnO exhibited somewhat more hydroxyl functional groups. A surface with this kind of functionality offers a more robust interface for the perovskite layer to adhere to [28]. However, the increased moisture absorption of such hydrophilic groups decreases the life of solar cells [29–31].

Characterizing the values of sheet resistance, conductivity, resistivity, and sheet carrier mobility was done using the Hall Effect swing system. Four probing approaches were employed to determine the conductivity characteristics of the nanofiber films. The resistivity, sheet resistance, carrier mobility, and conductivity of the two samples are contrasted in Table 1. The resistivity of pure TiO_2 is 7.6×10^3 Ω cm, but that of ZnO nanofibers is 9.8×10^3 Ω cm. Due to its slower electron transport compared to TiO_2, zinc oxide exhibits higher resistance. TiO_2 offers an even and straight channel for the movement of electrons. It decreases resistivity by accelerating the diffusion of photoelectrons. The conductivity of the pure ZnO sample is 1.12×10^{-4} 1/Ohm-cm, whereas pure TiO_2 has a conductivity of 1.28×10^{-4} 1/Ohm-cm.

Table 1. Conductivity, resistivity, sheet resistance, and sheet carrier mobility for both samples.

Sr	Conductivity 1/Ohm-cm	Resistivity Ohm-cm	Sheet Resistance Ohm-sq	Sheet Carrier Mobility
TiO_2	1.28×10^{-4}	7.6×10^3	4.3×10^8	1.52×10^{-2}
ZnO	1.12×10^{-4}	9.8×10^3	5.3×10^8	1.4×10^{-2}

TiO_2 is a more conductive substance than ZnO because conductivity is reciprocal to resistance. TiO_2 will help electrons move more quickly. As a result, the sheet resistance of the pure ZnO sample is 5.3×10^8 Ohm-sq. The sheet resistance of pure TiO_2 is 4.3×10^8 Ohm-sq. ZnO's photoactivity will be reduced since the electrons in this compound will encounter greater sheet resistance along their route. A carrier mobility of 1.4×10^{-2} cm^2/Vs was obtained from the pure ZnO sample. The carrier mobility of pure TiO_2 is 1.52×10^{-2} cm^2/Vs. This suggests that TiO_2 has slightly higher absorption of light and a tuned electrical band structure. These characteristics will help improve TiO_2's electron conduction and reduce resistance losses. Improved electron conduction leads to a higher current density, which raises solar cell efficiency. In general, TiO_2 exhibits more favorable electrical characteristics than ZnO, including effective electron transport [32].

Electrochemical impedance spectrum (EIS) measurements were characterized to understand the charge transfer behavior of the solar cells. Figure 6a shows the Nyquist plots of solar cells based on TiO_2 and ZnO nanofibers. The EIS show two arcs. The arc with higher frequency represents interface contact resistance, while low frequency generally represents recombination resistance (Rrec) and chemical capacitance (Cμ) of the device. Figure 6b shows an equivalent circuit for fitting the EIS. Table 2 also lists all the fitting values. The solar cells fabricated from TiO_2 nanofibers have smaller series resistance and larger recombination resistance compared to ZnO nanofibers. This shows that the charge transport ability was modified by TiO_2 nanofibers, thereby reducing carrier recombination. There are numerous bulk trap states as well as surface trap states due to oxygen vacancies in ZnO. TiO_2 can decrease the trap density, probably due to oxygen vacancy reduction. The smaller series resistance in the case of TiO2 nanofibers is attributed to the decreased traps density.

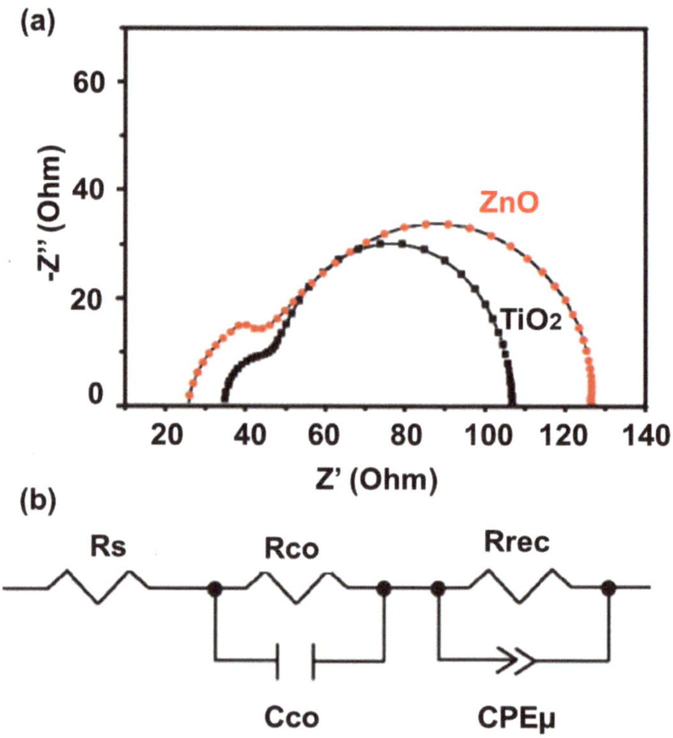

Figure 6. (**a**) Electrochemical Impedance spectra for TiO_2 and ZnO; (**b**) circuit diagram for EIS.

Table 2. EIS measurement values.

Sr	R_{rec}/Ohm	R_s/Ohm	R_{co}/Ohm	CPE-T/F
TiO_2	12.8	27.1	25.9	5.7×10^{-6}
ZnO	68.8	33.9	57.9	6.5×10^{-6}

As seen in Figure 7a, external quantum efficiency (EQE) was calculated for both TiO_2 and ZnO nanofiber photoanodes with perovskite in order to determine the cause of the Jsc boost. According to reports, the EQE is the result of charge injection/transfer, charge collecting, and light harvesting efficiency. In the range of 400 to 780 nm, the EQE spectrum of TiO_2 photoanodes exhibits a broad peak value of almost 90%, whereasthat of ZnO photoanodes shows a broad peak value below 80%. The creation of high-performance cells and their simple planar shape validates the superior potential of TiO_2 nanofibers over ZnO nanofibers in solar applications.

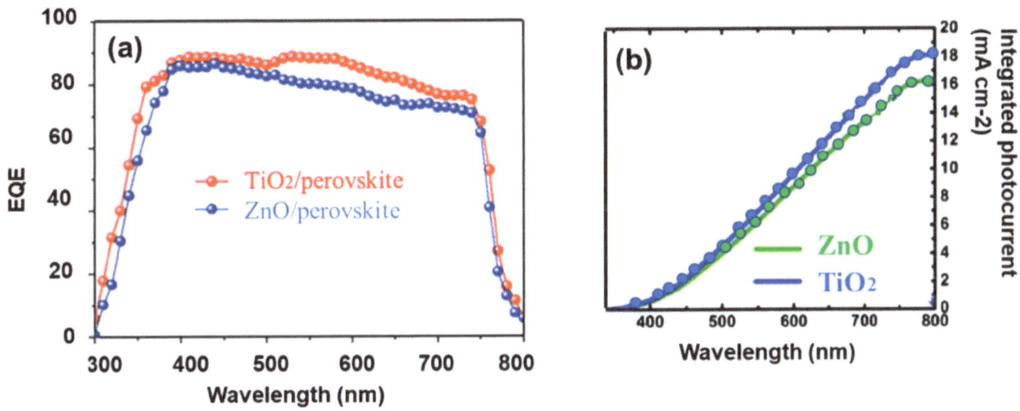

Figure 7. (**a**) EQE spectra for TiO_2, ZnO photoanodes; (**b**) integral current density for TiO_2 and ZnO photoanodes.

The integral of the current density from External Quantum Efficiency (EQE) curves is a key parameter for evaluating the performance of photovoltaic devices, such as the solar cells shown in Figure 7b. The EQE curve represents the fraction of incident photons that are converted into electrical current at each wavelength of light. In our devices, the integrated photocurrent density for the ZnO photoanode is 16 mA cm^{-2}, while for TiO_2, its value is 18 mA cm^{-2}. The integrated current density derived from the EQE spectra in Figure 7b is close to the Jsc measured under simulated sunlight.

A PL spectrophotometer was used for the characterization of charge recombination dynamics in perovskite solar cells. The electron and hole recombination of TiO_2 and ZnO were compared using photoluminescence (PL) spectra, as shown in Figure 8a. The high peak intensity of ZnO shows a higher direct recombination of electron-hole pairs. In contrast, TiO_2 nanofibers decrease the recombination of charges, thus resulting in better charge separation and associated photo-conversion efficiency. Charges have longer dwell times, hence charges can participate in photo-conversion efficiency for longer times.

Time-resolved photoluminescence spectra (TRPL) of undoped TiO_2/perovskite and ZnO/perovskite were recorded using a fluorometer, as shown in Figure 8b. The TRPL measurements were characterized by a bi-exponential decay function, with a full fast decay (τ1) component and a slow decay component (τ2). Free carrier transportation from perovskite to TiO_2 results in the fast decay component, while radiative decay results in the slow decay component. In the case of ZnO nanofibers, the fast decay time is 55.1 ns, and the slow decay time is 121.4 ns, while their weight fractions are 31.3 and 67.7%, respectively. In the case of TiO_2 nanofibers, the fast decay lifetime is decreased to 36.4 from 55.1 ns, and the slow decay lifetime to 109.5 from 121.4 ns, while the weight fraction of fast decay is increased to 35.1 from 31.3%. It is concluded that the TiO_2/perovskite interface presents a faster charge transfer and induced charge recombination than the ZnO/perovskite interface. The stability and efficiency of the solar cells based on TiO_2 and ZnO nanofibers could be due to the property change of the ETM, which affects the charge behavior at the interfaces.

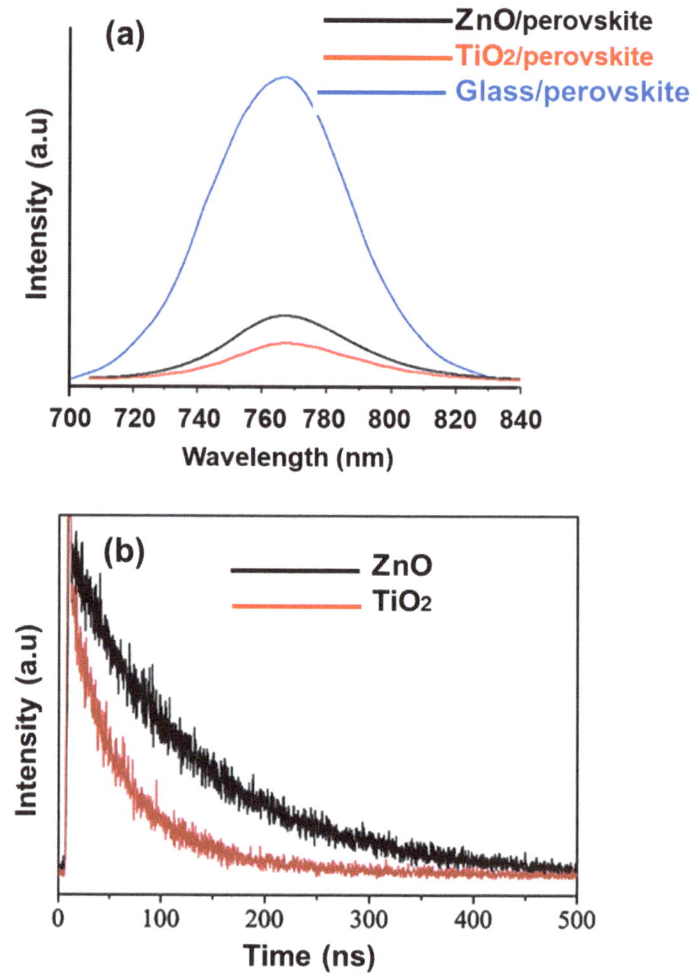

Figure 8. (**a**) PL spectra for TiO_2/perovskite, ZnO/perovskite, and glass/perovskite nanofibers; (**b**) TRPL spectra for TiO_2/perovskite and ZnO/perovskite nanofibers.

J-V measurements were performed to determine solar cell characteristics and to assess how TiO_2 and ZnO nanofibers affected the performance of solar cells. Figure 9 illustrates the JV performance of photoanodes built on titanium and zinc oxide nanofibers. Table 3 also provides a summary of all performance metrics. The current density (Jsc) of ZnO nanofibers was 17.77 mA/cm^2, while TiO_2 had a Jsc of 20.68 mA/cm^2. This increase is the result of TiO_2's improved conductivity of the ETL, which makes it possible for perovskite solar cells to efficiently use photons for energy harvesting [33]. As a result of TiO_2's superior charge transport characteristics over ZnO, more electrons are drawn to the external circuit, leading to a larger Jsc. In addition to effectively accelerating electron transport and lowering recombination rates to increase Jsc, band alignment with TiO_2 also produces effective ETLs. TiO_2's transparency contributes to increased Jsc values and improved light absorption by the perovskite layer. Jsc is much better with TiO_2 nanofibers ETL, which may be due to better band alignment. The electrons easily move from the perovskite conduction band to the TiO_2 conduction band due to appropriate band alignment.

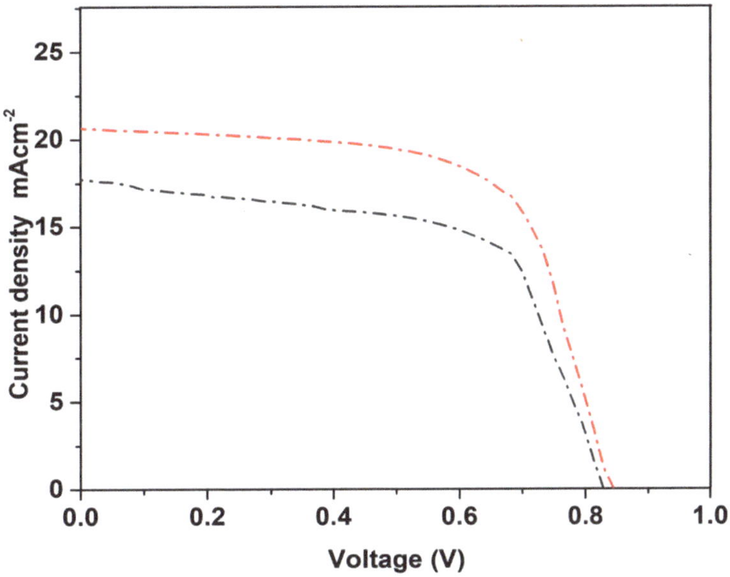

Figure 9. J-V characteristics for photoanodes based on TiO_2 (red) and ZnO (black).

Table 3. PSCs J-V parameters based on TiO_2 and ZnO nanofibers.

PSC	V_{oc} (V)	J_{sc} (mA/cm^2)	Shunt Resistance (Ohm)	Series Resistance (Ohm)	FF	η (%)
ZnO	0.82	17.77	52,278.68	205.423	0.59	8.5
TiO_2	0.83	20.68	53,469.8	197.553	0.61	10.4

Because holes have a higher mass than electrons, electrons have a higher mobility than holes. This is due to the fact that holes, which are often the space left by missing electrons, have a fixed orbit around atoms. The movement of electrons is favored by the electron transport structure, which raises conductivity and electron transport properties. Higher electron concentrations reduce charge recombination by increasing charge mobility. Furthermore, the conduction band and Fermi level tuning facilitate improved electron quenching to TiO_2 from the perovskite layer. Effective electron transport loops (ETLs) that aggressively suppress holes and recombination rates are produced by band alignment with TiO_2. Effective charge separation is possible because band alignment and conduction work together to limit electron-hole recombination and preserve their energy barrier [34]. Furthermore, TiO_2 nanofiber-based ETL proved successful in continuous electronic pathways; as a result, greater FF is achieved with improved electrical conductivity. One term for the increase in FF is "lowered series resistance" [11,35–37]. TiO_2 showed an efficiency of 10.4%, while ZnO demonstrated an efficiency of 8.5%. TiO_2 exhibits superior Jsc due to its efficient electron transport, which also results in less charge recombination. Increased electron transport from the valence band to the conduction band as a result of an optimized and tuned bandgap leads to increased efficiency. In conclusion, the TiO_2 nanofiber-fabricated ETL offers improved energy harvesting and perovskite solar cell efficiency [38].

The performance of our proposed ZnO and TiO_2 nanofibers was comparable to that of previously reported nanofibers, as shown in Table 4. Compared to other photoanodes, our materials are low-cost and earth-abundant, making them advantageous in commercial applications.

Table 4. Comparison of the performance of different nanofiber ETL-based PSCs.

PSC	Jsc (mA cm^{-2})	Voc (mV)	FF	η (%)	Ref
In-doped ZnO nanofibers	23.0	1000	70	16.10	[39]
Al doped Cu-ZnO	18.6	1080	70.77	14.18	[38]
TiO_2 fiber	4.02	1060	73.0	3.11	[35]
Pristine-TiO_2 NFs	23.32	1013	67.0	15.82	[3]
Ag doped CuO NFs	17.8	890	53.8	8.7	[40]
ZnO NFs	18.1	670	58.1	7.05	[41,42]
TiO_2 NFs	20.68	830	0.61	10.4	Our work
ZnO NFs	17.77	820	0.59	8.5	

In our optimized ZnO nanofibers, the current density is 17.77 mA cm^{-2}, possessing an efficiency 8.5%. For whole TiO_2 nanofibers, the current density is 20.68 mA cm^{-2}, possessing an efficiency 10.4%.

4. Conclusions

In conclusion, the electrospinning method was effectively used to synthesize ZnO and TiO_2 nanofibers on the same scale and under the same parameters, which were then annealed at 450 °C. TiO_2 and ZnO polycrystalline nanofibers were produced following the elimination of PVP during the annealing process, according to XRD spectra. The Raman spectra and the XRD data also revealed the polycrystalline nature of the nanofibers. It is evident from scanning electron microscopy images that ZnO and TiO_2 nanofibers are long, fibrous, and continuous in shape. Both zinc oxide as well as Titania nanofibers exhibit promising absorption spectra, with bandgaps of 3.2 and 3.3 eV, respectively, according to a preliminary characterization for photoactivity. According to Hall measurements, TiO_2 nanofibers have a greater conductivity of 1.28×10^{-4} Ohm/cm compared to ZnO nanofibers, which showed 1.12×10^{-4} Ohm/cm. TiO_2 decreases the trap density probably due to oxygen vacancy reduction. The smaller series resistance in the case of TiO_2 nanofibers is attributed to the decreased trap density. TiO_2 nanofibers had a higher conversion efficiency of 10.4% compared to 8.5% for ZnO, owing to their superior conductivity. TiO_2 will, therefore, have more photoactivity for energy harvesting if it is synthesized under the same processing conditions as ZnO.

Author Contributions: Methodology, A.Z.; Software, A.Z.; Validation, W.I.; Formal analysis, W.I.; Investigation, S.K.; Resources, M.F.; Writing—original draft, S.K. and M.F.; Writing—review and editing, A.A. All authors have read and agreed to the published version of the manuscript.

Funding: The researchers would like to thank the Deanship of Graduate studies and scientific research at Qassim University for financial support (QU-APC-2024-9/1).

Institutional Review Board Statement: Not applicable.

Informed Consent Statement: Not applicable.

Data Availability Statement: The original contributions presented in the study are included in the article; further inquiries can be directed to the corresponding author.

Conflicts of Interest: The authors declare no conflicts of interest.

References

1. Lotus, A.; Kang, Y.; Ramsier, R.; Chase, G.G. Investigation of the physical and electronic properties of indium doped zinc oxide nanofibers synthesized by electrospinning. *J. Vac. Sci. Technol. B* **2009**, *27*, 2331–2336. [CrossRef]
2. Wang, D.; Wright, M.; Elumalai, N.K.; Uddin, A. Stability of perovskite solar cells. *Sol. Energy Mater. Sol. Cells* **2016**, *147*, 255–275. [CrossRef]

3. Patil, J.V.; Mali, S.S.; Patil, A.P.; Patil, P.S.; Hong, C.K. Highly efficient mixed-halide mixed-cation perovskite solar cells based on rGO-TiO_2 composite nanofibers. *Energy* **2019**, *189*, 116396. [CrossRef]
4. Park, N.-G. Perovskite solar cells: An emerging photovoltaic technology. *Mater. Today* **2015**, *18*, 65–72. [CrossRef]
5. Mali, S.S.; Shim, C.S.; Kim, H.; Patil, P.S.; Hong, C.K. In situ processed gold nanoparticle-embedded TiO_2 nanofibers enabling plasmonic perovskite solar cells to exceed 14% conversion efficiency. *Nanoscale* **2016**, *8*, 2664–2677. [CrossRef]
6. Gonzalez-Pedro, V.; Juarez-Perez, E.J.; Arsyad, W.-S.; Barea, E.M.; Fabregat-Santiago, F.; Mora-Sero, I.; Bisquert, J. General working principles of $CH_3NH_3PbX_3$ perovskite solar cells. *Nano Lett.* **2014**, *14*, 888–893. [CrossRef]
7. Imran, M.; Haider, S.; Ahmad, K.; Mahmood, A.; Al-Masry, W.A. Fabrication and characterization of zinc oxide nanofibers for renewable energy applications. *Arab. J. Chem.* **2017**, *10*, S1067–S1072. [CrossRef]
8. Valadi, K.; Gharibi, S.; Taheri-Ledari, R.; Akin, S.; Maleki, A.; Shalan, A.E. Metal oxide electron transport materials for perovskite solar cells: A review. *Environ. Chem. Lett.* **2021**, *19*, 2185–2207. [CrossRef]
9. Son, D.-Y.; Im, J.-H.; Kim, H.-S.; Park, N.-G. 11% efficient perovskite solar cell based on ZnO nanorods: An effective charge collection system. *J. Phys. Chem. C* **2014**, *118*, 16567–16573. [CrossRef]
10. Amna, T.; Hassan, M.S.; Barakat, N.A.; Pandeya, D.R.; Hong, S.T.; Khil, M.-S.; Kim, H.Y. Antibacterial activity and interaction mechanism of electrospun zinc-doped titania nanofibers. *Appl. Microbiol. Biotechnol.* **2012**, *93*, 743–751. [CrossRef]
11. Zhen, C.; Wu, T.; Chen, R.; Wang, L.; Liu, G.; Cheng, H.-M. Strategies for modifying TiO_2 based electron transport layers to boost perovskite solar cells. *ACS Sustain. Chem. Eng.* **2019**, *7*, 4586–4618. [CrossRef]
12. Qiu, L.; Zhuang, Z.; Yang, S.; Chen, W.; Song, L.; Ding, M.; Xia, G.; Du, P.; Xiong, J. Fabrication of high efficiency perovskite solar cells based on mesoporous TiO_2 nanofibrous film under high humidity conditions. *Mater. Res. Bull.* **2018**, *106*, 439–445. [CrossRef]
13. Park, J.Y.; Kim, S.S. Growth of nanograins in electrospun ZnO nanofibers. *J. Am. Ceram. Soc.* **2009**, *92*, 1691–1694. [CrossRef]
14. Mali, S.S.; Su Shim, C.; Kook Hong, C. Highly porous Zinc Stannate (Zn_2SnO_4) nanofibers scaffold photoelectrodes for efficient methyl ammonium halide perovskite solar cells. *Sci. Rep.* **2015**, *5*, 11424. [CrossRef]
15. Erdogar, K.; Yucel, O.; Oruc, M.E. Investigation of Structural, Morphological, and Optical Properties of Novel Electrospun Mg-Doped TiO_2 Nanofibers as an Electron Transport Material for Perovskite Solar Cells. *Nanomaterials* **2023**, *13*, 2255. [CrossRef]
16. Aryal, S.; Kim, C.K.; Kim, K.-W.; Khil, M.S.; Kim, H.Y. Multi-walled carbon nanotubes/TiO_2 composite nanofiber by electrospinning. *Mater. Sci. Eng. C* **2008**, *28*, 75–79. [CrossRef]
17. Siddheswaran, R.; Sankar, R.; Ramesh Babu, M.; Rathnakumari, M.; Jayavel, R.; Murugakoothan, P.; Sureshkumar, P. Preparation and characterization of ZnO nanofibers by electrospinning. *Cryst. Res. Technol.* **2006**, *41*, 446–449. [CrossRef]
18. Xu, A.-W.; Gao, Y.; Liu, H.-Q. The preparation, characterization, and their photocatalytic activities of rare-earth-doped TiO_2 nanoparticles. *J. Catal.* **2002**, *207*, 151–157. [CrossRef]
19. Liao, D.; Liao, B. Shape, size and photocatalytic activity control of TiO_2 nanoparticles with surfactants. *J. Photochem. Photobiol. A Chem.* **2007**, *187*, 363–369. [CrossRef]
20. Liu, L.; Li, S.; Zhuang, J.; Wang, L.; Zhang, J.; Li, H.; Liu, Z.; Han, Y.; Jiang, X.; Zhang, P. Improved selective acetone sensing properties of Co-doped ZnO nanofibers by electrospinning. *Sens. Actuators B Chem.* **2011**, *155*, 782–788. [CrossRef]
21. Tai, M.; Zhao, X.; Shen, H.; Guo, Y.; Zhang, M.; Zhou, Y.; Li, X.; Yao, Z.; Yin, X.; Han, J. Ultrathin Zn_2SnO_4 (ZTO) passivated ZnO nanocone arrays for efficient and stable perovskite solar cells. *Chem. Eng. J.* **2019**, *361*, 60–66. [CrossRef]
22. Choi, H.C.; Jung, Y.M.; Kim, S.B. Size effects in the Raman spectra of TiO_2 nanoparticles. *Vib. Spectrosc.* **2005**, *37*, 33–38. [CrossRef]
23. Liu, Y.; Zhang, H.; An, X.; Gao, C.; Zhang, Z.; Zhou, J.; Zhou, M.; Xie, E. Effect of Al doping on the visible photoluminescence of ZnO nanofibers. *J. Alloys Compd.* **2010**, *506*, 772–776. [CrossRef]
24. Zhang, X.; Thavasi, V.; Mhaisalkar, S.; Ramakrishna, S. Novel hollow mesoporous 1D TiO_2 nanofibers as photovoltaic and photocatalytic materials. *Nanoscale* **2012**, *4*, 1707–1716. [CrossRef] [PubMed]
25. Wang, W.; Zhang, Z.; Cai, Y.; Chen, J.; Wang, J.; Huang, R.; Lu, X.; Gao, X.; Shui, L.; Wu, S. Enhanced performance of $CH_3NH_3PbI_3 - x Cl x$ perovskite solar cells by CH_3NH_3I modification of TiO_2-perovskite layer interface. *Nanoscale Res. Lett.* **2016**, *11*, 316. [CrossRef]
26. Mahmood, K.; Khalid, A.; Mehran, M.T. Nanostructured ZnO electron transporting materials for hysteresis-free perovskite solar cells. *Sol. Energy* **2018**, *173*, 496–503. [CrossRef]
27. Onozuka, K.; Ding, B.; Tsuge, Y.; Naka, T.; Yamazaki, M.; Sugi, S.; Ohno, S.; Yoshikawa, M.; Shiratori, S. Electrospinning processed nanofibrous TiO_2 membranes for photovoltaic applications. *Nanotechnology* **2006**, *17*, 1026. [CrossRef]
28. Zhang, Y.; Liu, X.; Li, P.; Duan, Y.; Hu, X.; Li, F.; Song, Y. Dopamine-crosslinked TiO_2/perovskite layer for efficient and photostable perovskite solar cells under full spectral continuous illumination. *Nano Energy* **2019**, *56*, 733–740. [CrossRef]
29. Duan, L.; Liu, S.; Wang, X.; Zhang, Z.; Luo, J. Interfacial Crosslinking for Efficient and Stable Planar TiO_2 Perovskite Solar Cells. *Adv. Sci.* **2024**, *11*, 2402796. [CrossRef]
30. Shakoor, A.; Nowsherwan, G.A.; Alam, W.; Bhatti, S.Y.; Bilal, A.; Nadeem, M.; Zaib, A.; Hussain, S.S. Fabrication and characterization of TiO_2: ZnO thin films as electron transport material in perovskite solar cell (PSC). *Phys. B Condens. Matter* **2023**, *654*, 414690. [CrossRef]
31. Hu, W.; Zhou, W.; Lei, X.; Zhou, P.; Zhang, M.; Chen, T.; Zeng, H.; Zhu, J.; Dai, S.; Yang, S. Low-temperature in situ amino functionalization of TiO_2 nanoparticles sharpens electron management achieving over 21% efficient planar perovskite solar cells. *Adv. Mater.* **2019**, *31*, 1806095. [CrossRef] [PubMed]

32. Yakuphanoglu, F. Electrical and photovoltaic properties of cobalt doped zinc oxide nanofiber/n-silicon diode. *J. Alloys Compd.* **2010**, *494*, 451–455. [CrossRef]
33. Son, D.-Y.; Bae, K.-H.; Kim, H.-S.; Park, N.-G. Effects of seed layer on growth of ZnO nanorod and performance of perovskite solar cell. *J. Phys. Chem. C* **2015**, *119*, 10321–10328. [CrossRef]
34. Laila, I.K.; Mufti, N.; Maryam, S.; Fuad, A.; Taufiq, A. Synthesis and characterization of ZnO nanorods by hydrothermal methods and its application on perovskite solar cells. *J. Phys. Conf. Ser.* **2018**, *1093*, 012012. [CrossRef]
35. Dharani, S.; Mulmudi, H.K.; Yantara, N.; Trang, P.T.T.; Park, N.G.; Graetzel, M.; Mhaisalkar, S.; Mathews, N.; Boix, P.P. High efficiency electrospun TiO_2 nanofiber based hybrid organic–inorganic perovskite solar cell. *Nanoscale* **2014**, *6*, 1675–1679. [CrossRef]
36. Cao, F.; Tian, W.; Gu, B.; Ma, Y.; Lu, H.; Li, L. High-performance UV–vis photodetectors based on electrospun ZnO nanofiber-solution processed perovskite hybrid structures. *Nano Res.* **2017**, *10*, 2244–2256. [CrossRef]
37. Chen, D.; Zhu, Y. Electrospun perovskite nanofibers. *Nanoscale Res. Lett.* **2017**, *12*, 114. [CrossRef]
38. Yang, H.; Kwon, H.-C.; Ma, S.; Kim, K.; Yun, S.-C.; Jang, G.; Park, J.; Lee, H.; Goh, S.; Moon, J. Energy level-graded Al-doped ZnO protection layers for copper nanowire-based window electrodes for efficient flexible perovskite solar cells. *ACS Appl. Mater. Interfaces* **2020**, *12*, 13824–13835. [CrossRef]
39. Mahmood, K.; Khalid, A.; Ahmad, S.W.; Mehran, M.T. Indium-doped ZnO mesoporous nanofibers as efficient electron transporting materials for perovskite solar cells. *Surf. Coat. Technol.* **2018**, *352*, 231–237. [CrossRef]
40. Zhuang, Z.; Qiu, L.; Dong, L.; Chen, Y.; Chu, Z.; Ma, X.; Du, P.; Xiong, J. Preparation of high-efficiency perovskite solar cells via doping Ag into CuO nanofibers as hole buffer layer. *Polym. Compos.* **2020**, *41*, 2145–2153. [CrossRef]
41. Mohtaram, F.; Borhani, S.; Ahmadpour, M.; Fojan, P.; Behjat, A.; Rubahn, H.-G.; Madsen, M. Electrospun ZnO nanofiber interlayers for enhanced performance of organic photovoltaic devices. *Sol. Energy* **2020**, *197*, 311–316. [CrossRef]
42. Abbas, A.; Shar, A.H.; Junejo, M.A. Why Brands Fail? Antecedents and Consequences of Brand Hate. A Study of Fashion Industry in Pakistan. *J. Manag. Sci.* **2023**, *17*, 1–26.

ceramics

Article

Effect of $(Ba_{1/3}Nb_{2/3})^{4+}$ Substitution on Microstructure, Bonding Properties and Microwave Dielectric Properties of $Ce_2Zr_3(MoO_4)_9$ Ceramics

Huamin Gao [1], Xiangyu Xu [2,3], Xinwei Liu [3], Xiaoyu Zhang [2], Mingling Li [1,*], Jialun Du [2] and Haitao Wu [2,*]

1 Engineering Technology Research Center of Preparation and Application of Industrial Ceramics of AnHui Province, Chaohu University, Hefei 238000, China
2 School of Environmental and Material Engineering, Yantai University, Yantai 264005, China
3 School of Materials Science and Engineering, University of Jinan, Jinan 250022, China
* Correspondence: 053014@chu.edu.cn (M.L.); wuhaitao@ytu.edu.cn (H.W.)

Abstract: In this study, $Ce_2[Zr_{1-x}(Ba_{1/3}Nb_{2/3})_x]_3(MoO_4)_9$ ($0.02 \leq x \leq 0.1$, $CZ_{1-x}N_x$) ceramics were sintered at 600 °C and 700 °C using the traditional solid-state method. An analysis conducted through XRD and Rietveld refinement confirmed that all the $CZ_{1-x}N_x$ ceramics displayed a single phase with a trigonal structure (space group R-3c). The observed increases in cell volume with increasing x values indicate the successful substitution of $(Ba_{1/3}Nb_{2/3})^{4+}$. The high densification of the synthesized phase was validated by the density and SEM results. Additionally, the P-V-L theory demonstrates a strong correlation between the Ce-O bond and ε_r, as well as τ_f, and between the Mo-O bond and $Q\times f$. Notably, the $CZ_{0.98}N_{0.02}$ ceramics demonstrated superior performance at 675 °C, exhibiting ε_r = 10.41, $Q\times f$ = 53,296 GHz, and τ_f = −23.45 ppm/°C. Finally, leveraging $CZ_{0.98}N_{0.02}$ ceramics as substrate materials enabled the design of a patch antenna suitable for the 5G communication band, demonstrating its significant potential in this field.

Keywords: $Ce_2[Zr_{1-x}(Ba_{1/3}Nb_{2/3})_x]_3(MoO_4)_9$; Rietveld refinement; P-V-L theory; patch antenna

Citation: Gao, H.; Xu, X.; Liu, X.; Zhang, X.; Li, M.; Du, J.; Wu, H. Effect of $(Ba_{1/3}Nb_{2/3})^{4+}$ Substitution on Microstructure, Bonding Properties and Microwave Dielectric Properties of $Ce_2Zr_3(MoO_4)_9$ Ceramics. *Ceramics* **2024**, *7*, 1172–1186. https://doi.org/10.3390/ceramics7030077

Academic Editors: Dawei Wang and Fayaz Hussain

Received: 27 June 2024
Revised: 5 August 2024
Accepted: 17 August 2024
Published: 29 August 2024

1. Introduction

The advent of 5G has precipitated the rapid advancement of novel technologies such as autonomous driving, intelligent transportation, and smart manufacturing, marking the onset of an era characterized by the intelligent interconnection of all things. Microwave dielectric ceramics play a pivotal role in the evolution of 5G technology due to their substantial market size. Furthermore, 5G technology necessitates lower signal latency, increased information-carrying capacity, and broader coverage, imposing heightened requirements on related microwave components [1–5]. It is noteworthy that materials with low dielectric constants (ε_r) offer advantages such as enhanced signal transmission speed, reduced coupling loss between ceramic substrates and metal electrodes, and a high-quality factor ($Q\times f$), rendering them the preferred choice for 5G technology. The nearly zero temperature coefficient of resonant frequency (τ_f) exhibited by ceramic substrates also constitutes an essential factor in maintaining signal stability within communication technology [6–9]. Furthermore, the utilization of low-temperature co-fired ceramics (LTCC) technology contributes to enhanced circuit density, facilitating device miniaturization and compliance with current environmental standards. This renders LTCC technology a highly promising multi-layer ceramic packaging solution. In LTCC technology, the firing temperature is typically maintained at a lower range of 800 °C to 900 °C to accommodate the sintering of ceramic materials with Au (melting temperature 1063 °C), Ag (melting temperature 961 °C), and Cu (melting temperature 1084 °C) electrode materials [10]. Notably, Mo-based ceramics have garnered attention from researchers due to their comparatively low sintering temperature and reduced losses [11,12].

In the double-molybdate system, $Re_2Zr_3(MoO_4)_9$ (Re = rare earth elements) has been reported to exhibit superior performance and can be synthesized at temperatures below 900 °C, showing great potential for development in LTCC applications [13–15]. Within this system, the $Eu_2Zr_3(MoO_4)_9$ ceramics demonstrate exceptional performance (ε_r = 10.75, $Q\times f$ = 74,900 GHz, τ_f = −8.88 ppm/°C), and they can be synthesized at 600 °C [15]. It is noteworthy that Tao et al. [12] and Shi et al. [16] have documented the low sintering temperature of $Ce_2Zr_3(MoO_4)_9$ (CZM) ceramics (575 °C and 750 °C respectively), along with their lowest τ_f value (−1.29 ppm/°C and −7.10 ppm/°C), indicating significant commercial application potential. However, the relatively low $Q\times f$ value (19,062 GHz and 24,720 GHz) of the CZM ceramics limits further scalability. Significant progress has been achieved in the doping and modification of CZM ceramics in recent years, representing an exciting development. Zheng et al. conducted studies on the substitution of Sn and Ti at the Zr site, resulting in $Q\times f$ value increases to 72,390 GHz and 84,200 GHz, respectively, while maintaining the τ_f value around −10 ppm/°C [17]. Furthermore, research on complex ion substitution at the Zr sites of CZM ceramics has expanded the range of substitution ions for performance enhancement. For instance, Xu et al. successfully elevated the $Q\times f$ value to 100,954 GHz at 750 °C by replacing the Zr site of CZM ceramics with $(Zn_{1/3}Sb_{2/3})^{4+}$. Xu et al. optimized the $Q\times f$ value to 59,381 GHz at 675 °C by substituting $(Zn_{1/3}Nb_{2/3})^{4+}$ for Zr in CZM ceramics [18]. Pan et al. also achieved an increased $Q\times f$ value of 82,696 GHz at 775 °C by replacing the Zr site with $(Sr_{1/3}Sb_{2/3})^{4+}$ [19]. It is evident that when using the pentavalent ion Nb for complex ion substitution in CZM ceramics, generally, a lower firing temperature is observed. Therefore, further research is warranted on $(Ba_{1/3}Nb_{2/3})^{4+}$ as a highly promising substitution ion in CZM ceramics.

This study presents a novel enhancement of the performance of CZM ceramics through the incorporation of $(Ba_{1/3}Nb_{2/3})^{4+}$ for the first time and, further, an examination of its impact on the microstructure and sintering characteristics of CZM ceramics. The Phillips-Van Vechten-Levine (P-V-L) valence bond theory is an analytical approach for investigating the correlation between properties and bonding characteristics. Therefore, the structure and properties of $Ce_2[Zr_{1-x}(Ba_{1/3}Nb_{2/3})_x]_3(MoO_4)_9$ ($0.02 \leq x \leq 0.1$, $CZ_{1-x}N_x$) ceramics were elucidated using the P-V-L theory. Additionally, a patch antenna tailored for the 5G communication band was developed based on the $CZ_{0.98}N_{0.02}$ ceramic substrate.

2. Experimental Section

Following the stoichiometric ratio required for $CZ_{1-x}N_x$ ceramics, high-purity CeO_2 (>99.9%, Macklin, Shanghai, China), ZrO_2 (>99.99%, Macklin, Shanghai, China), $BaCO_3$ (>99.99%, Macklin, Shanghai, China), Nb_2O_5 (>99.99%, Macklin, Shanghai, China), and MoO_3 (>99.5%, Macklin, Shanghai, China) powders were precisely weighed and synthesized via the conventional solid-phase reaction route. Initially, the weighed powders were a ball mill with zirconia balls and ethanol for 15 h of milling. Subsequently, the homogeneous slurry was extracted and subjected to drying at 80 °C in an oven. The resulting dried powders were then sieved through a 200-mesh screen and transferred to an alumina crucible. The precursor was formed by sintering the powder at 600 °C in a muffle furnace for 2 h. This precursor was then blended with 8 wt.% PVA before being pressed into cylindrical green bodies in a cylindrical mold under 200 MPa pressure. Finally, the cylindrical green bodies were placed in an alumina crucible and subjected to sintering at 600 to 700 °C in an air atmosphere for 4 h.

The phase of $CZ_{1-x}N_x$ ceramics was determined using an X-ray diffractometer (XRD) model BRUKER D8 ADVANCE in ambient air at room temperature, and the data were refined employing the Rietveld method. The apparent density of $CZ_{1-x}N_x$ ceramics was measured by the Archimedes displacement method at a test water temperature of 25.5 °C. The surface morphology of $CZ_{1-x}N_x$ ceramics was characterized using a scanning electron microscope (SEM) model JEOL JSM-IT200. The ε_r and τ_f values of $CZ_{1-x}N_x$ ceramics were measured utilizing a vector network analyzer, model N5234A, from Agilent for TE_{011} mode. The τ_f value was determined by measuring the resonant frequency of the sample at 25 °C

(f_{25}) and 85 °C (f_{85}) using a temperature chamber, and then calculated using Equation (1). The $Q{\times}f$ value of $CZ_{1-x}N_x$ ceramics was obtained through the closed-cavity method in $TE_{01\delta}$ mode [9]. Furthermore, the patch antenna of $CZ_{1-x}N_x$ ceramics was designed by initially calculating the values for radiation patch length (L_p), width (W_p), and feed position (X_f and Y_f) based on ε_r and the designed center frequency (f_0), followed by optimization using ANSYS HFSS software [20,21].

$$\tau_f = \frac{f_{85} - f_{25}}{60 f_{25}} \quad (1)$$

$$W_p = \frac{c}{2f_0}\left(\frac{\varepsilon_r + 1}{2}\right)^{-0.5} \quad (2)$$

$$L_p = \frac{c}{2f_0\sqrt{\varepsilon_e}} - 2\Delta L \quad (3)$$

$$\Delta L = 0.412h\frac{(\varepsilon_e + 0.3)(\frac{W_p}{h} + 0.264)}{(\varepsilon_e - 0.258)(\frac{W_p}{h} + 0.8)} \quad (4)$$

$$\varepsilon_e = \frac{\varepsilon_r + 1}{2} + \frac{\varepsilon_r - 1}{2}\left(1 + \frac{12h}{W_p}\right)^{-0.5} \quad (5)$$

$$\xi_{\mathrm{re}}(L) = \frac{\varepsilon_r + 1}{2} + \frac{\varepsilon_r - 1}{2}\left(1 + \frac{12h}{L_p}\right)^{-0.5} \quad (6)$$

$$X_f = \frac{L}{2\sqrt{\xi_{re}(L)}} \quad (7)$$

$$Y_f = 0 \quad (8)$$

where c represents the speed of light, ΔL denotes the corrected length, ε_e signifies the effective dielectric constant, and h stands for the thickness of the $CZ_{1-x}N_x$ ceramics substrate.

3. Results and Discussion

The XRD patterns of $CZ_{1-x}N_x$ ceramics sintered at the optimum temperature are depicted in Figure 1. All diffraction peaks of the $CZ_{1-x}N_x$ ceramics exhibit identical positions to those of $Pr_2Zr_3(MoO_4)_9$ (PDF# 51-1851), with no additional diffraction peaks observed [17,18]. This observation suggests that $(Ba_{1/3}Nb_{2/3})^{4+}$ was successfully incorporated into the crystal lattice of the CZM ceramics, forming a trigonal structure (space group R-3c) within the $CZ_{1-x}N_x$ solid solution. Furthermore, as the value of x increases, all diffraction peaks of $CZ_{1-x}N_x$ ceramics display a leftward shift, which is attributable to the larger ion radius of $(Ba_{1/3}Nb_{2/3})^{4+}$ (1.41 Å, CN = 6) compared to that of Zr^{4+} (0.72 Å, CN = 6) [22]. This finding further validates the successful substitution by $(Ba_{1/3}Nb_{2/3})^{4+}$ at Zr sites in the CZM ceramics. To investigate the influence of the $(Ba_{1/3}Nb_{2/3})^{4+}$ on the crystal structure of the CZM ceramics, the diffraction peaks of the $CZ_{1-x}N_x$ ceramics were refined using the Rietveld method, as illustrated in Figure 2 and Table 1. Figure 2 demonstrates a good match between the measured and fitted diffraction peaks. Furthermore, the R_p, R_{wp}, and χ^2 values obtained after the refinement are all within 10, indicating the reliability of the refined structure [23]. The crystal lattice parameters of the $CZ_{1-x}N_x$ ceramics are depicted in Figure 3. Owing to the disparity in ionic radii, the collective values of a, b, c, and V_m exhibit an increasing trend with higher substitution levels, corresponding to the shift of the peaks observed in the XRD pattern.

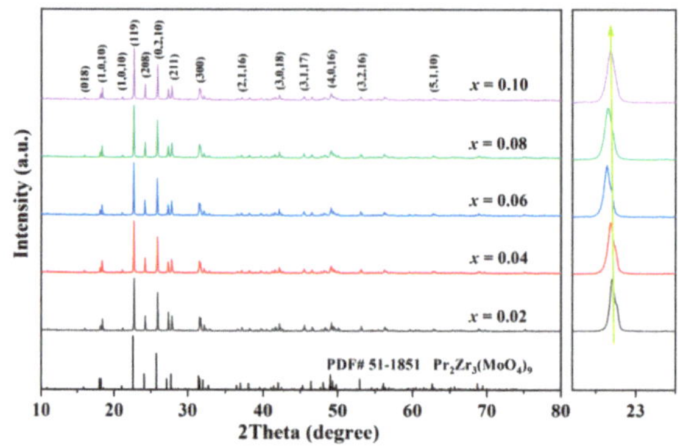

Figure 1. XRD patterns of $CZ_{1-x}N_x$ ceramics sintered at the optimal temperature with different x values.

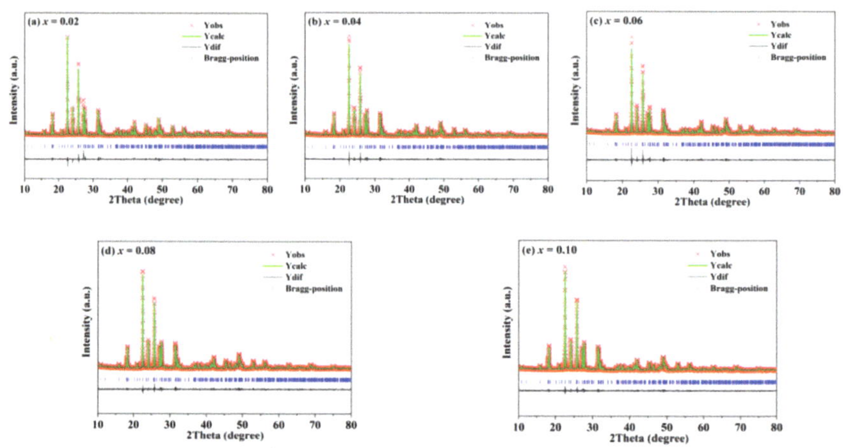

Figure 2. Rietveld refinement of $CZ_{1-x}N_x$ ceramics at the optimal sintering temperature with different x values.

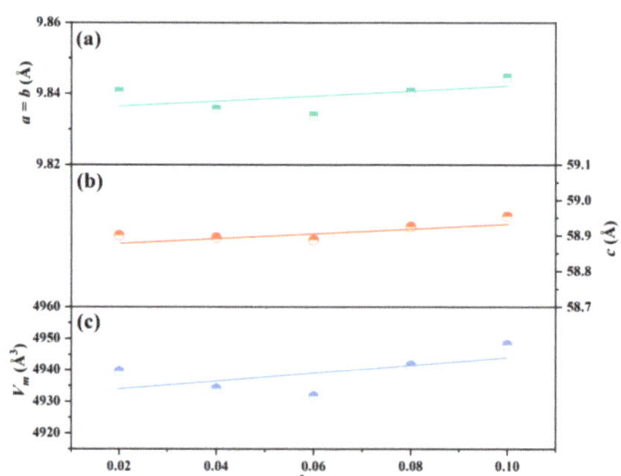

Figure 3. Lattice parameter changes (**a**) a and b, (**b**) c, and (**c**) V_m of $CZ_{1-x}N_x$ ceramics as a function of x values.

Table 1. The lattice parameters and reliability factors of $CZ_{1-x}N_x$ ceramics sintered at the optimal sintering temperatures.

x	Lattice Parameter					Reliability Factors		
	$a = b$ (Å)	c (Å)	$\alpha = \beta$ (°)	γ (°)	V_m (Å^3)	R_p (%)	R_{wp} (%)	χ^2
0.02	9.8405	58.9013	90	120	4939.56	6.17	8.07	2.52
0.04	9.8356	58.8950	90	120	4934.08	5.25	6.65	1.70
0.06	9.8337	58.8883	90	120	4931.71	5.60	7.06	1.85
0.08	9.8404	58.9261	90	120	4941.58	4.96	6.31	1.49
0.10	9.8445	58.9552	90	120	4948.08	4.76	6.08	1.49

The crystal structure of the $CZ_{1-x}N_x$ ceramics is illustrated in Figure 4. The $CZ_{1-x}N_x$ ceramics consist of [CeO_9] polyhedra, [ZrO_6] polyhedra, and [MoO_4] tetrahedra interconnected with each other. Ce atoms are coordinated with 9 × O atoms (3 × O(1), 3 × O(2), and 3 × O(6)) to form [CeO_9] polyhedra. Both Zr(1) and Zr(2) atoms are coordinated with 6 × O atoms (Zr(1): 6 × O(4), Zr(2): 3 × O(1) and 3 × O(5)) to form [ZrO6] polyhedra. Similarly, both Mo(1) and Mo(2) atoms are coordinated with four O atoms (Mo(1): O(1), O(2), O(3), O(4), Mo(2): 2 × O(5) and 2 × O(6)) to form [MoO_4] tetrahedra. In the $CZ_{1-x}N_x$ ceramics, the [CeO_9] polyhedra and [ZrO_6] polyhedra are not directly connected; they are linked through [MoO_4] tetrahedra serving as a bridge [18,24].

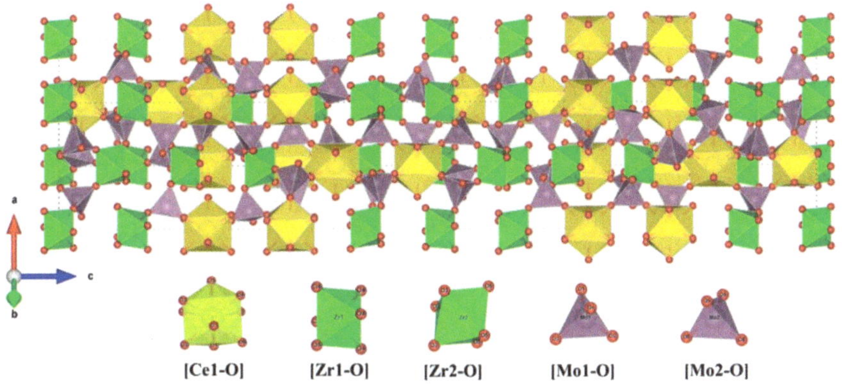

Figure 4. Crystal structure diagram of $CZ_{1-x}N_x$ ceramics.

The performance of ceramics is influenced by both intrinsic and extrinsic factors. Extrinsic factors, such as second phases and density, play a significant role. The XRD results confirmed the purity of the $CZ_{1-x}N_x$ ceramics phases, thus necessitating an investigation into the relationship between sintering behavior and the performance of $CZ_{1-x}N_x$ ceramics. Figure 5 illustrates the diameter shrinkage behavior and density changes of the $CZ_{1-x}N_x$ ceramics with temperature. With the increase in the sintering temperature, the shrinkage rate of the $CZ_{1-x}N_x$ ceramics with varying x values exhibited a pattern of initial augmentation followed by diminishment. This phenomenon is attributable to grain growth and pore elimination during sintering. However, excessive temperatures can induce secondary crystallization in the grains, resulting in reduced shrinkage rates. The trends observed in the apparent density align closely with those of the shrinkage rate. Furthermore, at the optimal sintering temperature, the relative density of the $CZ_{1-x}N_x$ ceramics consistently exceeded 90%, underscoring their favorable densification.

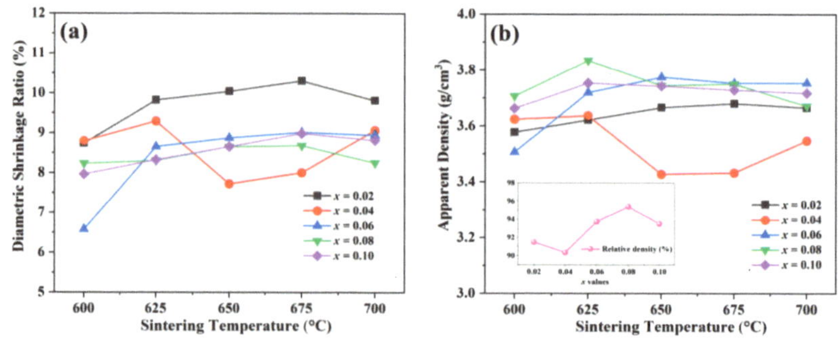

Figure 5. (**a**) Diameter shrinkage, (**b**) apparent density (the relative density at the optimal sintering temperature as a function of x values are shown in the inset) of $CZ_{1-x}N_x$ ceramics at 600 to 700 °C.

Figure 6 illustrates the microstructural images of the $CZ_{1-x}N_x$ ceramics at the optimal sintering temperature for each dopant content. The presence of atomic arrangement disorder and vacancies at grain boundaries has a detrimental effect on performance. Moreover, the well-developed regular grain morphology indicates favorable grain growth, contributing to performance enhancement [25,26]. Consequently, the $CZ_{0.98}N_{0.02}$ ceramics exhibit the largest grain size and most regular grain shape, positively impacting performance improvement. It is noteworthy that all the $CZ_{1-x}N_x$ ceramics demonstrated high densities without discernible pores, which is consistent with the relative density results.

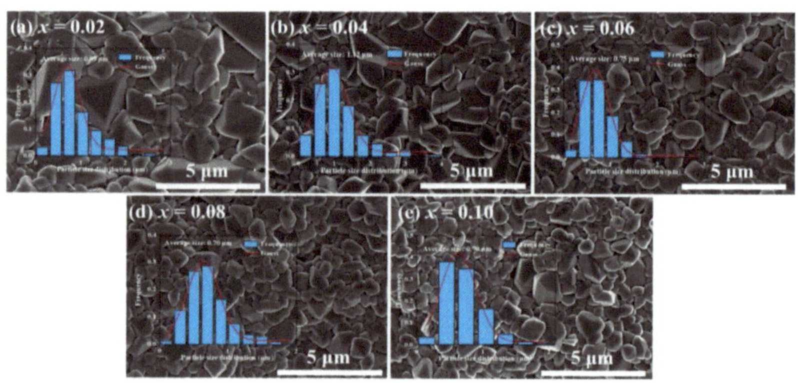

Figure 6. Microstructure of $CZ_{1-x}N_x$ ceramics at optimal sintering temperature with different x values.

Figure 7a illustrates the temperature-dependent variation in the ε_r for the $CZ_{1-x}N_x$ ceramics. The consistency between the ε_r and apparent density suggests that density is the primary factor influencing the dielectric constant of the ceramic. As evidenced by the presence of pores in the ceramic, where air has an ε_r value of 1, it was observed that the ε_r value of the $CZ_{1-x}N_x$ ceramics significantly exceeded this baseline, indicating that increased porosity leads to a reduction in dielectric constant [19]. Consequently, considering the presence of pores, Equation (9) was employed to correct for the ε_r ($\varepsilon_{corr.}$) of the $CZ_{1-x}N_x$ ceramics, as depicted in Figure 7c, where P ($P = 1 - \rho_{\text{relative}}$) is the porosity. Moreover, the relevant numerical values of the CZM ceramics matrix were derived from the reported findings of Tao et al. and Shi et al. [12,16]. The values of ε_r and $\varepsilon_{corr.}$ increased proportionally with x, a trend explicable by Equation (10) [27]. As per the Clausius–Mossotti equation, it is evident that ε_r exhibits a positive correlation with the polarizability (α). Additionally, based on the α values reported by Shannon et al. [28], the theoretical polarizability ($\alpha_{theo.}$) of the $CZ_{1-x}N_x$ ceramics was calculated using Equation (11).

$$\varepsilon_{corr.} = \varepsilon_r(1 + 1.5P) \tag{9}$$

$$\frac{\varepsilon_r - 1}{\varepsilon_r + 2} = \frac{4\pi\alpha}{3V_m} \tag{10}$$

$$\begin{aligned}\alpha_{theo.}(CZ_{1-x}N_x) = 2\alpha(Ce^{3+}) + 3(1-x)\alpha(Zr^{4+}) + x\alpha(Ba^{2+}) + 2x\alpha(Nb^{5+}) + 9\alpha(Mo^{6+}) + \\ 36\alpha(O^2) = 2 \times 6.15 + 3(1-x) \times 3.25 + x \times 6.04 + 2x \times 3.97 + 9 \times 3.28 + 36 \times 2.01\end{aligned} \tag{11}$$

Figure 7d illustrates the continuous increase in the $\alpha_{theo.}$ values of the $CZ_{1-x}N_x$ ceramics as the x value increases, aligning with the observed trend in the ε_r values. This observation suggests that α plays a significant role as an internal factor contributing to the enhancement of ε_r values. Furthermore, the P-V-L theory serves as a widely adopted approach for investigating the correlation between material structure and performance [29,30]. Within $CZ_{1-x}N_x$ ceramics, an analysis of the interna influence of $(Ba_{1/3}Nb_{2/3})^{4+}$ on the CZM ceramics was conducted through calculations of chemical bond variations using the P-V-L theory. Initially, complex chemical formulas were decomposed into binary bond formulas, as depicted in Equation (12).

$$\begin{gathered}Ce_2\left[(Zr_{1-x}(Ba_{1/3}Nb_{2/3})x\right]_3(MoO_4)_9 \\ Ce_{2/3}O(1)_3 + Ce_{2/3}O(2)_3 + Ce_{2/3}O(6)_3 + Zr/(Ba_{1/3}Nb_{2/3})(1)O(4)_3 + Zr/(Ba_{1/3}Nb_{2/3})(2)O(3)_3 \\ +Zr/(Ba_{1/3}Nb_{2/3})(2)O(5)_3 + Mo(1)_{3/2}O(1)_3 + Mo(1)_{3/2}O(2)_3 + Mo(1)_{3/2}O(3)_3 + Mo(1)_{3/2}O(4)_3 \\ +Mo(2)_{3/2}O(5)_3 + Mo(2)_{3/2}O(6)_3\end{gathered} \tag{12}$$

In the P-V-L theory, the correlation between ionicity (f_i) and ε_r was established, and the impact of $(Ba_{1/3}Nb_{2/3})^{4+}$ on the ε_r of the CZM ceramics was determined by evaluating the change in f_i for each chemical bond. As depicted in Equation (13), there was a positive correlation between f_i and ε_r. Furthermore, Equations (14)–(17) were utilized to compute the f_i of each chemical bond in the $CZ_{1-x}N_x$ ceramics [31,32].

$$\varepsilon_r = \frac{n_0^2 - 1}{1 - f_i} + 1 \tag{13}$$

$$f_i^\mu = \frac{(C^\mu)^2}{\left(E_g^\mu\right)^2} \tag{14}$$

$$\left(E_g^\mu\right)^2 = \left(E_h^\mu\right)^2 + (C^\mu)^2 \tag{15}$$

$$\left(E_h^\mu\right)^2 = \frac{39.74}{(d^\mu)^{2.48}} \tag{16}$$

$$\left(E_h^\mu\right)^2 = \frac{39.74}{(d^\mu)^{2.48}} C^\mu = 14.4 b^\mu \exp\left(-k_s^\mu r_0^\mu\right)\left[\left(Z_A^\mu\right)^* - \frac{n}{m}\left(Z_B^\mu\right)^*\right]/r_0^\mu \tag{17}$$

where n_0, d^μ, and b^μ denote the refractive index, bond length, and periodic correction factor, respectively. $(Z_A^\mu)^*$ and $(Z_B^\mu)^*$ represent the effective valence electron numbers of the cation and anion, respectively. Additionally, $\exp(-k_s^\mu r_0^\mu)$ represents the Thomas-Fermi factor. The calculated f_i values for each chemical bond in the $CZ_{1-x}N_x$ ceramics are summarized in Table 2. Notably, the f_i value of the Ce-O bond surpasses that of other chemical bonds. Moreover, the average f_i value ($f_{iave.(Ce\text{-}O)}$) for this bond was computed, as shown in Figure 7e. It was observed that the ε_r exhibited a similar upward trend to the $f_{iave.(Ce\text{-}O)}$, suggesting that the Ce-O bond exerts a dominant influence on the ε_r of $CZ_{1-x}N_x$ ceramics.

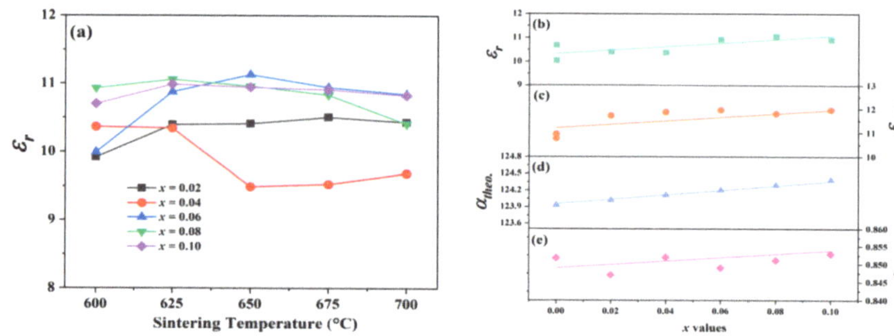

Figure 7. $CZ_{1-x}N_x$ ceramics (**a**) ε_r at a sintering temperature of 600 °C to 700 °C, (**b**) ε_r, (**c**) $\varepsilon_{corr.}$, (**d**) $\alpha_{theo.}$, and (**e**) $f_{iave.(Ce\text{-}O)}$ at the optimal sintering temperature as a function of x values.

Table 2. The f_i of the $CZ_{1-x}N_x$ ceramics sintered at optimal temperature.

Bond Type	$x = 0.02$	$x = 0.04$	$x = 0.06$	$x = 0.08$	$x = 0.10$
Ce-O(1) [a]	0.8393	0.8490	0.8476	0.8506	0.8517
Ce-O(1) [b]	0.8393	0.8490	0.8476	0.8506	0.8518
Ce-O(1) [c]	0.8393	0.8490	0.8476	0.8507	0.8518
Ce-O(2) [a]	0.8382	0.8509	0.8485	0.8493	0.8533
Ce-O(2) [b]	0.8382	0.8509	0.8485	0.8493	0.8533
Ce-O(2) [c]	0.8382	0.8510	0.8485	0.8494	0.8533
Ce-O(6) [a]	0.8638	0.8561	0.8511	0.8534	0.8538
Ce-O(6) [b]	0.8638	0.8561	0.8511	0.8534	0.8538
Ce-O(6) [c]	0.8639	0.8561	0.8511	0.8534	0.8539
Zr(BaNb)1-O(4) × 6	0.7865	0.7913	0.7952	0.7923	0.7913
Zr(BaNb)2-O(3) [a]	0.7666	0.7817	0.7781	0.7821	0.7835
Zr(BaNb)2-O(3) [b]	0.7666	0.7817	0.7781	0.7821	0.7836
Zr(BaNb)2-O(3) [c]	0.7666	0.7817	0.7782	0.7820	0.7836
Zr(BaNb)2-O(5) [a]	0.7643	0.7876	0.7762	0.7803	0.7836
Zr(BaNb)2-O(5) [b]	0.7644	0.7877	0.7762	0.7803	0.7836
Zr(BaNb)2-O(5) [c]	0.7644	0.7877	0.7763	0.7803	0.7837
Mo1-O(1)	0.7021	0.7170	0.7046	0.7103	0.7117
Mo1-O(2)	0.7095	0.7191	0.7062	0.7179	0.7198
Mo1-O(3)	0.7161	0.7370	0.7289	0.7312	0.7390
Mo1-O(4)	0.7048	0.7315	0.7222	0.7267	0.7280
Mo2-O(5) × 2	0.7104	0.7251	0.7268	0.7307	0.7317
Mo2-O(6) × 2	0.6281	0.7037	0.7006	0.7030	0.7104

[a–c] Three different bonds.

Figure 8a illustrates the variation in the $Q{\times}f$ values for the $CZ_{1-x}N_x$ ceramics at different temperatures. The trend in the $Q{\times}f$ values mirrors that of the apparent density, suggesting that density is the primary external factor influencing $Q{\times}f$ values. Lattice energy (U) is commonly linked to the $Q{\times}f$ values of ceramics. The magnitude of the U value reflects the stability of the ceramic, while higher U values indicate greater compound stability, thereby reducing nonharmonic vibration-induced losses and, consequently, enhancing the $Q{\times}f$ value of the ceramic. According to the P-V-L theory, the U values for the chemical bonds in the $CZ_{1-x}N_x$ ceramics were computed using Equations (18)–(21) [33].

$$U_{cal} = \sum_{\mu} U_b^{\mu} \tag{18}$$

$$U_b^{\mu} = U_{bc}^{\mu} + U_{bi}^{\mu} \tag{19}$$

$$U_{bc}^{\mu} = 2100m \frac{P_{A-B}^{1.64}}{d^{0.75}} f_c^{\mu} \tag{20}$$

$$U_{bi}^{\mu} = 1270\frac{(m+n)P_{A-B}P_{B-A}}{d}(1-\frac{0.4}{d})f_{\mathrm{i}}^{\mu} \quad (21)$$

where U_{bc}^{μ} and U_{bi}^{μ} represent the covalent and ionic bond energies of the material, respectively. f_c^{μ} and f_{i}^{μ} represent the bond covalency and ionicity, respectively. P_{A-B} and P_{B-A} represent the valence states of the cation and anion, respectively. Table 3 presents the specific U values for each chemical bond. The Mo-O bond exhibits a notably higher U value compared to the other two chemical bonds, indicating its significant influence on the $Q\times f$ values of the ceramic. Additionally, Figure 8b,c illustrates the $Q\times f$ and average U ($U_{ave.(\text{Mo-O})}$) of the $CZ_{1-x}N_x$ ceramics at optimal temperature. The sample with x = 0.02 demonstrates the highest $CZ_{1-x}N_x$ value (53,296 GHz), significantly surpassing that of the CZM ceramics, suggesting that $(Ba_{1/3}Nb_{2/3})^{4+}$ serves as an effective substituted complex ion. Furthermore, $U_{ave.(\text{Mo-O})}$ displays a similar trend to the $Q\times f$ value, underscoring the major influence of the Mo-O bond on the $Q\times f$ in the $CZ_{1-x}N_x$ ceramics.

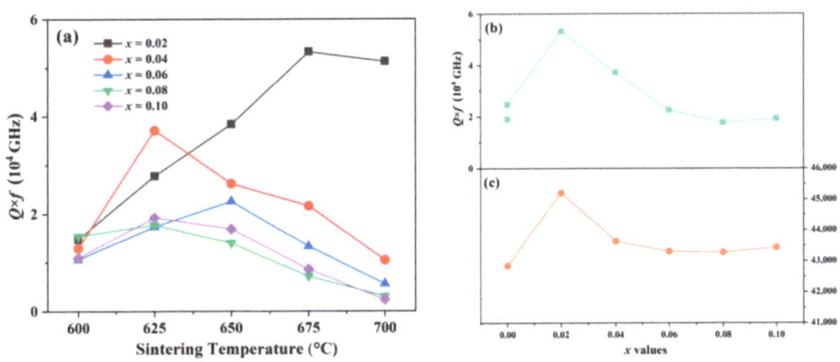

Figure 8. $CZ_{1-x}N_x$ ceramics (**a**) $Q\times f$ at a sintering temperature of 600 °C to 700 °C, (**b**) $Q\times f$, and (**c**) $U_{ave.(\text{Mo-O})}$ at the optimal sintering temperature as a function of x values.

Table 3. The U (kJ/mol) of the $CZ_{1-x}N_x$ ceramics sintered at optimal temperature.

Bond Type	x = 0.02	x = 0.04	x = 0.06	x = 0.08	x = 0.10
Ce-O(1) [a]	1095	1100	1080	1076	1085
Ce-O(1) [b]	1095	1100	1080	1075	1085
Ce-O(1) [c]	1095	1100	1080	1075	1085
Ce-O(2) [a]	1103	1086	1074	1084	1074
Ce-O(2) [b]	1102	1085	1074	1084	1074
Ce-O(2) [c]	1102	1085	1074	1084	1074
Ce-O(6) [a]	906	1048	1055	1056	1070
Ce-O(6) [b]	906	1048	1055	1056	1070
Ce-O(6) [c]	906	1048	1055	1056	1070
Zr(BaNb)1-O(4) × 6	10,161	10,653	10,145	10,476	10,702
Zr(BaNb)2-O(3) [a]	3726	3711	3676	3665	3696
Zr(BaNb)2-O(3) [b]	3726	3709	3676	3665	3696
Zr(BaNb)2-O(3) [c]	3726	3709	3675	3666	3695
Zr(BaNb)2-O(5) [a]	3761	3613	3707	3694	3695
Zr(BaNb)2-O(5) [b]	3761	3612	3707	3694	3695
Zr(BaNb)2-O(5) [c]	3760	3611	3706	3694	3694
Mo1-O(1)	43,587	44,245	45,021	44,810	45,295
Mo1-O(2)	42,314	43,919	44,773	43,585	44,009
Mo1-O(3)	41,121	40,777	40,864	41,257	40,646
Mo1-O(4)	43,133	41,786	42,088	42,075	42,641
Mo2-O(5) × 2	42,157	42,900	41,254	41,343	41,993
Mo2-O(6) × 2	53,631	46,310	45,640	45,941	45,491

[a–c] Three different bonds.

The τ_f of a ceramic material characterizes its stability in diverse environments, and the closer τ_f is to 0, the greater the stability of the ceramic. As per Equation (22), an inverse relationship exists between the thermal expansion coefficient (α) and the τ_f value of the ceramics. The α value of the $CZ_{1-x}N_x$ ceramics was determined using Equations (23)–(26) derived from the P-V-L theory [34,35].

$$\tau_f = -\left(\frac{\tau_\varepsilon}{2} + \alpha\right) \tag{22}$$

$$\alpha = \sum_\mu F^\mu_{mn}\alpha^\mu_{mn} \tag{23}$$

$$\alpha^\mu_{\mathrm{mn}} = -3.1685 + 0.8376\gamma_{mn} \tag{24}$$

$$\gamma_{mn} = \frac{kZ^\mu_A N^\mu_{CA}}{U^\mu_b \Delta_A}\beta_{mn} \tag{25}$$

$$\beta_{mn} = \frac{m(m+n)}{2n} \tag{26}$$

where τ_ε represents the temperature coefficient of the dielectric constant, N^μ_{CA} denotes the coordination number of the cation, Z^μ_A indicates the valence state of the cation, and k and Δ_A stand for the Boltzmann constant and the ionic periodicity constant, respectively. The α values of the chemical bonds in the $CZ_{1-x}N_x$ ceramics are detailed in Table 4. Among these ceramics, it was observed that the Ce-O bond exhibited the highest α value, thus exerting a significant influence on the τ_f value. As depicted in Figure 9, the average α value ($\alpha_{ave.(\text{Ce-O})}$) of the Ce-O bond demonstrated a similar trend to the τ_f. This suggests a strong correlation between the Ce-O bond and the τ_f value in the $CZ_{1-x}N_x$ ceramics.

Table 4. The α (10^{-6}/K) of the $CZ_{1-x}N_x$ ceramics sintered at optimal temperature.

Bond Type	$x = 0.02$	$x = 0.04$	$x = 0.06$	$x = 0.08$	$x = 0.10$
Ce-O(1) [a]	10.2227	10.1619	10.4087	10.4592	10.3461
Ce-O(1) [b]	10.2227	10.1619	10.4087	10.4719	10.3461
Ce-O(1) [c]	10.2227	10.1619	10.4087	10.4719	10.3461
Ce-O(2) [a]	10.1256	10.3337	10.4846	10.3586	10.4846
Ce-O(2) [b]	10.1377	10.3461	10.4846	10.3586	10.4846
Ce-O(2) [c]	10.1377	10.3461	10.4846	10.3586	10.4846
Ce-O(6) [a]	13.0163	10.8233	10.7304	10.7173	10.5356
Ce-O(6) [b]	13.0163	10.8233	10.7304	10.7173	10.5356
Ce-O(6) [c]	13.0163	10.8233	10.7304	10.7173	10.5356
Zr(BaNb)1-O(4) × 6	3.8445	3.5103	3.8340	3.6024	3.4493
Zr(BaNb)2-O(3) [a]	3.2064	3.2224	3.2734	3.2828	3.2189
Zr(BaNb)2-O(3) [b]	3.2064	3.2258	3.2734	3.2828	3.2189
Zr(BaNb)2-O(3) [c]	3.2064	3.2258	3.2751	3.2811	3.2207
Zr(BaNb)2-O(5) [a]	3.1471	3.3957	3.2195	3.2322	3.2207
Zr(BaNb)2-O(5) [b]	3.1471	3.3976	3.2195	3.2322	3.2207
Zr(BaNb)2-O(5) [c]	3.1488	3.3994	3.2212	3.2322	3.2224
Mo1-O(1)	−0.4054	−0.4465	−0.4934	−0.4808	−0.5096
Mo1-O(2)	−0.3223	−0.4263	−0.4786	−0.4053	−0.4319
Mo1-O(3)	−0.2397	−0.2150	−0.2213	−0.2494	−0.2055
Mo1-O(4)	−0.3763	−0.2863	−0.3070	−0.3061	−0.3441
Mo2-O(5) × 2	−0.3117	−0.3612	−0.2492	−0.2554	−0.3005
Mo2-O(6) × 2	−0.9229	−0.5679	−0.5297	−0.5470	−0.5211

[a–c] Three different bonds.

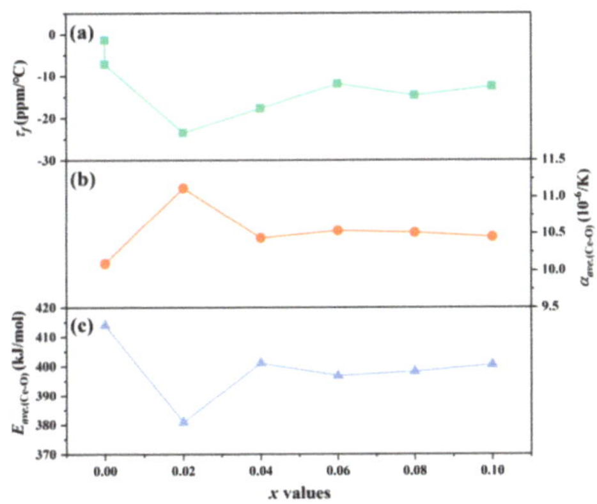

Figure 9. $CZ_{1-x}N_x$ ceramics (**a**) τ_f, (**b**) $\alpha_{ave.(Ce\text{-}O)}$, and (**c**) $E_{ave.(Mo\text{-}O)}$ at the optimal sintering temperature as a function of x values.

It is noteworthy that bond energy (E) is frequently employed to characterize the stability of materials. A higher E value indicates a stronger connection between chemical bonds. The E value of the $CZ_{1-x}N_x$ ceramics was determined using Equations (27)–(31) [36].

$$E^{\mu} = t_c E_c^{\mu} + t_i E_i^{\mu} \tag{27}$$

$$E_c^{\mu} = \frac{(r_{cA} + r_{cB})}{d^{\mu}} (E_{A-A} E_{B-B})^{1/2} \tag{28}$$

$$E_i^{\mu} = \frac{1389.088}{d^{\mu}} \tag{29}$$

$$t_i = \left| \frac{(S_A - S_B)}{6} \right| \tag{30}$$

$$(t_c + t_i) = 1 \tag{31}$$

Specifically, $E_{A\text{-}A}$ and $E_{B\text{-}B}$ represent the bond energies of atoms A and B, respectively, while r_{cA} and r_{cB} denote the covalent radii of ions A and B. The calculation results are presented in Figure 9c and Table 5. It is noteworthy that despite not displaying the highest E value, the Ce-O bond was indicated by the α results as linked to the τ_f value. Moreover, the average E value ($E_{ave.(Mo\text{-}O)}$) for the Ce-O bond aligned with the trend of the τ_f, suggesting the significant role played by the Ce-O bond in ceramic stability, which is consistent with the α calculation outcomes.

Microstrip antennas, which are widely utilized in wireless communication and radar fields due to their simple design, ease of production, relatively low cost, and very narrow bandwidth, were the focus of this study. A rectangular microstrip antenna with a center frequency of 3.47 GHz was designed using a coaxial feeding method and $CZ_{0.98}N_{0.02}$ ceramics as the basic materials for the antenna. It is noteworthy that 3.47 GHz falls within the S-band of the microwave radio frequency commonly employed in wireless communication and radar fields. Furthermore, it is situated within the frequency range of the 5G communication band (3.3 GHz~4.2 GHz), specifically the n77 and n78 frequency bands [37–39].

Figure 10a illustrates the optimized antenna model and corresponding design dimensions centered at 3.47 GHz, with radiation patches and ground parts designed using silver (Ag). The return loss parameters (S11) based on the designed model are depicted in Figure 10b. It is evident from the figure that the antenna's center frequency is 3.47 GHz, exhibiting a bandwidth of 75 MHz and an S11 value of −33.15 dB. Notably, when S11 falls below −10 dB, the minimal reflection impact on the transmission system allows for normal antenna operation. Furthermore, Figure 10c–e presents simulations of the antenna's radiation in both three-dimensional and electromagnetic field views. The distribution of the main lobe and side lobe radiation from the antenna is clearly delineated, with a maximum radiation gain of 4.32 dB observed. Additionally, unidirectional and symmetrical maximum electric field and magnetic field gain directions indicate favorable radiation characteristics for this antenna design. Consequently, $CZ_{0.98}N_{0.02}$ ceramics exhibit significant potential for application in 5G communication.

Table 5. The E (kJ/mol) of the $CZ_{1-x}N_x$ ceramics sintered at optimal temperature.

Bond Type	x = 0.02	x = 0.04	x = 0.06	x = 0.08	x = 0.10
Ce-O(1) [a]	406.8505	410.6544	401.4945	399.9166	404.4490
Ce-O(1) [b]	406.7539	410.5723	401.4004	399.8233	404.3694
Ce-O(1) [c]	406.6895	410.5067	401.3377	399.7611	404.3058
Ce-O(2) [a]	410.0645	404.6082	398.7994	403.8610	399.6523
Ce-O(2) [b]	410.0155	404.5445	398.7530	403.8133	399.6057
Ce-O(2) [c]	409.9500	404.4967	398.6911	403.7499	399.5435
Ce-O(6) [a]	326.0845	388.2832	390.5555	391.2096	397.8577
Ce-O(6) [b]	326.0742	388.2539	390.5109	391.1798	397.8269
Ce-O(6) [c]	326.0224	388.1952	390.4516	391.1053	397.7653
Zr(BaNb)1-O(4) × 6	459.2762	489.8926	461.1076	481.5940	496.2735
Zr(BaNb)2-O(3) [a]	518.9672	518.3938	512.9597	512.3371	519.4946
Zr(BaNb)2-O(3) [b]	518.8108	518.2381	512.9597	512.3623	519.3393
Zr(BaNb)2-O(3) [c]	518.7847	518.2122	512.7823	512.5390	519.3134
Zr(BaNb)2-O(5) [a]	525.3543	500.8407	518.6251	517.7693	519.3134
Zr(BaNb)2-O(5) [b]	525.2741	500.7680	518.5733	517.7177	519.2358
Zr(BaNb)2-O(5) [c]	525.1139	500.6227	518.4178	517.5632	519.1066
Mo1-O(1)	592.0846	602.8673	619.4640	614.6348	623.9548
Mo1-O(2)	567.3968	596.3924	614.4154	590.1899	598.0154
Mo1-O(3)	544.9440	536.5641	538.9721	545.7791	534.0395
Mo1-O(4)	583.1916	555.2862	561.8104	561.0780	571.3516
Mo2-O(5) × 2	564.4201	576.4860	546.1831	547.3699	559.0432
Mo2-O(6) × 2	820.3593	645.2379	632.2044	637.9052	627.9761

[a–c] Three different bonds.

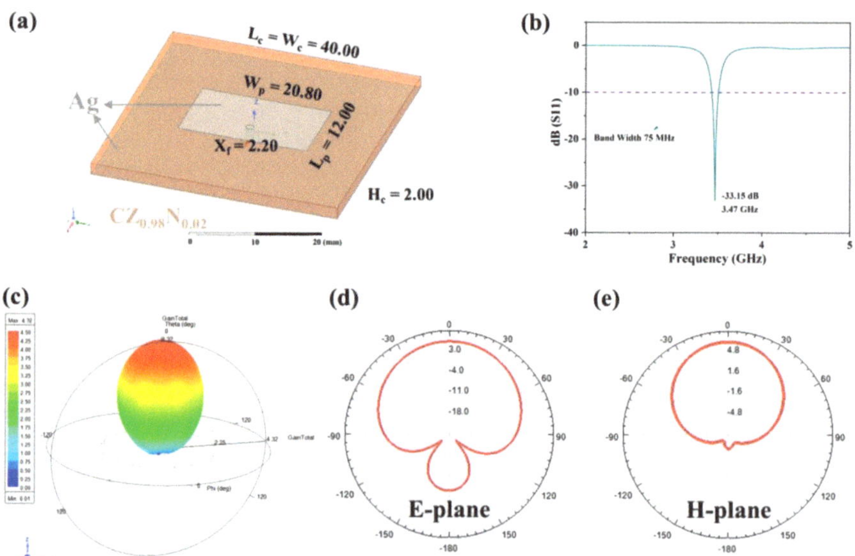

Figure 10. (**a**) The design model and dimensions, (**b**) simulated S11 parameters, (**c**) 3D radiation pattern, (**d**) E-plane, and (**e**) H-plane of the antenna.

4. Conclusions

In this work, $CZ_{1-x}N_x$ ($0.02 \leq x \leq 0.1$) ceramics were synthesized via the conventional solid-state method. The XRD and Rietveld refinement analyses confirmed that all the $CZ_{1-x}N_x$ ceramics exhibited a trigonal structure (space group R-3c) as a single phase. Furthermore, the increase in the unit cell volume with the increasing x value substantiated the successful substitution of $(Ba_{1/3}Nb_{2/3})^{4+}$. The density and SEM results validated the high density of the synthesized ceramics. Notably, the $CZ_{0.98}N_{0.02}$ ceramics exhibited the most uniform grain size, which enhances performance. Specifically, at 675 °C, the $CZ_{0.98}N_{0.02}$ ceramics demonstrated superior performance, with ε_r = 10.41, $Q\times f$ = 53,296 GHz, and τ_f = −23.45 ppm/°C. The P-V-L theory was employed to establish a correlation between structure and performance characteristics. The Ce-O bond predominantly influenced the ε_r and τ_f of the $CZ_{1-x}N_x$ ceramics. Furthermore, the Mo-O bond was strongly correlated with $Q\times f$. Finally, a patch antenna tailored for 5G communication bands (n77 and n78) was designed using the $CZ_{0.98}N_{0.02}$ ceramics as the substrate materials. This study represents an initial exploration of the practical applications of CZM ceramics.

Author Contributions: Conceptualization, H.G. and X.X.; methodology, H.G. and X.X.; software, H.G. and X.X.; validation, H.G., X.X., X.L. and X.Z.; formal analysis, H.G., X.X., X.L. and X.Z.; investigation, H.G., X.X., X.L. and X.Z.; resources, M.L., J.D. and H.W.; data curation, H.G.; writing—original draft preparation, H.G.; writing—review and editing, H.G., X.X., M.L., J.D. and H.W.; visualization, H.G. and X.X.; supervision, M.L., J.D. and H.W.; project administration, M.L., J.D. and H.W.; funding acquisition, M.L., J.D. and H.W. All authors have read and agreed to the published version of the manuscript.

Funding: The project was supported by the National Natural Science Foundation of China (52302116) and Graduate Innovation Foundation of Yantai University, GIFYTU. The authors are also thankful to Zeming Qi and Chuansheng Hu in the IR beamline workstation of the National Synchrotron Radiation Laboratory (NSRL) for the IR measurement. Thanks zkec (www.zkec.cc) for XRD and SEM examination.

Institutional Review Board Statement: Not applicable.

Informed Consent Statement: Not applicable.

Data Availability Statement: The data presented in this study are available in this article.

Conflicts of Interest: The authors declare no conflicts of interest.

References

1. Shafi, M.; Molisch, A.F.; Smith, P.J.; Haustein, T.; Zhu, P.; Silva, P.D.; Tufvesson, F.; Benjebbour, A.; Wunder, G. 5G: A tutorial overview of standards, trials, challenges, deployment, and practice. *IEEE J. Sel. Areas Commun.* **2017**, *35*, 1201–1221. [CrossRef]
2. Li, S.; Xu, L.D.; Zhao, S. 5G Internet of Things: A survey. *J. Ind. Inf. Integr.* **2018**, *10*, 1–9. [CrossRef]
3. Yao, G.; Yan, J.; Tan, J.; Pei, C.; Liu, P.; Zhang, H.; Wang, D. Structure, chemical bond and microwave dielectric characteristics of novel $Li_3Mg_4NbO_8$ ceramics. *J. Eur. Ceram. Soc.* **2021**, *41*, 6490–6494. [CrossRef]
4. Luo, W.; Yan, S.; Zhou, J. Ceramic-based dielectric metamaterials. *Interdiscip. Mater.* **2022**, *1*, 11–27. [CrossRef]
5. Tan, K.; Song, T.; Shen, T.; Yu, H.; Zhang, Y.; Cui, K.; Xu, X.; Li, W.; Wang, H. Research progress of low permittivity microwave dielectric ceramics. *Adv. Ceram.* **2022**, *43*, 11–29. [CrossRef]
6. Dai, Y.; Chen, J.; Tang, Y.; Xiang, H.; Li, J.; Fang, L. Relationship between bond characteristics and microwave dielectric properties of $REVO_4$ (RE = Yb, Ho) ceramics. *Ceram. Int.* **2023**, *49*, 875–881. [CrossRef]
7. Pei, C.; Hou, C.; Li, Y.; Yao, G.; Ren, Z.; Liu, P.; Zhang, H. A low ε_r and temperature-stable $Li_3Mg_2SbO_6$ microwave dielectric ceramics. *J. Alloys Compd.* **2019**, *792*, 46–49. [CrossRef]
8. Pei, C.; Tan, J.; Li, Y.; Yao, G.; Jia, Y.; Ren, Z.; Liu, P.; Zhang, H. Effect of Sb-site nonstoichiometry on the structure and microwave dielectric properties of $Li_3Mg_2Sb_{1-x}O_6$ ceramics. *J. Adv. Ceram.* **2020**, *9*, 588–594. [CrossRef]
9. Huang, X.; Guo, H.; Zhu, P.; Liu, L.; Xiao, J.; Tang, D.; Lin, C.; Wu, X.; Zheng, X. Microwave dielectric properties of $CaCu_3Ti_4O_{12}$ ceramics: A clue to its intrinsic dielectric response. *J. Adv. Dielectr.* **2023**, *13*, 2344001–2344006. [CrossRef]
10. An, Z.; Lv, J.; Wang, X.; Xu, Y.; Zhang, L.; Shi, F.; Guo, H.; Zhou, D.; Liu, B.; Song, K. Effects of LiF additive on crystal structures, lattice vibrational characteristics and dielectric properties of $CaWO_4$ microwave dielectric ceramics for LTCC applications. *Ceram. Int.* **2022**, *48*, 29929–29937. [CrossRef]

11. Zhou, D.; Pang, L.; Wang, D.; Guo, H.; Yang, F.; Qi, Z.; Li, C.; Jin, B.; Reaney, I.M. Crystal structure, impedance and broadband dielectric spectra of ordered scheelite-structured $Bi(Sc_{1/3}Mo_{2/3})O_4$ ceramic. *J. Eur. Ceram. Soc.* **2018**, *38*, 1556–1561. [CrossRef]
12. Tao, B.J.; Xing, C.F.; Wang, W.F.; Wu, H.T.; Zhou, Y.Y. A novel $Ce_2Zr_3(MoO_4)_9$ microwave dielectric ceramic with ultra-low firing temperature. *Ceram. Int.* **2019**, *45*, 24675–24683. [CrossRef]
13. Liu, W.; Zuo, R. Low temperature fired $Ln_2Zr_3(MoO_4)_9$ (Ln = Sm, Nd) microwave dielectric ceramics. *Ceram. Int.* **2017**, *43*, 17229–17232. [CrossRef]
14. Liu, W.; Zuo, R. A novel low-temperature firable $La_2Zr_3(MoO_4)_9$ microwave dielectric ceramic. *J. Eur. Ceram. Soc.* **2018**, *38*, 339–342. [CrossRef]
15. Zhang, Y.H.; Sun, J.J.; Dai, N.; Wu, Z.C.; Wu, H.T.; Yang, C.H. Crystal structure, infrared spectra and microwave dielectric properties of novel extra low-temperature fired $Eu_2Zr_3(MoO_4)_9$ ceramics. *J. Eur. Ceram. Soc.* **2019**, *39*, 1127–1131. [CrossRef]
16. Shi, L.; Liu, C.; Zhang, H.; Peng, R.; Wang, G.; Shi, X.; Wang, X.; Wang, W. Crystal structure, Raman spectroscopy, metal compatibility and microwave dielectric properties of $Ce_2Zr_3(MoO_4)_9$ ceramics. *Mater. Chem. Phys.* **2020**, *250*, 122954. [CrossRef]
17. Zheng, J.; Liu, Y.; Tao, B.; Zhang, Q.; Wu, H.; Zhang, X. Crystal structure and optimised microwave dielectric properties of $Ce_2(Zr_{1-x}Ti_x)_3(MoO_4)_9$ solid solutions. *Ceram. Int.* **2021**, *47*, 5624–5630. [CrossRef]
18. Xu, X.; Xi, Z.; Feng, Z.; Zhang, X.; Du, W.; Shan, L.; Du, J.; Wu, H.; Wangsuo, X. Microstructure, bonding characteristics, far-infrared spectra and microwave dielectric properties of co-substituted $Ce_2[(Zr_{1-x}(Zn_{1/3}Sb_{2/3})_x]_3(MoO_4)_9$ ceramics. *Ceram. Int.* **2024**, *50*, 24769–24780. [CrossRef]
19. Pan, H.; Yan, S.; Zhang, Y.; Du, J.; Zhang, X.; Gong, P.; Wu, H.; Tian, H.; Wübbenhorst, M. Crystal structure, bond characteristics and microwave dielectric properties of $Ce_2[Zr_{1-x}(Sr_{1/3}B_{2/3})_x]_3(MoO_4)_9$ (B = Ta, Sb) solid solutions. *Ceram. Int.* **2023**, *49*, 24038–24046. [CrossRef]
20. Du, K.; Zhou, M.; Li, C.; Yin, C.; Cai, Y.; Cheng, M.; Zhu, W.; Wei, G.; Wang, S.; Lei, W. Ultralow-permittivity and temperature-stable $Ba_{1-x}Ca_xMg_2Al_6Si_9O_{30}$ dielectric ceramics for C-Band patch antenna applications. *ACS Appl. Mater. Interfaces* **2024**, *16*, 23505–23516. [CrossRef]
21. Siragam, S. Synthesis and fabrication of proto-type microstrip patch antenna by using microwave dielectric ceramic nanocomposite material for C-band application. *Optik* **2024**, *306*, 171787. [CrossRef]
22. Shannon, R.D. Revised effective ionic radii and systematic studies of interatomic distances in halides and chalcogenides. *Found. Crystallogr.* **1976**, *32*, 751–767. [CrossRef]
23. McCusker, L.; Von Dreele, R.; Cox, D.; Louër, D.; Scardi, P. Rietveld refinement guidelines. *J. Appl. Crystallogr.* **1999**, *32*, 36–50. [CrossRef]
24. Yang, H.; Chai, L.; Liang, G.; Xing, M.; Fang, Z.; Zhang, X.; Qin, T.; Li, E. Structure, far-infrared spectroscopy, microwave dielectric properties, and improved low-temperature sintering characteristics of tri-rutile $Mg_{0.5}Ti_{0.5}TaO_4$ ceramics. *J. Adv. Ceram.* **2023**, *12*, 296–308. [CrossRef]
25. Bao, J.; Zhang, Y.; Kimura, H.; Wu, H.; Yue, Z. Crystal structure, chemical bond characteristics, infrared reflection spectrum, and microwave dielectric properties of $Nd_2(Zr_{1-x}Ti_x)_3(MoO_4)_9$ ceramics. *J. Adv. Ceram.* **2023**, *12*, 82–92. [CrossRef]
26. Wang, X.; Liu, T.; Cao, Z.; Li, Z.; Xu, Y.; Shi, F.; Zhang, L.; Qi, Z.-m. Lattice vibrational characteristics and structure-property relationships of $Ca(Mg_{1/2}W_{1/2})O_3$ microwave dielectric ceramics with different sintering temperatures. *Ceram. Int.* **2022**, *48*, 1415–1422. [CrossRef]
27. Li, J.; Wang, Z.; Guo, Y.; Ran, S. Influences of substituting of $(Ni_{1/3}Nb_{2/3})^{4+}$ for Ti^{4+} on the phase compositions, microstructures, and dielectric properties of $Li_2Zn[Ti_{1-x}(Ni_{1/3}Nb_{2/3})_x]_3O_8$ ($0 \leq x \leq 0.3$) microwave ceramics. *J. Adv. Ceram.* **2023**, *12*, 760–777. [CrossRef]
28. Shannon, R.D. Dielectric polarizabilities of ions in oxides and fluorides. *J. Appl. Phys.* **1993**, *73*, 348–366. [CrossRef]
29. Phillips, J.C.; Van Vechten, J.A. Charge redistribution and piezoelectric constants. *Phys. Rev. Lett.* **1969**, *23*, 1115–1117. [CrossRef]
30. Levine, B.F. Bond susceptibilities and ionicities in complex crystal structures. *J. Chem. Phys.* **1973**, *59*, 1463–1486. [CrossRef]
31. Wu, Z.J.; Meng, Q.B.; Zhang, S.Y. Semiempirical study on the valences of Cu and bond covalency in $Y_{1-x}Ca_xBa_2Cu_3O_{6+y}$. *Phys. Rev. B* **1998**, *58*, 958–962. [CrossRef]
32. Stepan, S.B. Dielectric methods of studying the chemical bond and the concept of electronegativity. *Russ. Chem. Rev.* **1982**, *51*, 684. [CrossRef]
33. Berkov, D.V. Evaluation of the energy barrier distribution in many-particle systems using the path integral approach. *J. Phys. Condens. Matter* **1998**, *10*, L89. [CrossRef]
34. Fang, W.; Chen, J.; Yang, Y.; Ao, L.; Tang, Y.; Li, J.; Fang, L. Anomalous microwave dielectric behaviour induced by the orthorhombic-tetragonal phase transition in $CaLaGaO_4$ ceramics. *J. Eur. Ceram. Soc.* **2022**, *42*, 1474–1479. [CrossRef]
35. Sun, Y.; Xiang, H.; Tang, Y.; Li, J.; Fang, L. Constructing the cationic rattling effect to realize the adjustability of the temperature coefficient in $Nd_{2-x}Sm_xO_3$ microwave dielectric ceramics. *J. Eur. Ceram. Soc.* **2024**, *44*, 2859–2865. [CrossRef]
36. Wang, G.; Zhang, D.; Li, J.; Gan, G.; Rao, Y.; Huang, X.; Yang, Y.; Shi, L.; Liao, Y.; Liu, C.; et al. Crystal structure, bond energy, Raman spectra, and microwave dielectric properties of Ti-doped $Li_3Mg_2NbO_6$ ceramics. *J. Am. Ceram. Soc.* **2020**, *103*, 4321–4332. [CrossRef]

37. Li, F.; Li, Y.; Li, S.; Luo, Y.; Lu, Y.; Tang, T.; Liao, Y.; Zhang, J.; Wen, Q. All-ceramic array patch for 5G signal enhancement based on B-site substituted zinc-cobalt molybdate low temperature co-fired ceramics. *Chem. Eng. J.* **2023**, *466*, 143325. [CrossRef]
38. Wu, F.F.; Zhou, D.; Du, C.; Xu, D.M.; Li, R.T.; Shi, Z.Q.; Darwish, M.; Zhou, T.; Jantunen, H. Design and fabrication of a satellite communication dielectric resonator antenna with novel low loss and temperature stabilized $(Sm_{1-x}Ca_x)(Nb_{1-x}Mo_x)O_4$ (x = 0.15–0.7) microwave ceramics. *Chem. Mater.* **2023**, *35*, 104–115. [CrossRef]
39. Tian, H.; Zhang, X.; Zhang, Z.; Liu, Y.; Wu, H. Low-permittivity $LiLn(PO_3)_4$ (Ln = La, Sm, Eu) dielectric ceramics for microwave/millimeter-wave communication. *J. Adv. Ceram.* **2024**, *13*, 602–620. [CrossRef]

ceramics

MDPI

Article

Tunable Optical Properties and Relaxor Behavior in Ni/Ba Co-Doped $NaNbO_3$ Ceramics: Pathways Toward Multifunctional Applications

Tawfik Chaabeni [1], Zohra Benzarti [1,2,*], Najmeddine Abdelmoula [1] and Slim Zghal [1,3,*]

1 Laboratory of Multifunctional Materials and Applications (LaMMA, LR16ES18), Faculty of Sciences of Sfax, University of Sfax, B.P. 1171, Sfax 3000, Tunisia; chaabeni.tawfik91@gmail.com (T.C.); najmeddine.abdelmoula@fss.rnu.tn (N.A.)

2 Department of Mechanical Engineering, ARISE, CEMMPRE, University of Coimbra, Rua Luís Reis Santos, 3030-788 Coimbra, Portugal

3 High School of Science and Technology of Hammam Sousse, University of Sousse, Sousse 4011, Tunisia

* Correspondence: zohra.benzarti@dem.uc.pt (Z.B.); slim.zghal70@gmail.com (S.Z.)

Abstract: In this study, Ni/Ba co-doped $NaNbO_3$ ceramics ($NBNNO_x$) were synthesized using a solid-state method to explore the effects of Ni^{2+} and Ba^{2+} ion substitution on the structural, optical, and dielectric properties of $NaNbO_3$. X-ray diffraction (XRD) confirmed that the ceramics retained an orthorhombic structure, with crystallinity improving as the doping content (x) increased. Significant lattice distortions induced by the Ni/Ba co-doping were observed, which were essential for preserving the perovskite structure. Raman spectroscopy revealed local structural distortions, influencing optical properties and promoting relaxor behavior. Diffuse reflectance measurements revealed a significant decrease in band gap energy from 3.34 eV for undoped $NaNbO_3$ to 1.08 eV at $x = 0.15$, highlighting the impact of co-doping on band gap tunability. Dielectric measurements indicated relaxor-like behavior at room temperature for $x = 0.15$, characterized by frequency-dependent anomalies in permittivity and dielectric loss, likely due to ionic disorder and structural distortions. These findings demonstrate the potential of Ni/Ba co-doped $NaNbO_3$ ceramics for lead-free perovskite solar cells and other functional devices, where tunable optical and dielectric properties are highly desirable.

Keywords: Ni/Ba co-doped $NaNbO_3$ ceramics; solid–solid method; structural distortions; band gap tunability; relaxor behavior; multifunctional applications

Citation: Chaabeni, T.; Benzarti, Z.; Abdelmoula, N.; Zghal, S. Tunable Optical Properties and Relaxor Behavior in Ni/Ba Co-Doped $NaNbO_3$ Ceramics: Pathways Toward Multifunctional Applications. *Ceramics* **2024**, *7*, 1670–1694. https://doi.org/10.3390/ceramics7040107

Academic Editors: Dawei Wang and Fayaz Hussain

Received: 28 September 2024
Revised: 26 October 2024
Accepted: 5 November 2024
Published: 8 November 2024

1. Introduction

Perovskite niobates, including $NaNbO_3$, have attracted considerable attention due to their environmentally friendly characteristics and wide-ranging technological applications [1–5]. Among the limited class of lead-free antiferroelectric (AFE) perovskites, $NaNbO_3$-based relaxor antiferroelectrics are particularly notable for their ultrahigh energy-storage density, positioning them as potential candidates for high-power electronics and energy-storage capacitors [6,7]. This exceptional energy-storage capability is largely attributed to their reversible phase transitions and high dielectric constants, which are crucial for capacitive energy storage. Additionally, the intrinsic high chemical stability of $NaNbO_3$ enhances its utility in sensor applications, particularly in humidity sensing, where it offers reliable performance even under severe environmental conditions [8]. Recent advances have demonstrated that lead-free $NaNbO_3$-based ceramics can exhibit relaxor-type dielectric behavior, characterized by diffuse phase transitions and a strong frequency dependence of the dielectric constant [9,10]. This makes them promising materials for applications in tunable microwave devices, actuators, and capacitors [11]. Sodium niobate, typically regarded as an AFE material, can transition to a ferroelectric state through suitable chemical doping [11,12] or the application of an external

DC electric field [2]. This tunable ferroelectricity, driven by modifications in the crystal structure, is pivotal for applications in high-density optical data storage and energy storage, where controlled polarization states are crucial [11].

Recently, $NaNbO_3$ has emerged as a promising photocatalytic material due to its greater stability compared to other Nb-based photocatalysts and its unique corner-shared $[NbO_6]$ octahedral crystal structure, which is reported to significantly enhance charge migration and reduce the recombination rates of photogenerated electron–hole pairs. Extensive studies have demonstrated that $NaNbO_3$ is highly efficient in photocatalytic applications, such as hydrogen production through water splitting [13], CO_2 reduction [13,14], decomposition of organic pollutants [15], and the degradation of organic contaminants in wastewater treatment [16]. These capabilities are largely attributed to its specific crystal structure and the efficient separation of charge carriers. However, the wide band gap, typically around 3.4 eV in pure $NaNbO_3$, restricts its absorption to the ultraviolet (UV) range, which covers a limited part of the solar spectrum. This limitation reduces its efficiency as a visible-light photocatalyst. To overcome this limitation and enhance the photocatalytic performance of UV-active materials like $NaNbO_3$, extending their light absorption capabilities into the visible range is crucial. Strategies such as metal decoration, including noble metals like Pt [17] and Au [18], and coupling with other narrow band gap semiconductors like AgO_2 [19] and g-C_3N_4 [20] have shown some success in this regard. Moreover, doping with transition metals such as Ag [21] or nonmetals like N [22] has been investigated to reduce the band gap and enhance the absorption of visible light. More recently, research has shifted towards developing new photocatalysts with tunable band gaps and band-edge positions through precise chemical doping [23]. In particular, the high susceptibility of ABO_3 perovskites to partial cation substitution at both the A and B sites presents a promising approach for continuous band energy modulation, facilitating the development of high-performance, visible-light-sensitive $NaNbO_3$-based photocatalysts. This approach not only tailors the electronic structure but also improves the photocatalytic activity by optimizing the redox potential and enhancing charge carrier mobility.

Moreover, when these materials exhibit ferroelectric properties, their spontaneous polarization positions them as promising candidates for photovoltaic applications and other multifunctional devices [24,25]. This polarization can induce a bulk photovoltaic effect, where charge carriers, particularly electrons and holes, are efficiently separated, thereby reducing recombination rates [26]. This effect is particularly advantageous in ferroelectric materials, where the intrinsic electric field can drive charge separation without the need for an external p-n junction, leading to potential applications in self-powered devices. Several materials, such as $KNbO_3$ [27], $LiNbO_3$ [28], $BaTiO_3$ [29], and $BiFeO_3$ [30], have been recognized as potential candidates for photovoltaic applications due to their strong ferroelectric characteristics. These materials have been extensively studied for their ability to generate photocurrents under light illumination. Among them, $BiFeO_3$ stands out for its relatively narrow band gap and notable ferroelectric properties, which significantly improve its photocatalytic and photovoltaic performances [31,32]. However, a primary challenge in utilizing perovskite-type ferroelectrics for photovoltaic applications is their inherently wide band gap. This wide band gap, influenced by the nature of the A-O and B-O metal–oxygen bonds, restricts light absorption primarily to the UV region. Extending light absorption into the visible wavelength range is therefore crucial for improving the photovoltaic efficiency of these materials. Recent research has focused on various strategies to narrow the band gap of these ferroelectrics, such as doping with transition metals or creating solid solutions with other perovskites [33,34]. These approaches aim to enhance visible light absorption and optimize the electronic structure for better photovoltaic performance, making these materials more competitive with other emerging photovoltaic technologies.

Identifying methods to reduce the band gap of ferroelectrics while preserving their beneficial ferroelectric properties could lead to significant advancements in ABO_3 oxide materials. Recent theoretical work based on density functional theory (DFT) calculations suggests that the substitution of the B site by Group 10 metals such as Ni, Pd, and Pt, as well as stabilization of these substitutions by the introduction of oxygen vacancies, can significantly reduce the band gap while preserving or even increasing the spontaneous polarization of the parent phase [35]. These theoretical predictions have been experimentally validated by Grinberg et al., who demonstrated that Ni-containing $(K_{1-x}Ba_x)(Nb_{1-x/2}Ni_{x/2})O_{3-\delta}$ (KBNNO) solid solutions exhibit both ferroelectricity and band gap tenability up to 2.7 eV [36]. Such doping strategies effectively narrow the band gap and enable these ferroelectric perovskites to absorb visible light, which opens promising pathways for the development of semiconductor-based ferroelectric perovskite solar cells and other advanced applications.

In this context, our approach involves developing $NaNbO_3$-based materials by co-doping with Ni^{2+} and Ba^{2+} ions, selected specifically for their ability to adjust optical energy bands and modify dielectric properties. The substitution of the A-site (Na^+) with the larger cation (Ba^{2+}) is expected to further enhance solubility and promote the formation of vacancies. This study aims to understand how varying the concentration of these dopants affects the structural, dielectric, and optical properties of $[NaNbO_3]_{1-x}[BaNi_{1/2}Nb_{1/2}O_{3-\delta}]_x$ ($NBNNO_x$) ceramics, synthesized via solid-state reactions. Structural characterization was performed using X-ray diffraction and Raman spectroscopy. The dielectric properties were conducted by measuring the dielectric constant and loss over a range of temperatures and frequencies, while optical properties were examined at room temperature using diffuse reflectance spectroscopy. The goal is to explore the potential applications of these materials in advanced functional devices, potentially advancing technologies for photocatalysis, photovoltaics, and other solar-light-sensitive applications.

2. Materials and Methods

$[NaNbO_3]_{1-x}[BaNi_{1/2}Nb_{1/2}O_{3-\delta}]_x$ ceramic samples with compositions of x = 0, 0.05, 0.10, and 0.15 were synthesized using the solid-state reaction method. These ceramic samples are subsequently designated as NNO, $NBNNO_{0.05}$, $NBNNO_{0.10}$, and $NBNNO_{0.15}$, respectively. The starting materials, which included highly pure powders of Na_2CO_3, $BaCO_3$, Nb_2O_5, and NiO (Sigma-Aldrich 99.9%, St. Louis, MO, USA), were dried at 473 K for 2 h. These powders were then weighed in stoichiometric ratios and mixed for 2 h in an agate mortar. The resulting mixtures were calcined at 1273 K for 12 h, followed by an additional 2 h mixing before being pressed into pellets measuring 8 mm in diameter and 1 mm in thickness. Finally, the pellets were sintered at 1473 K for 3 h. The compactness values (C), defined as the ratio of experimental density (d_{exp}) to theoretical density (d_{theo}), for the sintered specimens increased from 93% for the undoped NNO (x = 0) to 96% for the $NBNNO_{0.15}$-doped sample. The diameter shrinkages of the ceramic disks ($\Delta\Phi/\Phi$) were calculated as $(\Phi_{in} - \Phi_{fin})/\Phi_{in}$, where Φ_{in} and Φ_{fin} represent the initial and final diameters, respectively. These values were found to range between 0.13 and 0.16. To form electrodes for dielectric measurements, silver paste was applied to both sides of the sintered ceramic pellets.

The crystal structure of the ceramic samples was analyzed at room temperature using X-ray diffraction (XRD). The XRD patterns were recorded on a Philips diffractometer (Philips, Eindhoven, The Netherlands) with CuKα radiation (λ = 1.5418 Å) over the angular range of $20° < 2\theta < 70°$, employing a step size of 0.02° and a counting time of 1 s per step. The surface morphology of the as-prepared ceramics was examined using an FEI XL-30 Scanning Electron Microscope (FEI, Oregon, OR, USA), which offers a typical resolution of 1.5 nm at 10 kV or higher. Raman scattering data were collected at room temperature in the wavenumber range of 50–1000 cm^{-1} using a Horiba HR800 Raman spectrometer (Horiba Scientific, Kyoto, Japan) equipped with a 632 nm red laser. Diffuse reflectance measurements were carried out at room temperature in the wavelength range of 350–900 nm, using an Ocean Optics HR2000 monochromator, a DH-2000-Bal lamp, a UV–Vis fiber, and a WS-1 reflectance standard featuring a PTFE optical diffuser (Ocean Optics, Dunedin, FL, USA). The temperature dependence of the complex relative permittivity and dielectric loss were measured using an HP 4192A impedance analyzer (HP, Santa Clara, CA, USA) within the temperature range of 100–800 K and at frequencies from 1 kHz to 1 MHz.

3. Results

3.1. Structural and Microstructure Analysis

3.1.1. X-Ray Diffraction

Figure 1a shows the XRD patterns of the sintered $[NaNbO_3]_{1-x}[BaNi_{1/2}Nb_{1/2}O_{3-\delta}]_x$ ceramics ($NBNNO_x$), exhibiting typical diffraction peaks of an ABO_3 perovskite structure. Room-temperature XRD analysis confirmed that the $NBNNO_x$ ceramics crystallized in a single-phase, with no traces of residual precursors detected. The indexing of Bragg reflections indicates that the $NBNNO_x$ ceramics crystallize in an orthorhombic structure, with progressive changes in crystal symmetry as the doping rate (x) increases. The identified space groups are Pbma (for x = 0), Pnma (for x = 0.05), and Amm2 (for both x = 0.10 and 0.15). Notably, the irregular shifts of the (020) diffraction peak observed in Figure 1b can be attributed to structural changes and space group transitions induced by doping. Transitions within orthorhombic phases (e.g., Pbma → Pnma or Pbma → Amm2) affect peak positions, even when lattice parameters change systematically. Factors such as octahedral tilting, atomic displacements, and internal stress likely contribute to these shifts. For $NBNNO_{0.10}$ and $NBNNO_{0.15}$ (both with the Amm2 space group), the observed leftward shift of the (020) and (111) peaks corresponds to an increase in unit cell volume, confirming that when the space group remains consistent, peak shifts align with the trends in lattice parameter expansion.

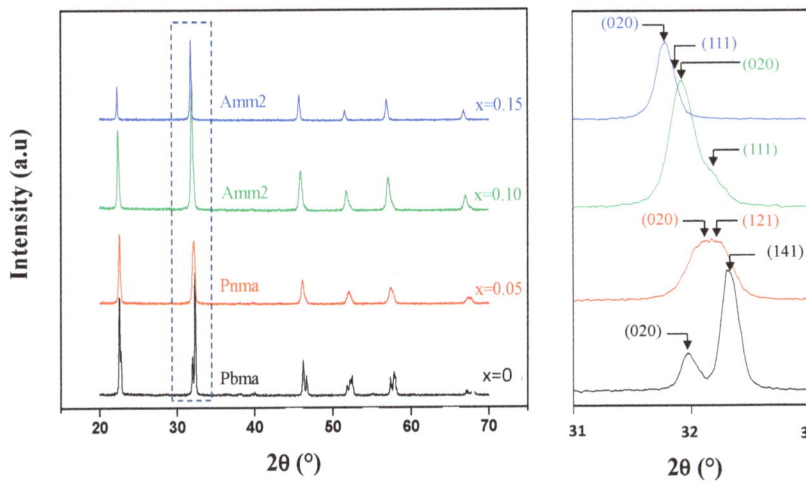

Figure 1. (**a**) X-ray diffraction patterns of sintered $NBNNO_x$ ceramics recorded at room temperature; (**b**) Enlarged view of the XRD patterns in the rectangular window between 31° and 33°, focusing on the principal diffraction peaks.

Structural refinement of the XRD patterns was conducted using the Fullprof program. The experimental and calculated Rietveld-refined XRD patterns for $NBNNO_x$ ceramics are presented in Figure 2. To evaluate the quality of the refinement, various agreement factors between the observed and calculated profiles were defined in Fullprof (2020 version). Satisfactory agreement factors were achieved after several refinement steps. The refinement confirms the presence of an orthorhombic structure for all phases, with space groups Pbma for NNO, Pnma for $NBNNO_{0.05}$, and Amm2 for $NBNNO_{0.10}$ and $NBNNO_{0.15}$ ceramics. Table 1 summarizes the Rietveld-refinement data, including the lattice parameters, unit cell volume, and adjustment parameters. The profile residual (R_p), weighted profile residual (R_{wp}), and goodness of fit (χ^2) parameters are used to evaluate the quality of the refinement [37], all of which confirm the reliability of the XRD results.

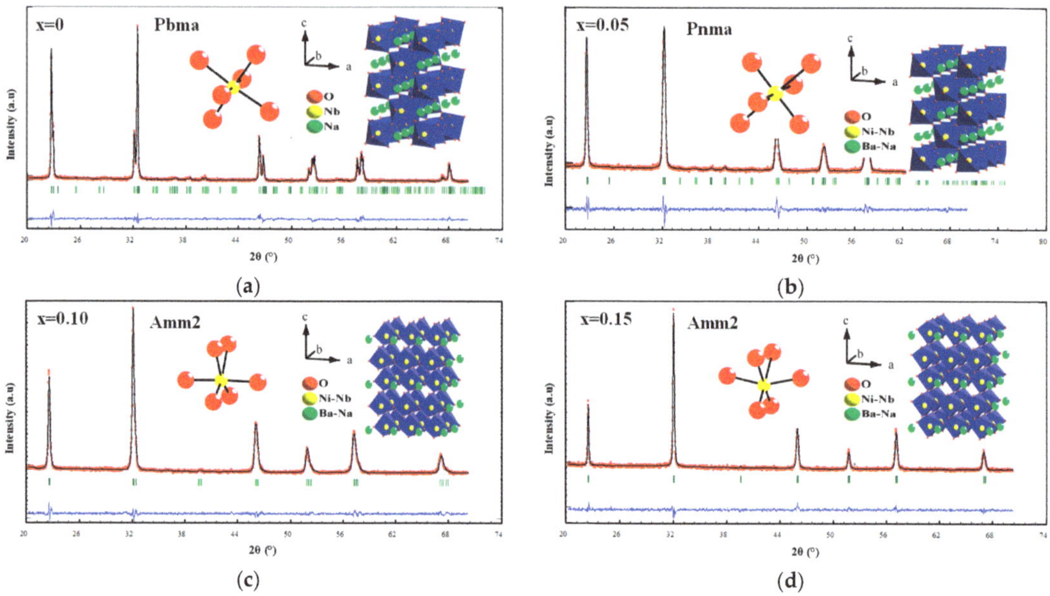

Figure 2. The Rietveld pattern obtained from X-ray powder diffraction of the $NBNNO_x$ ceramics. (**a**) NNO; (**b**) $NBNNO_{0.05}$; (**c**) $NBNNO_{0.10}$; (**d**) $NBNNO_{0.15}$. The points represent the experimental profile, and the red solid line corresponds to the calculated profile. Positions of the Bragg reflections are indicated by green vertical bars. The blue curve at the bottom of the figure represents the discrepancy between the experimental and calculated profiles. The crystal structures were plotted inside using the "Diamond" program.

Table 1. Rietveld-refined lattice parameters of $NBNNO_x$ compounds, bond lengths and bond angles, and discrepancy factors: profile residual (R_p), weighted profile residual (R_{wp}), and goodness of fit (χ^2).

Compounds	NNO	$NBNNO_{0.05}$	$NBNNO_{0.10}$	$NBNNO_{0.15}$
Structure	Orthorhombic	Orthorhombic	Orthorhombic	Orthorhombic
Space group	Pbma	Pnma	Amm2	Amm2
a (Å)	5.56427	5.57398	3.94260	3.95321
b (Å)	5.50223	5.54044	5.52592	5.57900
c (Å)	15.50840	7.82424	5.57834	5.58175
V (Å^3)	474.80352	241.63054	121.53246	123.10526
Bond lengths (Å) and bond angles (°)				
$\langle d(Nb/Ni)–O\rangle_{axial}$	1.95050	2.14500	1.97200	1.97300
$\langle d(Nb/Ni)–O_{pl}\rangle_{apical}$	1.99925	1.90500	2.00500	1.99725
$\langle d(Nb/Ni)–O_{pl}\rangle$	1.98300	1.98500	1.99400	1.98916
$\langle \theta_1(Nb/Ni)–O–(Nb/Ni)\rangle$	165.17	163.99	156.37	179.26
$\langle \theta_2(Nb/Ni)–O–(Nb/Ni)\rangle$	168.29	163.99	156.37	179.26
Discrepancy factors				
R_p	7.46	7.71	6.07	7.77
R_{wp}	9.61	9.74	7.78	9.87
R_{exp}	7.74	7.92	6.47	8.94
χ^2	1.54	1.51	1.44	1.22

Figure 3 illustrates the variation in the unit cell volume and lattice parameters with respect to the Ni/Ba substitution content (x) in the NNO parent phase. It is evident that the change in space group leads to a significant decrease in the unit cell volume with an increased doping rate (x) (Figure 3a), with the volume for $NBNNO_{0.10}$ being up to four times smaller compared to that of NNO. This reduction is mainly attributed to decreases in the lattice parameters (a) and (c), while the (b) parameter remains almost unchanged (Figure 3b). The observed lattice contraction in the (010) planes is likely due to the ionic character of the Ni-O bonds. However, when comparing $NBNNO_{0.15}$ with $NBNNO_{0.10}$, which are both crystallized in the same space group (Amm2) within the orthorhombic structure, a slight increase in cell volume (1.3%) is noted. This increase corresponds to the shift of diffraction peaks toward lower angles, as shown in Figure 1b. The larger ionic radii of Ba^{2+} (1.35 Å) and Ni^{2+} (0.69 Å) compared to Na^{+} (1.02 Å) and Nb^{5+} (0.64 Å) lead to an expansion of the unit cell volume, resulting in a shift of peak positions towards lower angles.

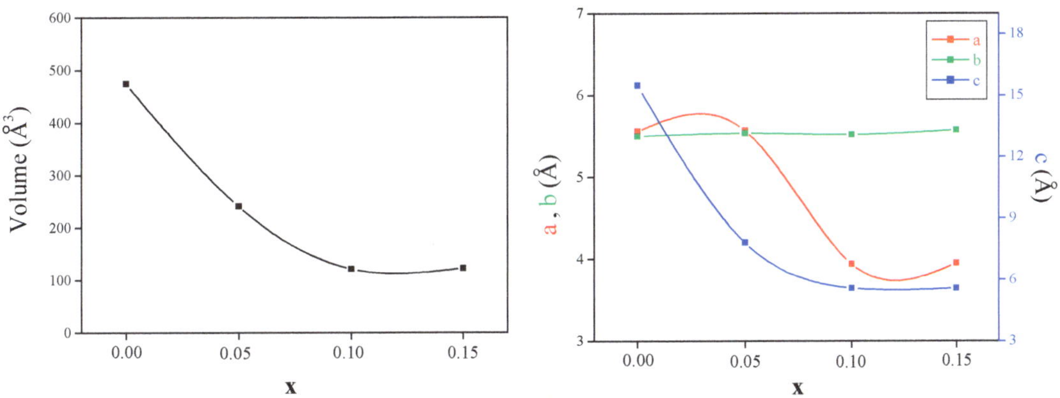

(a) (b)

Figure 3. (**a**) Variation in the unit cell volume of the $NBNNO_x$ orthorhombic phases; (**b**) Lattice parameters a and b shown on the left-hand side ordinate, and c on the right-hand side ordinate as a function of the doping content (x).

For a more detailed understanding of the structural changes induced by Ni/Ba substitution, the crystal structure and BO_6 octahedra of the $NBNNO_x$ phases were visualized using the "Diamond" program, utilizing atomic positions derived from the Rietveld-refinement data. As shown in Figure 4, the BO_6 octahedra form structural chains with periodically modulated tilt angles denoted as θ_1 and θ_2, within the orthorhombic structures of $NBNNO_x$. The octahedra alternate between these angles, where pairs of neighboring octahedra are tilted by θ_1 within each chain, and adjacent pairs are tilted by θ_2 relative to each other. This alternating pattern of tilt angles influences the bonding environment and contributes to the structural stability of the material [38]. The values of the O–B–O angles θ_1 and θ_2 as a function of the doping rate (x) are measured and plotted in Figure 5. The O–B–O tilt angles change significantly with increasing content (x), which may lead to varying levels of structural ordering and distortions in the $NBNNO_x$ ceramics. The orthorhombic structure of the NNO parent phase with the Pbma space group exhibits two distinct tilt angles (θ_1 and θ_2). However, for compositions with $x \geq 0.05$, these angles converge, signaling a transition toward a more symmetric configuration. This shift is accompanied by significant variations in the O–B–O angle. As shown in Figure 5a, the sequential change in space groups, from Pbma (x = 0) to Pnma (x = 0.05) and Amm2 (x = 0.10), is accompanied by decreasing tilt angles, which reflect increased misalignment between octahedra. The misalignment, defined as a 180° tilt angle, indicates that smaller tilt angles lead to greater structural distortion. Consequently, $NBNNO_{0.05}$ and $NBNNO_{0.10}$ show higher distortion compared to the NNO parent phase. Although both $NBNNO_{0.10}$ and $NBNNO_{0.15}$ adopt the Amm2 space group, the significant increase in the O–B–O tilt angle from 130° to 180° suggests reduced misalignment between adjacent octahedra, indicating enhanced structural stability and a decrease in the degree of crystal structure distortion, as discussed in the following section.

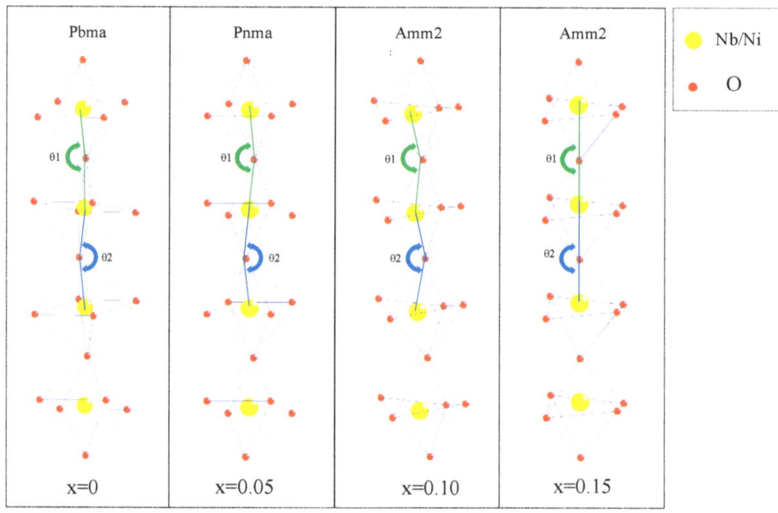

Figure 4. Arrangements of the BO_6 octahedra along the c-axis of the $NBNNO_x$ structures. θ_1 and θ_2 represent the periodically modulated tilt angles of the structural chains. The oxygen ions are represented by red spheres, and the Nb/Ni ions are represented by yellow spheres.

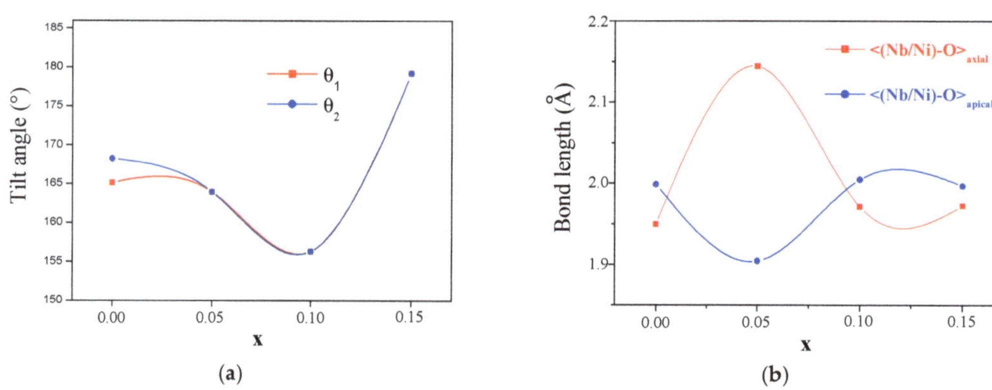

Figure 5. (**a**) Variation in the tilt angles within the structural octahedral chain; (**b**) Variation in the axial, apical and average <(Nb/Ni)–O> bond lengths as a function of doping content (x).

Regarding the geometric aspects of the BO_6 octahedron, this configuration typically exhibits distortions, giving rise to two types of nonequivalent B–O bonds: axial bonds and apical bonds. The axial bonds, aligned along the c-axis, are often either longer or shorter than the apical bonds, depending on the nature of the distortion and the surrounding bonding environment. The apical bonds, located within the (a, b) plane, form two pairs that may differ in length and angle relative to the axial bonds, further contributing to the overall distortion of the octahedron.

The variation in B–O bond lengths and angles usually influences the material's physical properties and stability. A clear understanding of the differences between axial and apical bonds is essential for analyzing the electronic structure, bonding characteristics, and potential applications of materials with orthorhombic BO_6 octahedra. In this context, the average bond lengths, <B–O>$_{axial}$ and <B–O>$_{apical}$, representing the axial and apical bonds, respectively, were measured and are presented in Figure 5b. This can be useful in providing information on the lattice distortions induced by Ni/Ba substitution. The variation in <(Nb/Ni)–O> bond lengths as a function of the doping rate (x) shows a specific change in <(Nb/Ni)–O>$_{axial}$ and <(Nb/Ni)–O>$_{apical}$ bonding lengths for x = 0.05 compared to the other compositions. This change, characterized by an axial expansion accompanied by a planar shortening of the BO_6 octahedra, highlights the presence of significant lattice distortion in the $NBNNO_{0.05}$ compound. When examining the overall change by averaging the axial and apical bond lengths, as shown in Figure 5b, the substitution effect reveals an increase of about 2% in the <(Nb/Ni)–O> bond length for the $NBNNO_{0.05}$ compound compared to the other compositions.

These observations confirm the presence of structural distortions in the crystal lattice, which are essential for preserving the perovskite structure of the doped compounds. This suggests that distortions in the BO_6 octahedra, driven by variations in B–O bond lengths, contribute to an increased degree of local structural disorder, mainly induced by Ba and Ni substitutions at the A and B sites, respectively.

3.1.2. Raman Analysis

To obtain additional information on the structural features of the crystal lattice, such as order–disorder, crystallinity, and bond distortion, Raman analysis was carried out on $NBNNO_x$ ceramics. Figure 6 shows the Raman spectra recorded at room temperature for $NBNNO_x$ samples with different Ni/Ba contents, identifying various vibration modes. These modes are classified into the translational mode of an isolated cation at the A-site and six internal modes of the coordination polyhedra (BO_6), as schematically illustrated in Figure 7. The internal vibration modes of the BO_6 octahedra include three stretching modes ($^1A_{1g}$ (ϑ_1), ^1E_g (ϑ_2), and $^1F_{1u}$ (ϑ_3)) and three bending modes ($^1F_{1u}$ (ϑ_4), $^1F_{2g}$ (ϑ_5), and $^1F_{2u}$ (ϑ_6)). It should be noted that the modes ϑ_1, ϑ_2, and ϑ_5 are symmetrical, while the modes ϑ_3, ϑ_4, and ϑ_6 exhibit non-symmetrical vibrations [39].

Three main frequency regions can be distinguished from the deconvoluted Raman spectra (Figure 6). The low-frequency region, below 150 cm^{-1}, displays an intense peak at 71 cm^{-1}, associated with translational movements of Na^+ and/or Ba^{2+} cations. The intermediate region, extending over 150 to 450 cm^{-1}, contains several fine peaks associated with the internal vibratory modes of BO_6 (F_{2g} (ϑ_5) and F_{2u} (ϑ_6)) and a low band linked to the combination mode of stretching and bending vibrations F_{1u} (ϑ_4). The high-frequency region between 450 and 900 cm^{-1} displays three peaks corresponding to the stretching vibrations $\vartheta 2$, $\vartheta 1$, and $\vartheta 3$, located at 553, 598, and 665 cm^{-1}, respectively [40].

Table 2 summarizes the vibration modes related to each $NBNNO_x$ sample, highlighting a significant substitution effect. As seen in Figure 6, the number of peaks related to the ϑ_6 vibration mode decreases as the doping rate (x) increases, indicating a notable structural change in the crystal structure. Additionally, a broad peak at about 850 cm^{-1} is specifically observed in doped ceramics, associated with the combination of ϑ_1 and ϑ_5 modes [39,40]. This peak is attributed to the valence vibrations of the B–O–B bridges, where B designates the elements occupying site B (Nb^{5+}, Ni^{2+}). Its intensity increases with the Ni-doping rate, suggesting that the concentration of Ni–O–Nb bridges increases while that of Nb–O–Nb decreases as the Ni content increases. This vibrational mode may be associated with local cation ordering/disordering within the structure, which can be correlated with changes in the NbO_6 octahedron parameters, such as volume, bond angles, and bond lengths, as previously confirmed by Rietveld refinement.

The relative shifts in vibrational modes observed in the doped ceramics are closely associated with the significant lattice distortions induced by the incorporation of Ba^{2+} and Ni^{2+} ions into the crystal lattice of the NNO parent phase. These distortions introduce an imbalance in bond lengths and strengths between the B-site ions (Nb^{5+}, Ni^{2+}) and their coordinated oxygen atoms, resulting in shifts in the vibrational modes [41]. Specifically, the incorporation of these dopants modifies the local bonding environment by introducing lattice strain, octahedral tilting, and structural disorder.

Table 2. Vibration modes of $NBNNO_x$ ceramics, collected in the wavenumber range of 50–1000 cm^{-1}.

Composition	Wave Number (cm^{-1})																		
x	Na^+/Ba^{2+}			$\vartheta6$								$\vartheta5$	$\vartheta4$			$\vartheta2$	$\vartheta1$	$\vartheta3$	$\vartheta1 + \vartheta5$
0	57	69	88	118	139	150	173	179	197	219	253	272	293	375	429	555	597	660	
0.05	60	73		110	136			176	196	216		246	276	290	428	569	609	655	867
0.10	60	88			135			174		213		245	275	312	426	563	610	662	863
0.15	62	88			137							228		335	425	556	597	666	859

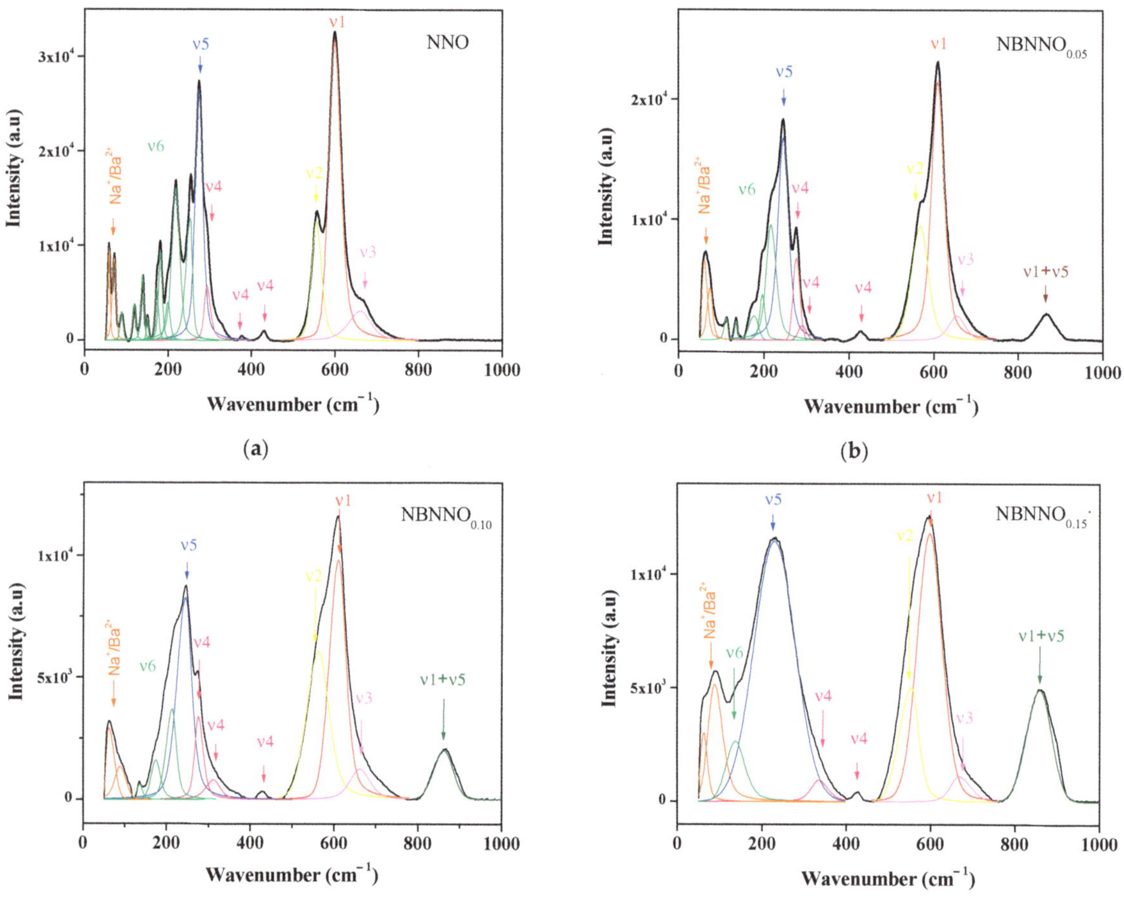

Figure 6. Raman spectra recorded at room temperature for $NBNNO_x$ ceramics with different Ni/Ba contents. (**a**) NNO; (**b**) $NBNNO_{0.05}$; (**c**) $NBNNO_{0.10}$; (**d**) $NBNNO_{0.15}$. Various vibration modes are identified from the deconvoluted Raman spectra, represented in different colors.

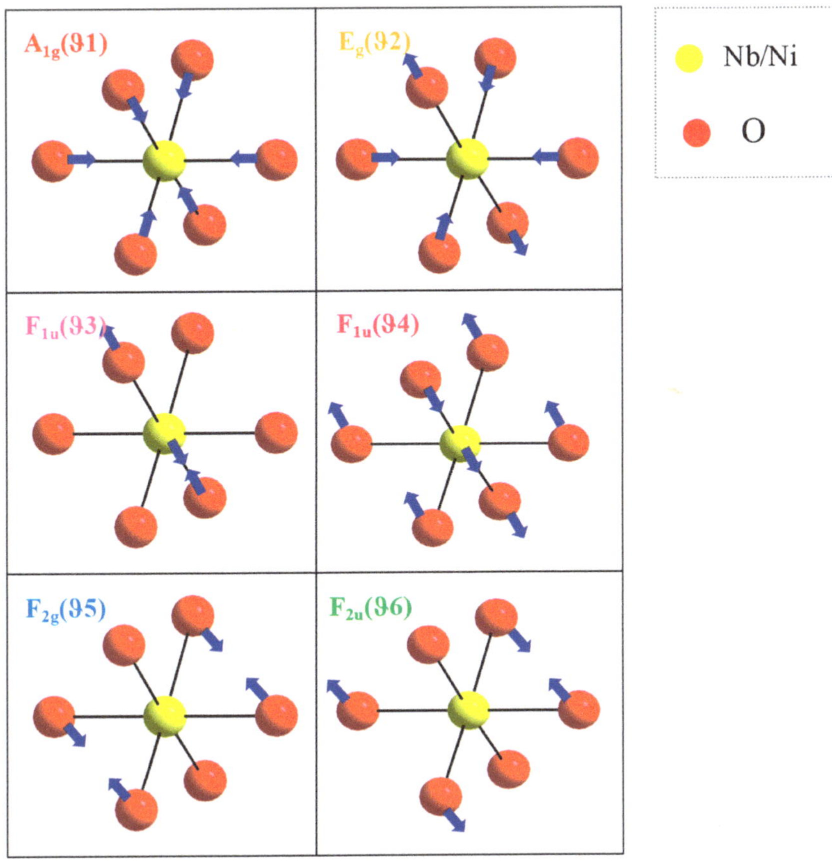

Figure 7. Schematic representations of internal vibrational modes of BO_6 octahedra, showing that the ϑ_1, ϑ_2, and ϑ_5 modes are symmetric; the ϑ_3, ϑ_4, and ϑ_6 modes are non-symmetric; $^1A_{1g}$ (ϑ_1), ^1E_g (ϑ_2), and $^1F_{1u}$ (ϑ_3) are stretching modes; and $^1F_{1u}$ (ϑ_4), $^1F_{2g}$ (ϑ_5), and $^1F_{2u}$ (ϑ_6) are bending modes. The blue arrows indicate the direction of atom vibrations. The oxygen ions are represented by red spheres, and the Nb/Ni ions are represented by yellow spheres.

To further illustrate these changes, Figure 8 shows the shifts in the internal vibration modes of the BO_6 octahedra, with an emphasis on the bending mode ϑ_1 and the stretching mode ϑ_5, plotted as a function of doping content. A progressive red shift (the peak moves to lower wavenumbers) is observed for the ϑ_5 mode with increasing Ni/Ba doping, indicating bond weakening due to lattice expansion and local distortions. In contrast, the ϑ_1 mode exhibits a blue shift (peak moves to higher wavenumbers) at $x = 0.05$ and $x = 0.10$, suggesting bond strengthening or lattice compression. Notably, for the $x = 0.15$ composition, almost no shift is observed compared to the NNO parent phase ($x = 0$), indicating a structural stabilization at higher doping levels.

These mode shifts highlight the complex interplay between strengthened and weakened bonds, driven by lattice expansion, compression, and octahedral tilting. The results reflect the extensive structural distortions within the BO_6 octahedra and the overall lattice disorder introduced by Ba and Ni co-doping, which plays a crucial role in modifying the material's structural and dielectric behavior.

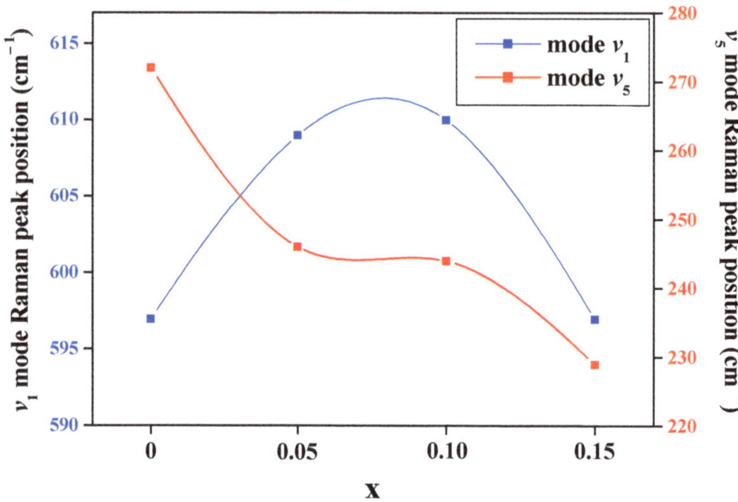

Figure 8. Variation in the Raman peak position of the principal internal vibration modes of the BO_6 octahedra, ϑ_1 (bending mode) and ϑ_5 (stretching mode), as a function of the doping content.

3.1.3. Microstructure Analysis

To investigate the microstructure changes in $NBNNO_x$ ceramics, the microstructural parameters, microstrain, and crystallite size are evaluated based on X-ray diffraction peak broadening. The integral breadth (β) of diffraction peaks is mainly associated with two contributions: β_G, due to the presence of microstrain, and β_L, related to crystallite size. These two effects were deconvoluted from the overall peak broadening using the Halder and Wagner formalism [42,43]. In this approach, the crystallite size contribution is modeled with a Lorentzian profile, while the microstrain contribution follows a Gaussian profile. The integral breadths β_G and β_L are related to the microstrain (ε) and crystallite size (D) via the Stokes and Wilson formula [44] and Scherrer's equation [45], respectively:

$$\beta_G = 4\varepsilon \, \tan\theta \tag{1}$$

$$\beta_L = \frac{K\lambda}{D\cos\theta}, \tag{2}$$

where λ represents the wavelength of CuK_α radiation, θ denotes the Bragg's diffraction angle of the peak, and k is a constant, taken as 4/3 for spherical crystallites [46]. By applying these equations, the microstrain and crystallite size can be accurately determined, providing insight into the microstructural changes occurring in the $NBNNO_x$ ceramics.

The two contributions β_G and β_L are extracted from the integral broadening of the most intense diffraction peak using the "Winplotr" program. The microstrain (ε) and crystallite size (D) are then determined from Equations (1) and (2), respectively. The obtained microstructural parameters D and ε are plotted as a function of the substitution rate (x). The pronounced variation in microstructural parameters illustrated in Figure 9 underlines the notable influence of Ni/Ba substitution in the NNO parent phase. A sharp decrease in crystallite size, from 140 nm to 71 nm, is observed when the Ni/Ba content increases from x = 0 to x = 0.10. This effect is accompanied by an increase in microstrain which reaches its maximum level of about 0.3% for x = 0.05 and x = 0.10 compared to the other compositions. In contrast, the $NBNNO_{0.15}$ ceramic exhibits a slight relaxation of microstrain, associated with a relative increase in crystallite size up to 86 nm. Such an observation indicates that the crystallinity of the $NBNNO_{0.15}$ phase is relatively improved compared to the $NBNNO_{0.10}$ phase.

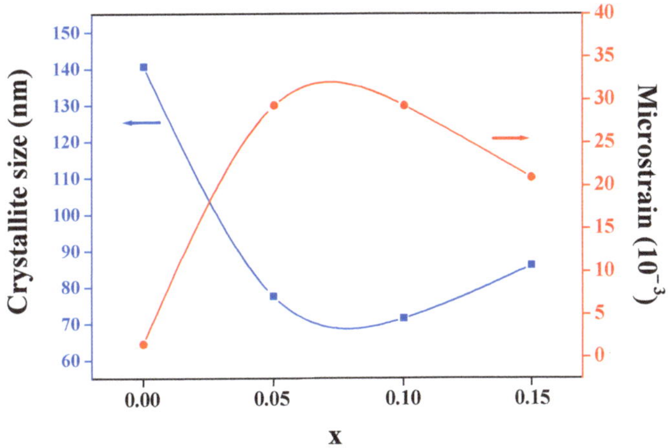

Figure 9. Variation in the microstructural parameters: crystallite size (D) and microstrain (ε) as a function of the doping content (x), determined through X-ray diffraction peak broadening and derived from the Halder–Wagner approach.

For further microstructural analysis, the surface morphology of the $NBNNO_{0.15}$-doped ceramic was examined and compared to that of the NNO parent phase. The SEM micrographs in Figure 10 reveal significant changes in grain size and porosity between the two samples. The NNO parent phase consists of large particles averaging about 3 µm in size, with clearly visible pores identified by dark contrasts as depicted in Figure 10a. In contrast, the $NBNNO_{0.15}$-doped ceramic, shown in Figure 10b, exhibits a surface morphology with low porosity and well-faceted, uniformly distributed fine grains, averaging about 0.75 µm in size. This marked reduction in grain size is consistent with various studies reported in the literature [47–49]. It is suggested that oxygen vacancies influence grain growth differently depending on the material system, doping levels, and synthesis conditions. On one hand, they can enhance ion transport by providing vacant sites for diffusion, promoting grain growth, particularly in high-temperature sintering processes. On the other hand, they can accumulate at grain boundaries, forming space charge layers that inhibit grain boundary migration and thereby reduce grain size. Additionally, the substitution of Ni^{2+} and Ba^{2+} ions introduces chemical disorder, which creates lattice distortions, oxygen vacancies, and internal stresses that further impede grain boundary movement. This combined effect of oxygen vacancies and chemical disorder likely accounts for the smaller grain size observed in the doped sample compared to the larger grains of the undoped NNO ceramic.

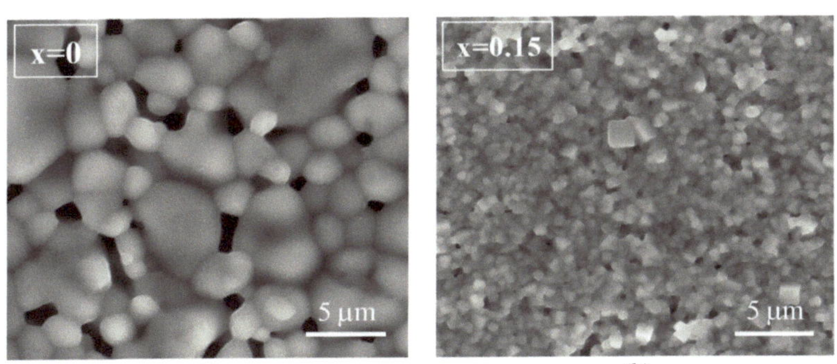

(**a**) (**b**)

Figure 10. SEM micrographs using secondary electron imaging of (**a**) parent phase NNO; (**b**) doped $NBNNO_{0.15}$ ceramic.

In conclusion, the substitution of Ba and Ni substitution at the A and B sites induces significant structural distortions in the crystal lattice, as evidenced by structural, microstructural, and Raman analysis. These distortions contribute to increasing the local structural disorder in the doped compounds, which are expected to influence their optical and dielectric properties.

3.2. Diffuse Reflectance Study

The diffuse reflectance spectra (DRS) at room temperature for $NBNNO_x$ ceramics, measured in the 350–900 nm range, are shown in Figure 11. A comparison with the parent NNO phase reveals significant changes in the DRS profiles of the doped $NBNNO_x$ ceramics. While the NNO phase shows a stable reflectance plateau over the 450–900 nm spectral range, the doped $NBNNO_x$ ceramics exhibit an enhanced absorption band in the visible region, particularly in the red-light spectrum. This suggests that the incorporation of Ni^{2+} and Ba^{2+} ions alters the optical properties, likely due to modifications in the electronic structure [35,36].

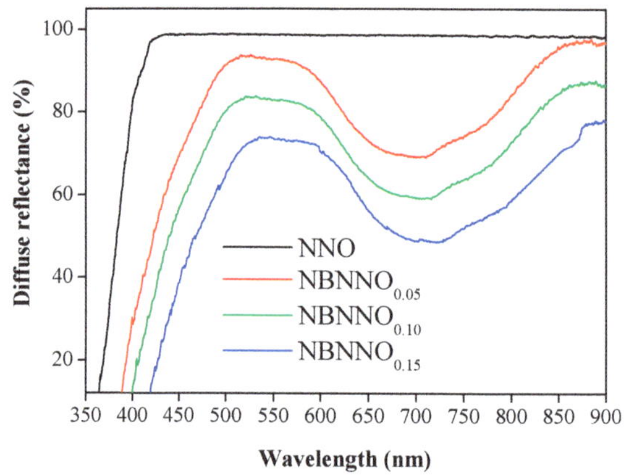

Figure 11. Diffuse reflectance spectra of $NBNNO_x$ ceramics measured at room temperature in the 350–900 nm range.

Moreover, the diffuse reflectance of the doped ceramics decreases significantly compared to the parent NNO, which maintains a reflectance as high as 96%. The reduction in reflectance becomes increasingly pronounced with higher doping contents (x), indicating a clear correlation between doping levels and optical response. This change in reflectance and absorption is likely influenced by multiple factors, including surface roughness, crystallinity, and grain size [50–52]. Surface roughness, in particular, plays a key role, as it has been widely reported that rougher surfaces tend to produce more diffuse reflection [53]. Additionally, variations in grain size and the crystallinity of the materials may contribute to the observed optical changes, affecting light scattering and absorption behavior within the ceramic matrix.

3.2.1. Band Gap Energy

The optical band gap energy (Eg) of $NBNNO_x$ compounds was estimated through diffuse reflectance measurements using the Kubelka–Munk method [54], which allows for precise extraction of the Eg value. The reflectance data were first converted into the Kubelka–Munk function following the expression below [55].

$$F(R_\infty) = \frac{K}{S} = \frac{(1 - R_\infty)^2}{2R_\infty}, \tag{3}$$

where $F(R_\infty)$ denotes the Kubelka–Munk function, R_∞ indicates the relative diffuse reflectance of the layer with respect to a standard, K represents the absorption coefficient of the sample, and S is the scattering coefficient.

When materials scatter in a perfectly diffuse manner, the Kubelka–Munk absorption coefficient K varies proportional to the linear absorption coefficient α [55]. The optical band gap E_g is related to this coefficient by the following equation [56]:

$$(\alpha\, hv) = C_1\, (hv - E_g)^n, \tag{4}$$

where hv is the photon energy, C_1 is a proportionality constant, and n an exponent that reflects the type of electronic transition, with $n = 1/2$ indicating direct allowed transitions and $n = 2$ indicating indirect allowed transitions.

Assuming K is equal to 2α for perfect scattering and incorporating this relation into Equation (4), the Kubelka–Munk expression is reformulated as follows:

$$(F(R)\, hv) = C_1\, (hv - E_g)^n \tag{5}$$

The n value is evaluated following the method outlined in [57]. By plotting ln($F(R)hv$) against ln($hv - E_g$) for an estimated E_g value, the slope of the linear region near the band edge provides the n value. This value is found to be equal to 2, affirming that the optical transition is indirectly allowed for the undoped and doped compounds. In this case, the photon absorption process promotes an indirect electronic transition between the valence and conduction bands. The E_g values for $NBNNO_x$ ceramics are determined by plotting $[F(R_\infty)hv]^{1/2}$ versus (hv) and extrapolating the linear region of the curve to its intercept with the abscissa.

As shown in Figure 12, the extrapolation of the linear part of the Kubelka–Munk curves, derived from the absorption data, reveals a significant reduction in the band gap of the doped $NBNNO_x$ compounds. Additionally, Figure 13 illustrates a nearly linear correlation between the band gap and the doping content. The E_g value decreases significantly from 3.34 eV to 1.08 eV as the doping content increases from 0 to 0.15, corresponding to a band gap tunability of 2.26 eV. A similar extensive band gap tunability has also been reported in KBNNO solid solutions [36,58].

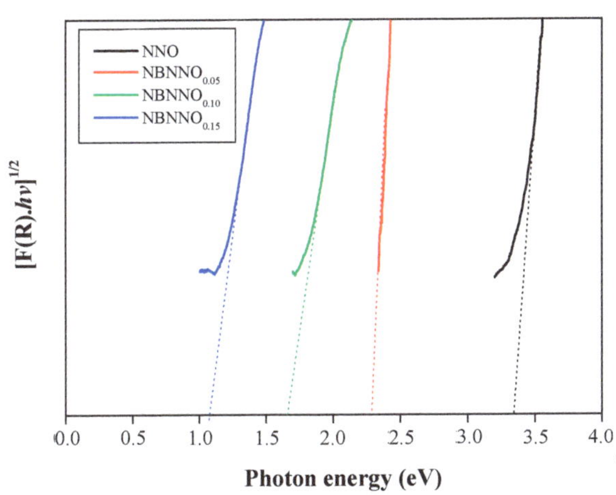

Figure 12. Kubelka–Munk transformed reflectance spectra of $NBNNO_x$ ceramics at room temperature (solid lines), showing the determination of the band gap energy (E_g) through the extrapolation of the linear portion of the spectra (dotted lines).

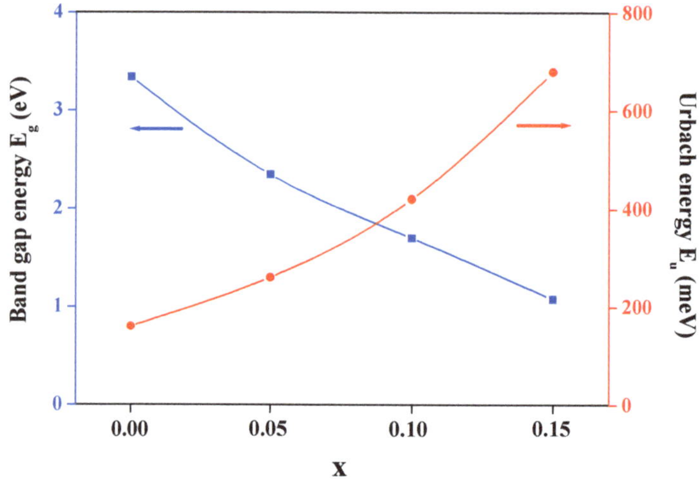

Figure 13. Dependence of the band gap and Urbach energy on doping content. The Urbach energy (E_u) is extracted from the slope of the linear fit in the plot of ln(*F(R)*) versus photon energy (hν).

These findings highlight the critical role of Ni, as well as the generated oxygen vacancies, in the significant lowering of the band gap, consistent with theoretical expectations for Ni-doped $PbTiO_3$ ceramics [35]. Specifically, it has been proven that the Ni–VO–Nb configuration tailored by the oxygen vacancies generates a higher density of Ni-3d states in the valence band. In this case, the valence band maximum consists of hybridized Ni-3d and O-2p states, while the conduction band minimum is governed by Nb-4d states. The presence of filled Ni-3d gap states within the NNO host is thus expected to play a crucial role in the observed reduction in the E_g band gap. In contrast, the wide band gap of the parent NNO phase arises from the intrinsic properties of the metal–oxygen (Nb–O) bonds, where the excitation across the band gap mainly involves charge transfer from the O-2p states at the valence band maximum to the Nb-4d states at the conduction band minimum.

The pronounced band gap reduction observed in $NBNNO_x$ ceramics highlights the significant influence of transition metal doping on the electronic structure of perovskite materials. The quasi-linear relationship between band gap energy and doping content indicates that the degree of band gap tunability can be precisely controlled by adjusting the dopant concentration, making these materials highly versatile for various applications. These findings confirm the crucial role of Ni and oxygen vacancies in modulating the material's optical properties, potentially enabling their use in visible-light-driven photocatalysis or other applications where light absorption is critical. Additionally, comparisons with KBNNO solid solutions suggest that such band gap engineering strategies may be widely applicable to similar perovskite systems, paving the way for the design of functional materials with tailored properties [36,58,59].

3.2.2. Urbach Energy

In general, numerous factors associated with structural disorder in the lattice can significantly affect the energy band structure [13,60]. Specifically, structural defects introduced within the band gap can contribute prominently to reducing the E_g value by generating intermediate levels within the band gap [61]. The extent of these defects can be quantified through the Urbach energy (E_u), which measures the magnitude of defect-related energy states [62,63]. The Urbach energy is determined using the following equation [64]:

$$\alpha = \alpha_0 + exp\left(\frac{hv}{E_u}\right), \quad (6)$$

where α is the absorption coefficient and α_0 is a constant. Since the absorption coefficient α is proportional to the Kubelka–Munk function *F(R)*, Equation (6) can be rewritten in logarithmic form as follows:

$$ln\ (F(R)) = C + \frac{h\nu}{E_u}, \tag{7}$$

where C is a constant. The Urbach energy E_u is determined by plotting *ln(F(R))* against photon energy ($h\nu$). The reciprocal of the slope of the linear fit provides the E_u value. The obtained E_u values, along with the corresponding band gap energy (E_g), are depicted in Figure 13. The graph clearly shows that the Urbach energy increases significantly with higher Ni/Ba content, rising from 170 meV for x = 0 to 700 meV for x = 0.15. This suggests that as the defect energy increases, the band gap decreases. This observation highlights how structural defects contribute to the narrowing of the band gap and emphasizes the role of defect states in modifying the material's optical properties, as commonly reported in several works [49,61,65].

It can be concluded that the substantial increase in Urbach energy with higher Ni/Ba doping confirms that the introduction of these elements leads to a higher density of defect states. This increased defect density contributes to the effective narrowing of the band gap. Understanding this phenomenon is crucial for elucidating the electronic and optical behavior of doped $NBNNO_x$ ceramics and is valuable for applications that require precise band gap tuning.

3.3. Dielectric Study

Figure 14 shows the temperature dependence of the real part of the dielectric permittivity (ε_r') and dielectric loss (δ) of the $NBNNO_x$ ceramics, measured at various frequencies.

The pure NNO ceramic, featuring a Pbma antiferroelectric structure, known as the P phase, displays three distinct dielectric peaks, each associated with a phase transition during heating, as shown in Figure 14a. The first dielectric peak at 686 K indicates the transition from the P phase to the antiferroelectric R phase, which has an orthorhombic Pmmn structure. The second peak, observed near 765 K, marks the transition from the R phase to the paraelectric S phase, also adopting an orthorhombic Pmmn structure. The third peak, observed at 796 K, is attributed to the paraelectric–paraelectric transition from the S phase to the *T(1)* phase, characterized by an orthorhombic Ccmm structure. At higher temperatures, further transitions are expected to occur from the *T(1)* phase to the *T(2)* phase, and subsequently to the C phase [66]. These subsequent transitions involve paraelectric phases that do not exhibit spontaneous polarization. Instabilities in the tilting of the oxygen octahedra and the displacement of the Nb^{5+} ions from the centers of their octahedral sites drive these temperature-induced phase transitions.

Upon heating, the material undergoes successive phase transitions, progressively reducing tilts and displacements until it achieves the cubic aristotype structure (Pm3m). Both the permittivity and dielectric loss peaks show significant frequency dispersion, while the corresponding T_{P-R}, T_{R-S}, and $T_{R-T(1)}$ phase transition temperatures remain invariant with frequency. The high values of ε_r' at low frequencies can be attributed to the presence of various types of polarizations (interfacial, dipolar, atomic, ionic, and electronic) within the ceramic. However, at high frequencies, some of these polarizations contribute less to ε_r', which is consistent with normal dielectric behavior.

A significant change in dielectric behavior is observed in $NBNNO_x$ ceramics, characterized by a noticeable shift of the dielectric peaks toward lower temperatures at specific frequencies, as shown in Figure 14b–d. In doped $NBNNO_x$ ceramics, two primary dielectric transition peaks are observed. We define Region I as the temperature range corresponding to the first transition, occurring at the temperature T_{1m}. Region II is defined by a higher temperature range, where the second phase transition occurs at the temperature T_{2m}, indicated by a second peak.

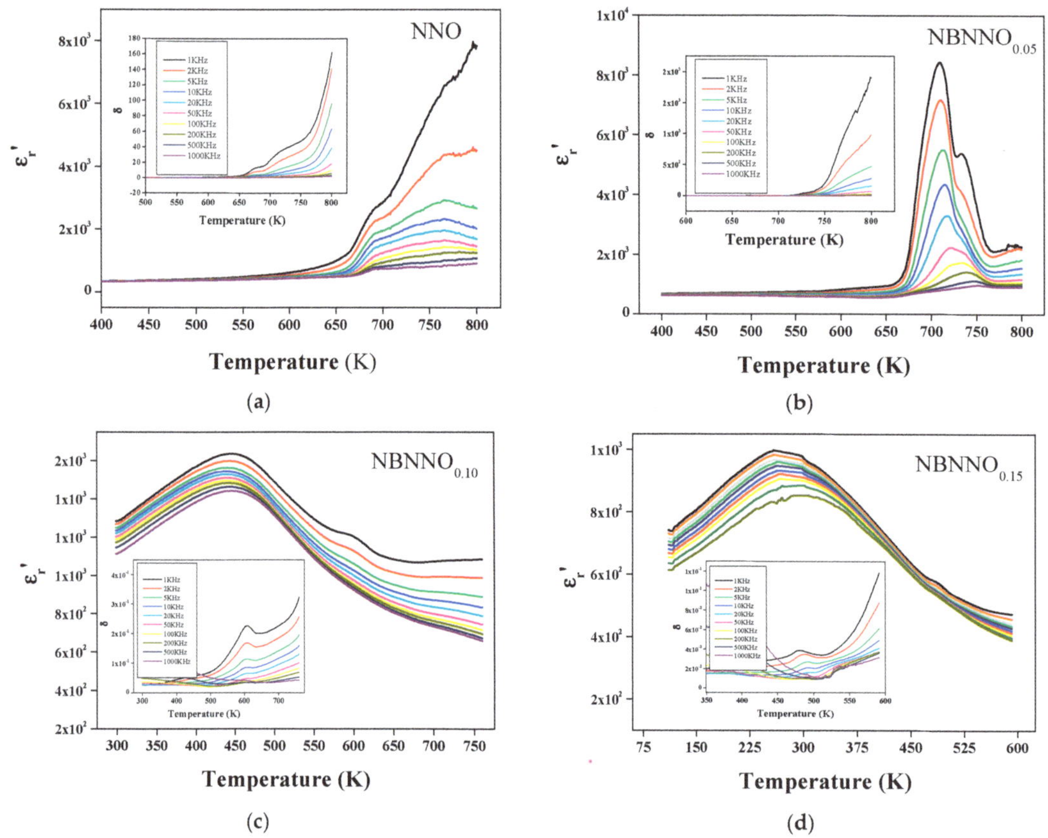

Figure 14. Temperature dependence of the real part of the dielectric permittivity (ε_r') and dielectric loss (δ) of the NBNNO$_x$ ceramics: (**a**) NNO; (**b**) NBNNO$_{0.05}$; (**c**) NBNNO$_{0.10}$; (**d**) NBNNO$_{0.15}$. The dielectric loss (δ) curves are presented in the inset of each figure.

At a lower doping level, the NBNNO$_{0.05}$ ceramic sample exhibits two distinct dielectric transitions, characterized by sharp and intense permittivity peaks at 710 K and 734 K, as shown in Figure 14b. The peak maxima shift significantly toward higher temperatures with increasing frequency, indicating a significant frequency dependence of T_m, which suggests relaxor-like behavior in the material. As the doping level increases, an interesting feature emerges in Figure 14c,d. The permittivity of NBNNO$_{0.10}$ and NBNNO$_{0.15}$ ceramics exhibits a broad main peak at temperatures T_{1m} of 445 K and 270 K for 1 kHz, respectively. These dielectric transitions are marked by a diffuse nature, with frequency-dependent variations in both permittivity and dielectric loss around the peak maxima. Specifically, in the NBNNO$_{0.15}$ sample, the temperature T_{1m} of the ε_r' maximum shifts to higher values with increasing frequency, while the intensity of the ε_r' maximum decreases (Figure 14d). This temperature shift and significant frequency dispersion observed for $T < T_{1m}$ suggest relaxor-like behavior for this transition. At higher temperatures, both NBNNO$_{0.10}$ and NBNNO$_{0.15}$ ceramics exhibit a second phase transition, as clearly shown by the dielectric loss curves in the inset of Figure 14c,d. The T_{2m} temperature decreases from 605 K to 480 K with increasing doping, along with a reduction in the intensity of the corresponding permittivity and dielectric loss peaks. In contrast to NBNNO$_{0.10}$, the second transition in NBNNO$_{0.15}$ ceramic shows a slight shift to higher temperatures with increasing frequency, confirming the relaxor-like behavior of this transition. These dielectric changes in doped

$NBNNO_x$ ceramics are generally attributed to the inhomogeneous distribution of incorporated ions within the NNO host. It can be concluded that substituting Ni^{2+} for Nb^{5+} and Ba^{2+} for Na^+, along with the creation of oxygen vacancies, primarily influences the dielectric properties of $NaNbO_3$ by lowering the transition temperature and promoting relaxor-like behavior. Similar dielectric behavior has been observed in other $NaNbO_3$-based lead-free solid solutions, including $Na_{1-x}Ba_xNb_xTi_{1-x}O_3$ [4,67] and Zr/Ln co-doped $Na_{0.8}Ba_{0.2}Nb_{0.8}Ti_{0.2}O_3$ [10].

Based on the temperature dependence of dielectric permittivity, several empirical parameters can be defined to thoroughly evaluate the nature of the dielectric response in $NBNNO_x$ ceramics. Among these parameters, ΔT_m is commonly used to indicate the degree of deviation from the Curie–Weiss law, while the coefficient γ reflects the degree of diffuseness in dielectric transitions. Additionally, the T_{relax} parameter, which describes the frequency dispersion of T_m, mainly reflects the degree of relaxor behavior [68–70].

The deviation of the observed dielectric transitions in the doped $NBNNO_x$ ceramics from the Curie–Weiss law, which is typically associated with normal dielectric behavior, is evaluated by plotting the inverse dielectric permittivity against temperature at 1 kHz, as illustrated in Figure 15. Extrapolating the linear region at higher temperatures clearly shows a deviation from the Curie–Weiss law (Equation (8)) for all compounds [71].

$$\frac{1}{\varepsilon_r{}'} = \frac{T - T_0}{C}, \ (T \succ T_0), \tag{8}$$

where C is the Curie–Weiss constant and T_0 is the Curie–Weiss temperature. The degree of deviation ΔT_m from the Curie–Weiss law is defined as follows [72]:

$$\Delta T_m = T_{dev} - T_m, \tag{9}$$

where T_{dev} indicates the temperature at which the permittivity begins to deviate from the Curie–Weiss law, while T_m corresponds to the temperature of the dielectric maximum, as illustrated in Figure 15. The calculated ΔT_m values and the fitting parameters T_0 and C, derived for each dielectric transition using Equation (8), are shown in Table 3. The observed deviation of dielectric transitions in the doped $NBNNO_x$ ceramics from the Curie–Weiss law offers significant insights into the complex dielectric behavior of these materials. The increase in ΔT_m with higher levels of Ni and Ba doping, particularly evident in the first transition of the $NBNNO_{0.10}$ and $NBNNO_{0.15}$ ceramics, suggests the emergence of diffuse phase transition behavior. This diffuse transition is characterized by a broadening of the dielectric peak and a more gradual change in permittivity with temperature, rather than a sharp, well-defined transition typical of standard dielectric materials. Similar dielectric behavior has been noted in various lead-free compounds, including $Ba(Zr_yTi_{1-y})O_3$ [69], $(1-x)BaTiO_3$–$xBi(Zn_{0.5}Ti_{0.5})O_3$ [73], and $Na_{1-x}Ba_xNb_xTi_{1-x}O_3$ [4,67].

Table 3. The Curie–Weiss temperature (T_0), Curie–Weiss constant (C), temperature of the dielectric permittivity constant maximum (T_m), the temperature above which the dielectric constant follows the Curie–Weiss law (T_{dev}), and the degree of deviation from the Curie–Weiss law parameter (ΔT_m) for the doped $NBNNO_x$ ceramics at 1 kHz.

		$NBNNO_{0.05}$	$NBNNO_{0.10}$	$NBNNO_{0.15}$
First transition	T_0 (K)	700	173	178
	C ($\times 10^5$ K)	1.39	4.92	2.09
	T_m K)	710	444.5	276
	T_{dev} (K)	720	486	420
	$\Delta T_m = T_{dev} - T_m$ (K)	10	41.5	44
Second transition	T_0 (K)	729	256	182
	C ($\times 10^5$ K)	0.99	3.97	1.48
	T_m (K)	735	598.6	486
	T_{dev} (K)	746	612	505
	$\Delta T_m = T_{dev} - T_m$ (K)	11	13,4	19

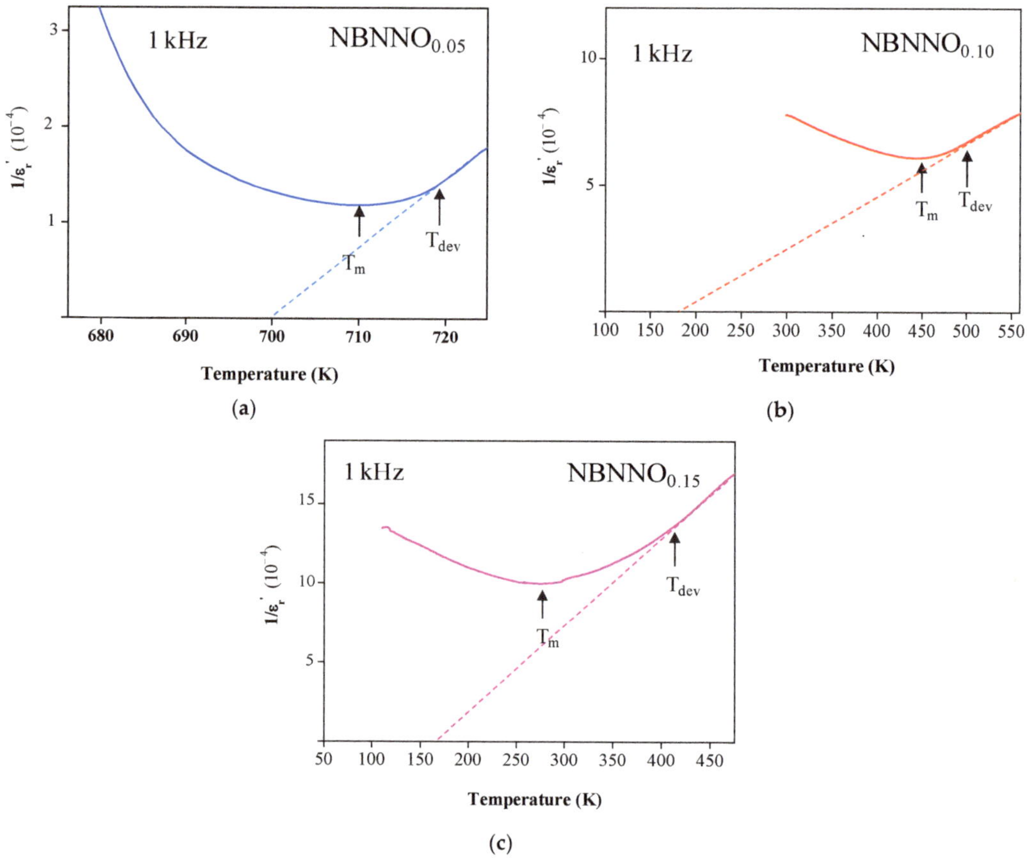

Figure 15. Temperature dependence of the inverse dielectric constant for the doped $NBNNO_x$ ceramics at 1 kHz: (**a**) $NBNNO_{0.05}$; (**b**) $NBNNO_{0.10}$; (**c**) $NBNNO_{0.15}$. Solid lines represent the experimental data, while dotted lines indicate the extrapolation of the linear fit to the Curie–Weiss law. T_m marks the temperature of the dielectric constant maximum, and T_{dev} denotes the temperature above which the dielectric constant follows the Curie–Weiss law. The deviation from the Curie–Weiss law is illustrated for the first dielectric transition.

The diffuseness of the dielectric transition can be effectively modeled by the modified Curie–Weiss law introduced by Uchino and Nomura, expressed as follows [74]:

$$\frac{1}{\varepsilon_r{}'} - \frac{1}{\varepsilon_r{}'_m} = \frac{(T - T_m)^{\gamma}}{C'}, (T \succ T_m), \tag{10}$$

where γ reflects the diffuseness degree indicator ($1 \leq \gamma \leq 2$) and C' is the modified Curie–Weiss constant. The value of γ provides insight into the nature of the phase transition: $\gamma = 1$ indicates respect for the normal Curie–Weiss law, while $\gamma = 2$ signifies a complete diffuse phase transition. According to Equation (10), the plots of $\ln(1/\varepsilon_r{}' - 1/\varepsilon_r{}'_m)$ versus $\ln(T - T_m)$ at 1 kHz are presented in Figure 16. A clear linear relationship is evident for all studied compositions, suggesting the effectiveness of the modified Curie–Weiss law.

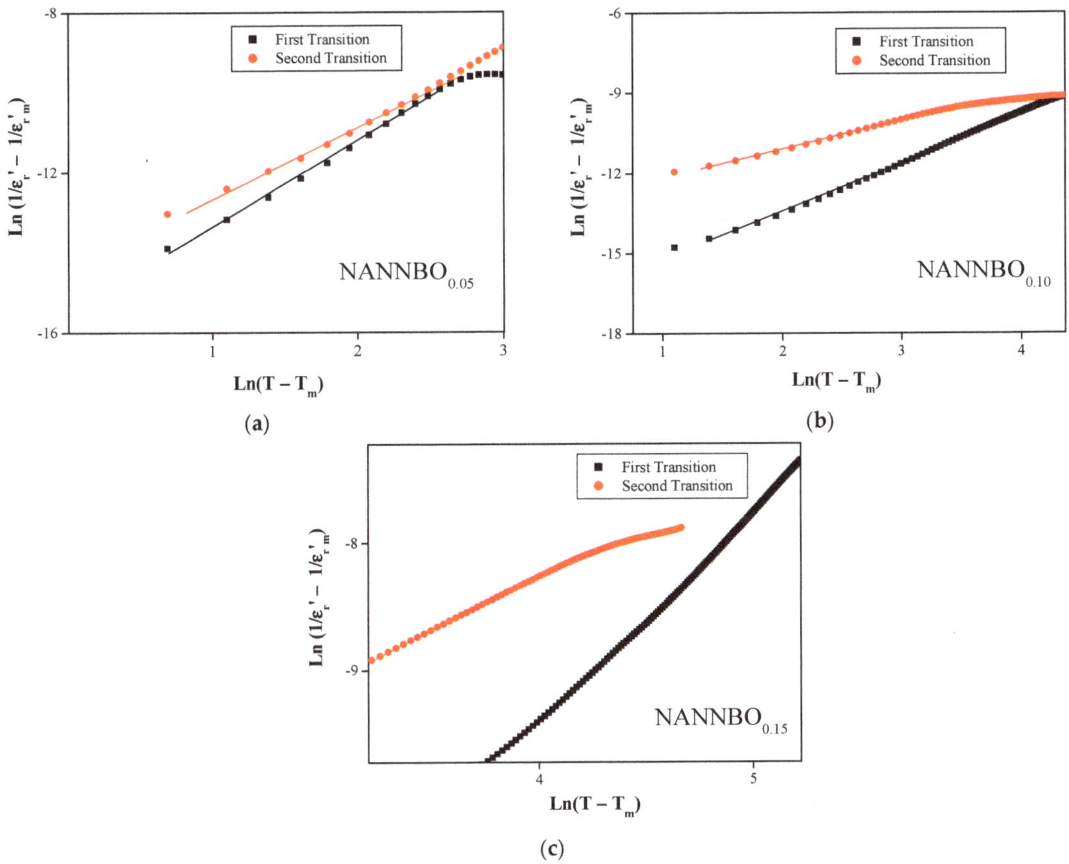

Figure 16. Plot of $\ln(1/\varepsilon_r' - 1/\varepsilon_r'_m)$ versus $\ln(T - T_m)$ for the doped $NBNNO_x$ ceramics at 1 kHz: (**a**) $NBNNO_{0.05}$; (**b**) $NBNNO_{0.10}$; (**c**) $NBNNO_{0.15}$. The filled squares represent the experimental data, while the dotted lines indicate the extrapolation of the linear fit to the modified Curie–Weiss law. The slope of the fitting curves reflects the degree of diffuseness, γ.

The slope of the fitting curves determines the γ value. The fitting parameters γ and C', obtained using Equation (10), are listed in Table 4. The high γ values clearly indicate that the doped $NBNNO_x$ ceramics display strong diffuse phase transition behavior, with a near-complete diffuse transition observed specifically in the $NBNNO_{0.05}$ and $NBNNO_{0.10}$ ceramics. This pronounced diffuseness implies that the dielectric response becomes increasingly complex and less characteristic of normal dielectric behavior. This suggests that the introduction of Ni^{2+} and Ba^{2+} ions induces greater structural disorder, leading to a more gradual transition and a broader temperature range over which the dielectric constant changes.

The diffuse phase behavior can thus be related to the disruption of long-range ordering and lattice distortion induced by the substitution of Ni^{2+} and Ba^{2+} ions in the NNO host lattice. The incorporation of these ions creates a complex local environment, introducing strain and disorder due to variations in ionic radii and valence states. Consequently, this results in a broadening of the phase transition, where different regions within the material undergo the transition over a range of temperatures.

Table 4. The dielectric permittivity maximum ($\varepsilon_r{'}_m$), the temperature of the dielectric permittivity maximum (T_m), the diffuseness parameter (γ), and the modified Curie–Weiss constant (C') for the doped NBNNOx ceramics at 1 kHz.

		NBNNO$_{0.05}$	**NBNNO$_{0.10}$**	**NBNNO$_{0.15}$**
First transition	$\varepsilon_r{'}_m$ (K)	8430	1638	1002
	T_m (K)	710	444.5	276
	γ	2.1	1.9	1.7
	C' ($\times 10^5$ K)	56	369	99
Second transition	$\varepsilon_r{'}_m$ (K)	5405	1210	574
	T_m (K)	734	598.6	486
	γ	2.0	1.2	1.6
	C' ($\times 10^5$ K)	34	7	10

The degree of relaxor behavior parameter δT_{relax}, which characterizes the frequency dispersion of the temperature at the maximum dielectric constant, T_m, is expressed as follows [10]:

$$\delta T_{relax} = T_m \text{ (1 MHz)} - T_m \text{ (1 kHz)} \tag{11}$$

This parameter, derived from the dielectric measurements of NBNNO$_x$ ceramics, quantifies the material's tendency toward relaxor-like behavior. Higher δT_{relax} values indicate a stronger relaxor characteristic. The δT_{relax} values listed in Table 5 reveal a distinct evolution in the dielectric behavior of the NBNNO$_x$ ceramics. For the NBNN$_{0.05}$ and NBNNO$_{0.15}$ compositions, these values suggest a clear transition from normal dielectric behavior to relaxor-like characteristics, indicative of the increasing complexity and disorder within the material's structure. This effect is particularly evident in the NBNNO$_{0.15}$ ceramic, where a strong relaxor behavior is observed near ambient temperature, marking a notable shift in the material's dielectric properties. The enhanced relaxor behavior observed in the NBNNO$_{0.05}$ and NBNNO$_{0.15}$ ceramics can be attributed to the significant lattice distortion and chemical disorder resulting from the Ni^{2+} and Ba^{2+} ion incorporation into the NNO parent phase. The partial substitution of Na^+ by Ba^{2+} and Nb^{5+} by Ni^{2+} results in significant local compositional fluctuations, which disrupt the long-range Coulombic interactions critical for the stabilization of ferroelectric microdomains. This disruption prevents the formation of long-range ordered ferroelectric domains, instead giving rise to nano-polar regions characterized by short-range order [75–77]. Within these regions, each nano-domain undergoes a phase transition at a distinct temperature, contributing to the diffuse nature of the phase transition and the pronounced relaxor behavior observed in the NBNNO$_{0.05}$ and NBNNO$_{0.15}$ ceramics. In contrast, the NBNNO$_{0.10}$ ceramic exhibits only a pronounced diffuseness in the dielectric transition, without fully developing the relaxor-like behavior seen in the other compositions. This suggests that while NBNNO$_{0.10}$ does experience a disruption in long-range ferroelectric order, it does not reach the same level of disorder and local heterogeneity necessary to fully manifest relaxor properties. The differences in dielectric response among these compositions highlight the critical role of doping concentration in tuning the material's phase transition behavior, offering valuable insights for the design of materials with tailored dielectric properties.

Table 5. The temperature of the dielectric permittivity maximum (T_m) and the degree of relaxor behavior parameter (δT_{relax}) for the doped NBNNO$_x$ ceramics.

		NBNNO$_{0.05}$	**NBNNO$_{0.10}$**	**NBNNO$_{0.15}$**
First transition	T_m (K) (1 kHz)	710	444	272
	T_m (K) (1 MHz)	729	444	309
	δT_{relax} (K)	19	0	37
Second transition	T_m (K) (1 kHz)	734	598	486
	T_m (K) (1 MHz)	750	598	497
	δT_{relax} (K)	16	0	11

4. Conclusions

The incorporation of Ni and Ba into $NaNbO_3$ antiferroelectric ceramics has profoundly influenced their structural, dielectric, and optical properties, with promising implications for advanced technologies. Doping introduced significant distortions in the crystal lattice and BO_6 octahedra, preserving the perovskite structure in $[NaNbO_3]_{1-x}[BaNi_{1/2}Nb_{1/2}O_{3-\delta}]_x$ ceramics and heightening local structural disorder, which in turn impacted the material's dielectric and optical behaviors. Ni and Ba doping led to lead-free ceramics with strong relaxor behavior near room temperature, particularly at x = 0.15, making them ideal for applications requiring stable dielectric properties. Additionally, the optical band gap energy was reduced from 3.34 eV to 1.08 eV, demonstrating remarkable tunability, which underscores the role of Ni and oxygen vacancies in modulating the material's electronic structure, making it suitable for photocatalysis and photovoltaic applications. The combination of room-temperature relaxor behavior and significant band gap tunability positions $NBNNO_x$ ceramics as excellent candidates for multifunctional applications, such as cellular solar energy, photocatalysis, and energy storage. These structural, dielectric, and optical advancements expand the potential of perovskite materials in high-performance, environmentally friendly devices and mark a major step forward in the development of lead-free materials for energy and electronic technologies.

Author Contributions: Conceptualization, T.C., Z.B., N.A. and S.Z.; methodology, T.C., Z.B., N.A. and S.Z.; formal analysis, T.C. and S.Z.; investigation T.C., Z.B., N.A. and S.Z.; resources, Z.B., N.A. and S.Z.; writing—original draft preparation, T.C.; review, and editing, S.Z. and Z.B.; visualization, T.C., Z.B., N.A. and S.Z.; supervision, S.Z. All authors have read and agreed to the published version of the manuscript.

Funding: Tunisian Ministry of Higher Education.

Institutional Review Board Statement: Not applicable.

Informed Consent Statement: Not applicable.

Data Availability Statement: The original contributions presented in the study are included in the article, further inquiries can be directed to the corresponding authors.

Acknowledgments: Zohra Benzarti gratefully acknowledges her support by Project No. 7225-ILLIANCE. Projeto apoiado pelo PRR-Plano de Recuperaçaõ e Resiliência e pelos Fundos Europeus Next Generation EU, na sequência do AVISO No. 02/C05-i01/2022, Componente 5-Capitalizaçaõ e Inovaçaõ Empresarial-Agendas Mobilizadores para a Inovaçao Empresarial. She also acknowledges her support by national funds through FCT-Fundaçaõ para a Ciência e a Tecnologia under the project UID/EMS/00285/2020, ARISE-LA/P/0112/2020.

Conflicts of Interest: The authors declare no conflicts of interest.

References

1. Saito, Y.; Takao, H.; Tani, T.; Nonoyama, T.; Takatori, K.; Homma, T.; Nagaya, T.; Nakamura, M. Lead-free piezoceramics. *Nature* **2004**, *432*, 84–87. [CrossRef] [PubMed]
2. Cross, E. Lead-free at last. *Nature* **2004**, *432*, 24–25. [CrossRef] [PubMed]
3. Shrout, T.R.; Zhang, S.J. Lead-free piezoelectric ceramics: Alternatives for PZT. *J. Electroceramics* **2007**, *19*, 113–126. [CrossRef]
4. Zuo, R.; Qi, H.; Fu, J.; Li, J.; Shi, M.; Xu, Y. Giant electrostrictive effects of $NaNbO_3$-$BaTiO_3$ lead-free relaxor ferroelectrics. *Appl. Phys. Lett.* **2016**, *108*, 232904. [CrossRef]
5. Rödel, J.; Jo, W.; Seifert, K.T.P.; Anton, E.M.; Granzow, T.; Damjanovic, D. Perspective on the Development of Lead-Free Piezoceramics. *J. Am. Ceram. Soc.* **2009**, *92*, 1153–1177. [CrossRef]
6. Jiang, J.; Li, X.; Li, L.; Guo, S.; Zhang, J.; Wang, J.; Zhu, H.; Wang, Y.; Zhang, S.-T. Novel lead-free $NaNbO_3$-based relaxor antiferroelectric ceramics with ultrahigh energy storage density and high efficiency. *J. Materiomics* **2022**, *8*, 295–301. [CrossRef]
7. Qi, H.; Zuo, R.; Xie, A.; Tian, A.; Fu, J.; Zhang, Y.; Zhang, S. Ultrahigh energy-storage density in $NaNbO_3$-based lead-free relaxor antiferroelectric ceramics with nanoscale domains. *Adv. Funct. Mater.* **2019**, *29*, 1903877. [CrossRef]
8. Gu, L.; Zhou, D.; Cao, J.C. Piezoelectric Active Humidity Sensors Based on Lead-Free $NaNbO_3$ Piezoelectric Nanofibers. *Sensors* **2016**, *16*, 833. [CrossRef] [PubMed]
9. Khemakhem, H.; Simon, A.; Von Der Mühll, R.; Ravez, J. Relaxor or classical ferroelectric behaviour in ceramics with composition $Ba_{1-x}Na_xTi_{1-x}Nb_xO_3$. *J. Phys. Condens. Matter.* **2000**, *12*, 5951–5959. [CrossRef]

10. Ben Mrad, M.; Hannachi, R.; Dammak, M.; Abdelmoula, N.; Zghal, S.; Khemakhem, H. Zirconium and lanthanide effects on the structural, dielectric and optical properties of $Na_{0.8}Ba_{0.2}Nb_{0.8}Ti_{0.2}O_3$ ferroelectric ceramic. *Mater. Today Commun.* **2020**, *24*, 101223. [CrossRef]
11. Xie, A.; Zuo, R.; Qiao, Z.; Fu, Z.; Hu, T.; Fei, L. $NaNbO_3$-$(Bi_{0.5}Li_{0.5})TiO_3$ Lead-Free Relaxor Ferroelectric Capacitors with Superior Energy-Storage Performances via Multiple Synergistic Design. *Adv. Energy Mater.* **2021**, *11*, 2101378. [CrossRef]
12. Shirane, G.; Newnham, R.; Pepinsky, R. Dielectric Properties and Phase Transitions of $NaNbO_3$ and (Na, K)NbO_3. *Phys. Rev.* **1954**, *96*, 581–588. [CrossRef]
13. Li, P.; Ouyang, S.; Xi, G.; Kako, T.; Ye, J. The effects of crystal structure and electronic structure on photocatalytic H_2 evolution and CO_2 reduction over two phases of perovskite-structured $NaNbO_3$. *J. Phys. Chem. C* **2012**, *116*, 7621–7628. [CrossRef]
14. Fresno, F.; Jana, P.; Reñones, P.; Coronado, J.M.; Serrano, D.P.; De La Peña O'Shea, V.A. CO_2 Reduction over $NaNbO_3$ and $NaTaO_3$ Perovskite Photocatalysts. *Photochem. Photobiol. Sci.* **2017**, *16*, 17–23. [CrossRef]
15. Li, G. Photocatalytic properties of $NaNbO_3$ and $Na_{0.6}Ag_{0.4}NbO_3$ synthesized by polymerized complex method. *Mater. Chem. Phys.* **2010**, *121*, 42–46. [CrossRef]
16. Sharma, A.; Bhardwaj, U.; Jain, D.; Kushwaha, H.S. $NaNbO_3$ Nanorods: Photopiezocatalysts for Elevated Bacterial Disinfection and Wastewater Treatment. *ACS Omega* **2022**, *7*, 7595–7605. [CrossRef]
17. Liu, Q.; Chai, Y.; Zhang, L.; Ren, J.; Dai, W.-L. Highly efficient Pt/$NaNbO_3$ nanowire photocatalyst: Its morphology effect and application in water purification and H_2 production. *Appl. Catal. B Environ.* **2017**, *205*, 505–513. [CrossRef]
18. Baeissa, E.S. Photocatalytic degradation of malachite green dye using Au/$NaNbO_3$ nanoparticles. *J. Alloys Compd.* **2016**, *672*, 564–570. [CrossRef]
19. Zhang, B.; Zhang, D.; Xi, Z.; Wang, P.; Pu, X.; Shao, X.; Yao, S. Synthesis of Ag_2O/$NaNbO_3$ p-n junction photocatalysts with improved visible light photocatalytic activities. *Sep. Purif. Technol.* **2017**, *178*, 130–137. [CrossRef]
20. Shi, H.; Chen, G.; Zhang, C.; Zou, Z. Polymeric g-C_3N_4 coupled with $NaNbO_3$ nanowires toward enhanced photocatalytic reduction of CO_2 into renewable fuel. *ACS Catal.* **2014**, *4*, 3637–3643. [CrossRef]
21. Li, G.; Kako, T.; Wang, D.; Zou, Z.; Ye, J. Composition dependence of the photophysical and photocatalytic properties of $(AgNbO_3)_{1-x}(NaNbO_3)_x$ solid solutions. *J. Solid State Chem.* **2007**, *180*, 2845–2850. [CrossRef]
22. Shi, H.; Li, X.; Iwai, H.; Zou, Z.; Ye, J. 2-Propanol photodegradation over nitrogen-doped $NaNbO_3$ powders under visible-light irradiation. *J. Phys. Chem. Solids* **2009**, *70*, 931–935. [CrossRef]
23. Tong, H.; Ouyang, S.; Bi, Y.; Umezawa, N.; Oshikiri, M.; Ye, J. Nano-photocatalytic Materials: Possibilities and Challenges. *Adv. Mater.* **2012**, *24*, 229–251. [CrossRef]
24. Choi, T.; Lee, S.; Choi, Y.J.; Kiryukhin, V.; Cheong, S.W. Switchable ferroelectric diode and photovoltaic effect in $BiFeO_3$. *Science* **2009**, *324*, 63–66. [CrossRef]
25. Brody, P.S. High voltage photovoltaic effect in barium titanate and lead titanate-lead zirconate ceramics. *J. Solid State Chem.* **1975**, *12*, 193–200. [CrossRef]
26. Inoue, Y.; Sato, K.; Sato, K.; Miyama, H. Photoassisted water decomposition by ferroelectric lead zirconate titanate ceramics with anomalous photovoltaic effects. *Phys. Chem.* **1986**, *90*, 2809–2810. [CrossRef]
27. Tiwari, R.P.; Birajdar, B.; Ghosh, R.K. Strain engineering of ferroelectric $KNbO_3$ for bulk photovoltaic applications: An insight from density functional theory calculations. *J. Phys. Condens. Matter* **2019**, *31*, 505502. [CrossRef]
28. Simon, M.; Wevering, S.; Buse, K.; Krätzig, E. The bulk photovoltaic effect of photorefractive $LiNbO_3$: Fe crystals at high light intensities. *J. Phys. D Appl. Phys.* **1997**, *30*, 144–149. [CrossRef]
29. Zenkevich, A.; Matveyev, Y.; Maksimova, K.; Gaynutdinov, R.; Tolstikhina, A.; Fridkin, V. Giant bulk photovoltaic effect in thin ferroelectric $BaTiO_3$ films. *Phys. Rev. B* **2014**, *90*, 161409. [CrossRef]
30. Ji, W.; Yao, K.; Liang, Y.C. Bulk Photovoltaic Effect at Visible Wavelength in Epitaxial Ferroelectric $BiFeO_3$ Thin Films. *Adv. Mater.* **2010**, *22*, 1763–1766. [CrossRef]
31. Haruna, A.; Abdulkadir, I.; Idris, S.O. Photocatalytic activity and doping effects of $BiFeO_3$ nanoparticles in model organic dyes. *Heliyon* **2020**, *6*, e03237. [CrossRef] [PubMed]
32. Tablero, C. An evaluation of $BiFeO_3$ as a photovoltaic material. *Sol. Energy Mater. Sol. Cells.* **2017**, *171*, 161–165. [CrossRef]
33. Yang, F.; Lin, S.; Yang, L.; Liao, J.; Chen, Y.; Wang, C.-Z. First-principles investigation of metal-doped cubic $BaTiO_3$. *Mater. Res. Bull.* **2017**, *96*, 4372–4378. [CrossRef]
34. Wang, J.; Peng, Z.; Wang, J.; Wu, D.; Yang, Z.; Chao, X. Band gap tunning to enhance photovoltaic response in $NaNbO_3$-based bulk ferroelectrics. *Scr. Mater.* **2022**, *221*, 114976. [CrossRef]
35. Bennett, J.W.; Grinberg, I.; Rappe, A.M. New Highly Polar Semiconductor Ferroelectrics through d^8 Cation-O Vacancy Substitution into $PbTiO_3$: A Theoretical Study. *J. Am. Chem. Soc.* **2008**, *130*, 17409–17412. [CrossRef] [PubMed]
36. Grinberg, I.; West, D.V.; Torres, M.; Gou, G.; Stein, D.M.; Wu, L.; Chen, G.; Gallo, E.M.; Akbashev, A.R.; Davies, P.K.; et al. Perovskite oxides for visible-light-absorbing ferroelectric and photovoltaic materials. *Nature* **2013**, *503*, 509–512. [CrossRef]
37. Kichanov, S.E.; Kozlenko, D.P.; Belozerova, N.M.; Jabarov, S.H.; Mehdiyeva, R.Z.; Lukin, E.V.; Mammadov, A.I.; Liermann, H.-P.; Morgenroth, W.; Dubrovinsky, L.S.; et al. An intermediate antipolar phase in $NaNbO_3$ under compression. *Ferroelectrics* **2017**, *520*, 22–33. [CrossRef]
38. Rietveld, H. A profile refinement method or nuclear and magnetic structures. *J. Appl. Crystallogr.* **1969**, *2*, 65–71. [CrossRef]

39. Jarupoom, P.; Pengpat, K.; Pisitpipathsin, N.; Eitssayeam, S.; Intatha, U.; Rujijanagul, G.; Tunkasiri, T. Development of electrical properties in lead-free bismuth sodium lanthanum titanate–barium titanate ceramic near the morphotropic phase boundary. *Current Appl. Phys.* **2008**, *8*, 253–257. [CrossRef]
40. Kakimoto, K.-I.; Akao, K.; Guo, Y.; Ohsato, H. Raman Scattering Study of Piezoelectric $(Na_{0.5}K_{0.5})NbO_3$-$LiNbO_3$ Ceramics. *Jpn. J. Appl. Phys.* **2005**, *44*, 7064–7067. [CrossRef]
41. Shen, Z.X.; Wang, X.B.; Kuok, M.H.; Tang, S.H. Raman Scattering Investigations of the Antiferroelectric–Ferroelectric Phase Transition of $NaNbO_3$. *J. Raman Spectrosc.* **1998**, *29*, 379–384. [CrossRef]
42. Halder, N.C.; Wagner, C.N.J. Separation of particle size and lattice strain in integral breadth measurements. *Acta Crystallogr.* **1966**, *20*, 312–313. [CrossRef]
43. Halder, N.C.; Wagner, C.N.J. Analysis of the broadening of powder pattern peaks using variance, integral breadth, and Fourier coefficients of the line profile. *Adv. X-Ray Anal.* **1966**, *9*, 91–102.
44. Stokes, A.R.; Wilson, A.J.C. The diffraction of X rays by distorted crystal aggregates—I. *Proc. Phys. Soc.* **1944**, *56*, 174. [CrossRef]
45. Scherrer, P. Bestimmung der Größe und der inneren Struktur von Kolloidteilchen mittels Röntgenstrahlen. In *Nachrichten von der Gesellschaft der Wissenschaften zu Göttingen, Mathematisch-Physikalische Klasse*; Forgotten Books: London, UK, 1918; pp. 96–100.
46. Ida, T.; Shimazaki, S.; Hibino, H.; Toraya, H. Diffraction peak profiles from spherical crystallites with lognormal size distribution. *J. Appl. Crystallogr.* **2003**, *36*, 1107–1115. [CrossRef]
47. Chou, X.; Zhai, J.; Jiang, H.; Yao, X. Dielectric properties and relaxor behavior of rare-earth (La, Sm, Eu, Dy, Y) substituted barium zirconium titanate ceramics. *J. Appl. Phys.* **2007**, *102*, 084106. [CrossRef]
48. Reddy, S.B.; Rao, K.P.; Rao, M.S.R. Effect of La substitution on the structural and dielectric properties of $BaZr_{0.1}Ti_{0.9}O_3$ ceramics. *J. Alloys Compd.* **2009**, *481*, 692–696. [CrossRef]
49. Chalfouh, C.; Lahmar, A.; Zghal, S.; Hannachi, R.; Abdelmoula, N.; Khemakhem, H. Effects of lanthanide amphoteric incorporation on structural, electrical, and photoluminescence properties of $BaTi_{0.925}(Yb_{0.5}Nb_{0.5})_{0.075}O_3$ ceramic. *J. Alloys Compd.* **2017**, *711*, 205–214. [CrossRef]
50. Teixeira, G.F.; Silva Junior, E.; Simoes, A.Z.; Longo, E.; Zaghete, M.A. Unveiling the correlation between structural order-disorder character and photoluminescence emissions of $NaNbO_3$. *Cryst. Eng. Comm.* **2017**, *19*, 4378–4392. [CrossRef]
51. Kortum, G. *Reflectance Spectroscopy: Principles, Methods, Applications*; Springer: Berlin/Heidelberg, Germany, 1969.
52. Alkallas, F.H.; Elshokrofy, K.M.; Mansour, S.A. Structural and diffuse reflectance characterization of cobalt-doped titanium dioxide nanostructured powder prepared via facile sonochemical hydrolysis technique. *Nanomater. Nanotechnol.* **2019**, *9*, 1–7. [CrossRef]
53. Tian, P.; Chen, X.; Jin, J.; Lu, J.Q.; Liang, X.; Hu, X.-H. Stochastic model for quantifying effect of surface roughness on light reflection by diffuse reflectance standards. *Opt. Eng.* **2018**, *57*, 094104. [CrossRef]
54. Kubelka, P.; Munk, F. Ein Beitrag Zur Optik Der Farbanstriche. *Z. Tech. Phys.* **1931**, *12*, 593–601.
55. Morales, A.E.; Mora, E.S.; Pal, U. Use of diffuse reflectance spectroscopy for optical characterization of un-supported nanostructures. *Rev. Mex. Fis.* **2007**, *53*, 18–22.
56. Smith, R.A. *Semiconductors*, 2nd ed.; Cambridge University Press: London, UK, 1978; pp. 434–436.
57. Tang, J.W.; Zou, Z.G.; Ye, J.H. Photophysical and Photocatalytic Properties of $AgInW_2O_8$. *J. Phys. Chem. B* **2003**, *107*, 14265–14269. [CrossRef]
58. Zhou, W.; Deng, H.; Yang, P.; Chu, J. Structural phase transition, narrow band gap, and room-temperature ferromagnetism in $[KNbO_3]_{1-x}[BaNi_{1/2}Nb_{1/2}O_{3-\delta}]_x$ ferroelectrics. *Appl. Phys. Lett.* **2014**, *105*, 111904. [CrossRef]
59. Song, B.; Wang, X.; Xin, C.; Zhang, L.; Song, B.; Zhang, Y.; Wang, Y.; Wang, J.; Liu, Z.; Sui, Y.; et al. Multiferroic properties of Ba/Ni co-doped $KNbO_3$ with narrow band-gap. *J. Alloys Compd.* **2017**, *703*, 67–72. [CrossRef]
60. Vlazan, P.; Rus, S.F.; Poienar, M.; Sfirloaga, P. The effect of dopant concentrations on the structural, morphological and optical properties of $NaNbO_3$ semiconductor materials. *Mater. Sci. Semicond. Proc.* **2019**, *102*, 104602. [CrossRef]
61. Parida, S.; Satapathy, A.; Sinha, E.; Bisen, A.; Rout, S.K. Effect of Neodymium on Optical Band Gap and Microwave Dielectric Properties of Barium Zirconate Ceramic. *Met. Mat. Trans. A* **2015**, *46*, 1277–1286. [CrossRef]
62. Saadi, H.; Khaldi, O.; Dahri, A.; Abdelmoula, N.; Hammami, I.; Graça, M.P.F.; Benzarti, Z. Influence of (Co+Al) Co-Doping on Structural, Micro-Structural, Optical and Electrical Properties of Nanostructured Zinc Oxide. *Ceram. Int.* 2024; *in press*.
63. Borah, M.; Mohanta, D. Structural and optoelectronic properties of Eu^{2+} doped nanoscale barium titanates of pseudo-cubic form. *J. Appl. Phys.* **2012**, *112*, 124321. [CrossRef]
64. Urbach, F. The long-wavelength edge of photographic sensitivity and of the electronic absorption of solids. *Phys. Rev.* **1953**, *92*, 1324. [CrossRef]
65. Parida, S.; Rout, S.K. Effect of zirconia on the structural and optical properties of strontium titanate ceramic. *Ferroelectrics* **2017**, *517*, 81–89. [CrossRef]
66. Mishra, S.K.; Choudhury, N.; Chaplot, S.L.; Krishna, P.S.R.; Mittal, R. Competing antiferroelectric and ferroelectric interactions in $NaNbO_3$: Neutron diffraction and theoretical studies. *Phys. Rev. B* **2007**, *76*, 024110. [CrossRef]
67. Lu, X.; Hou, L.; Jin, L.; Wang, L.; Tian, Y.; Yu, K.; Hu, Q.; Zhang, L.; Wei, X. Structure evolution and exceptionally ultra-low hysteresis unipolar electric field-induced strain in $(1-x)NaNbO_3$-$xBaTiO_3$ lead-free ferroelectrics. *Ceram. Int.* **2018**, *44*, 5492–5499. [CrossRef]

68. Gaied, A.I.; Dahri, A.; Perrin, V.; Seveyrat, L.; Sassi, Z.; Khemakhem, H.; Abdelmoula, N.; Benzarti, Z. Synergistic effects of Zn B-site substitution in lead-free $Ba_{0.95}Ca_{0.05}Ti_{0.92}Sn_{0.08}O_3$ ferroelectric ceramics for enhancing piezoelectric properties in energy harvesting applications. *J. Alloys Compd.* **2023**, *958*, 170419. [CrossRef]
69. Tang, X.G.; Chew, K.-H.; Chan, H.L.W. Diffuse phase transition and dielectric tunability of $Ba(Zr_yTi_{1-y})O_3$ relaxor ferroelectric ceramics. *Acta Mater.* **2004**, *52*, 5177–5183. [CrossRef]
70. Yu, Z.; Ang, C.; Guo, R.; Bhalla, A.S. Dielectric properties of $Ba(Ti_{1-x}Zr_x)O_3$ solid solutions. *Mater. Lett.* **2007**, *61*, 326–329. [CrossRef]
71. Sun, E.; Qi, X.; Yuan, Z.; Sang, S.; Zhang, R.; Yang, B.; Cao, W.; Zhao, L. Relaxation behavior in $0.24Pb(In_{1/2}Nb_{1/2})O_3$–$0.49Pb(Mg_{1/3}Nb_{2/3})O_3$–$0.27PbTiO_3$ ferroelectric single crystal. *Ceram. Int.* **2016**, *42*, 4893–4898. [CrossRef]
72. Bourguiba, F.; Dhahri, A.; Tahri, T.; Dhahri, J.; Abdelmoula, N.; Taibi, K.; Hlil, E.K. Structure Properties and Relaxor Characteristics of the Phases Transformation in $BaTi_{0.5}(Fe_{0.33}Mo_{0.17})O_3$ Perovskite Ceramic. *J. Alloys Compd.* **2016**, *675*, 174–182. [CrossRef]
73. Ostos, C.; Mestres, L.; Martinez-Sarrion, M.L.; Garcia, J.E.; Albareda, A.; Perez, R. The influence of A-site rare-earth for barium substitution on the chemical structure and ferroelectric properties of BZT thin films. *Solid State Sci.* **2009**, *11*, 1016–1022. [CrossRef]
74. Uchino, K.; Nomura, S. Critical exponents of the dielectric constants in diffused-phase-transition crystals. *Ferroelectrics Lett.* **1982**, *44*, 55–61. [CrossRef]
75. Cross, L.E. Relaxor ferroelectrics. *Ferroelectrics* **1987**, *76*, 241–267. [CrossRef]
76. Cross, L.E. Relaxor ferroelectrics: An overview. *Ferroelectrics* **1994**, *151*, 305–320. [CrossRef]
77. Ravez, J.; Broustera, C.; Simon, A. Lead-free ferroelectric relaxor ceramics in the $BaTiO_3$-$BaZrO_3$-$CaTiO_3$ system. *J. Mater. Chem.* **1999**, *9*, 1609–1613. [CrossRef]

ceramics

MDPI

Article

Excellent Energy Storage and Photovoltaic Performances in $Bi_{0.45}Na_{0.45}Ba_{0.1}TiO_3$-Based Lead-Free Ferroelectricity Thin Film

Jianhua Wu [1,2], Tiantian Zhang [1], Xing Gao [1], Lei Ning [1], Yanhua Hu [3], Xiaojie Lou [4], Yunying Liu [1,*], Ningning Sun [1,*] and Yong Li [1,*]

1 Inner Mongolia Key Laboratory of Ferroelectric-Related New Energy Materials and Devices, School of Materials Science and Engineering, Inner Mongolia University of Science and Technology, Baotou 014010, China; wjh2219333820@126.com (J.W.); ztt935810698@163.com (T.Z.); gx15247821582@163.com (X.G.); n2189956990@163.com (L.N.)

2 School of Mechanical Engineering, Inner Mongolia University of Science and Technology, Baotou 014010, China

3 Department of Chemical Engineering, Ordos Institute of Technology, Erdos 017000, China; aoqibaby080410@163.com

4 State Key Laboratory for Mechanical Behavior of Materials, Frontier Institute of Science and Technology, Xi'an Jiaotong University, Xi'an 710049, China; xlou03@mail.xjtu.edu.cn

* Correspondence: liu_yunying@126.com (Y.L.); sunning@imust.edu.cn (N.S.); liyong3062545@126.com (Y.L.)

Abstract: Inorganic dielectric films have attracted extensive attention in the field of microelectronic and electrical devices because of their wide operating temperature range, small size, and easy integration. Here, we designed and prepared eco-friendly (1-x)$Bi_{0.45}Na_{0.45}Ba_{0.1}TiO_3$-x$Bi(Mg_{1/3}Nb_{2/3})O_3$ multifunctional ferroelectric thin films for energy storage and photovoltaic. The results show that $Bi(Mg_{1/3}Nb_{2/3})O_3$ can effectively improve the energy storage performance. At x = 0.05, the energy storage density and efficiency are as high as 73.1 J/cm^3 and 86.2%, respectively, and can operate stably in a wide temperature range. The breakdown field strength of the thin films increased significantly, and the analysis showed that the addition of $Bi(Mg_{1/3}Nb_{2/3})O_3$ caused a change in the internal conduction mechanism. At the same time, the generation of polar nanoregions increases the relaxation characteristics, thus improving the energy storage properties. In addition, the thin film material also has excellent ferroelectric photovoltaic properties. This work represents a new design paradigm that can serve as an effective strategy for developing advanced multi-functional materials.

Keywords: thin film; energy storage; photovoltaic; ceramics

Citation: Wu, J.; Zhang, T.; Gao, X.; Ning, L.; Hu, Y.; Lou, X.; Liu, Y.; Sun, N.; Li, Y. Excellent Energy Storage and Photovoltaic Performances in $Bi_{0.45}Na_{0.45}Ba_{0.1}TiO_3$-Based Lead-Free Ferroelectricity Thin Film. *Ceramics* **2024**, 7, 1043–1052. https://doi.org/10.3390/ceramics7030068

Academic Editors: Francesco Scotognella and Gilbert Fantozzi

Received: 23 May 2024
Revised: 4 July 2024
Accepted: 30 July 2024
Published: 1 August 2024

1. Introduction

With the increasing demand for renewable energy and the rapid advancements in electrical and electronic technology, the research focus in the energy field has shifted towards developing electrical energy storage devices with high power density, high voltage, low cost, and potential for large-scale applications [1–5]. Dielectric capacitors offer advantages such as higher power density, faster charge and discharge rates, exceptional cycle life, and improved safety, making them suitable for various applications, including green energy storage, pulse power systems, and military facilities. Despite these benefits, dielectric capacitors have limitations in energy storage density, and the evolving trend towards smaller, lighter, and more integrated electronic devices poses challenges for their energy storage performance [6–9]. Addressing these gaps to enhance energy storage density remains a key research area. The energy storage performance of dielectric capacitors is typically assessed using the following formula [10–12]:

$$\text{Energy storage density } (W) = \int_0^{P_{\max}} E dP \quad (1)$$

$$\text{Recoverable energy storage density } (W_{\text{rec}}) = \int_{P_r}^{P_{\max}} EdP \tag{2}$$

$$\text{Energy storage efficiency } (\eta) = \frac{W_{\text{rec}}}{W} \times 100\% \tag{3}$$

wherein $P_{\max}$, P_r, P, and E are the maximum field polarization, residual polarization, polarization, and external electric field, respectively. Increasing the gap between $P_{\max}$ and P_r and enhancing the breakdown strength (BDS) is an effective strategy to maximize energy storage density.

At present, commercially available dielectric energy storage devices are mainly made of organic polymer materials, but such capacitors are often not resistant to high temperatures [13]. In contrast, inorganic ceramic-based dielectric materials have better temperature stability, making them suitable for a wide temperature range [14,15]. Additionally, ceramic-based dielectric materials have high dielectric constant and polarization strength, which can achieve significant energy storage density even under low electric field conditions [16–18]. Among them, ceramic films have a small thickness and can achieve high electric field strength and energy storage density at very low voltage levels. These materials are lightweight, highly integrated, and show great potential for various applications. Lead-containing ceramics generally exhibit good energy storage properties by reducing grain size and constructing field-induced phase transition with higher releasable energy density (W_{rec}) and energy storage efficiency (η) [19]. However, lead can cause environmental pollution problems and damage human health, so it is very necessary to develop alternative lead-free energy storage ceramics. Lead-free ferroelectric thin film materials, such as sodium bismuth titanate-based ($Na_{0.5}Bi_{0.5}TiO_3$, NBT) ferroelectric thin films, have garnered attention due to their unique diffusion phase transition behavior and relaxation ferroelectric properties [20–22]. Despite advantages like high dielectric constant, large $P_{\max}$ (>40 μC/cm^2), and high Curie temperature (approximately 320 °C), pure NBT faces limitations such as high P_r (approximately 38 μC/cm^2), large coercive field (approximately 70 kV/cm), and low BDS (<100 kV/cm), hindering its energy storage density and efficiency improvements. These constraints restrict its widespread application in energy storage and development. In response to the above problems, researchers have improved the energy storage performance of NBT films by inducing or enhancing relaxation and designing material structure. For example, Liu et al. prepared lead-free thin film (0.94-x)$Bi_{0.5}Na_{0.5}TiO_3$-0.06$BaTiO_3$-x$SrTiO_3$ (referred to as BNT-BT-xST, x = 0, 0.05, 0.10, 0.15, 0.20) using a sol-gel/spin-coating method and studied the effect of the introduction of strontium on the microstructure, dielectric properties, and energy storage density of the film [23]. The results show that the addition of strontium improves the maximum polarization and dielectric constant of the film. The film has a high maximum recyclable energy storage density of 22.5 J/cm^3 and a dielectric constant of 1120 at 1 kHz. Chen et al. conducted component exploration of (1-x)$Bi_{0.5}(Na_{0.8}K_{0.2})_{0.5}TiO_3$-x$SrZrO_3$ (BNKT-100xSZ) thin film [24]. When x = 0.15, the BNKT-SZ thin film has an energy storage density of 34.69 J/cm^3. In addition, the BNKT-15SZ film has good thermal stability over a wide temperature range of 30–100 °C. Therefore, it is effective to modify NBT-based thin film materials using specific components of doped solid solutions.

In this work, based on $(Bi_{0.5}Na_{0.5})TiO_3$ ceramic film, Ba^{2+} ions were doped at the A-site, and a solid solution was performed with the second component $Bi(Mg_{1/3}Nb_{2/3})O_3$ to optimize the properties of the films. We systematically studied the phase structure, microstructure, dielectric properties, and energy storage properties of $Bi_{0.45}Na_{0.45}Ba_{0.1}TiO_3$-x$Bi(Mg_{1/3}Nb_{2/3})O_3$ films. The results show that the introduction of $Bi(Mg_{1/3}Nb_{2/3})O_3$ significantly improves the breakdown strength and relaxation properties of films, thereby enhancing its energy storage properties. In addition, the film material not only demonstrates outstanding temperature stability but also showcases excellent photovoltaic and photodetection properties. These qualities are essential to ensure the reliable operation of the equipment in practical applications. For energy storage applications, this material can be further made into multi-layer ceramic capacitors (MLCC). MLCC is ubiquitous in modern electronic devices, such as hybrid cars and advanced medical equipment. In photovoltaic applications, photodetectors made mainly of such materials can be used in a variety of fields, such as sensing, imaging, and night vision.

2. Materials and Methods

The (1-x)$Bi_{0.45}Na_{0.45}Ba_{0.1}TiO_3$-xBi($Mg_{1/3}Nb_{2/3}$)$O_3$ (BNBT-xBMN, x = 0, 0.025, 0.050, and 0.075) film were prepared on Pt/Si substrate via a sol-gel method. PVD technique was used to deposit Pt on Si substrate to obtain the bottom electrode. Bismuth nitrate pentahydrate (99%, Aladdin reagent Co., Ltd., Shanghai, China) sodium acetate trihydrate (99%, Sinopharm Chemical Reagent Co., Ltd., Shanghai, China), tetrabutyl titanate (98%, Sinopharm Chemical Reagent Co., Ltd., Shanghai, China), barium acetate (99%, Aladdin reagent Co., Ltd., Shanghai, China) magnesium acetate tetrahydrate (99%, Aladdin reagent Co., Ltd., Shanghai, China) ammonium niobate oxalate hydrate (99.99%, Aladdin reagent Co., Ltd., Shanghai, China) were selected as the raw materials. Glacial acetic acid (99.5%, Sinopharm Chemical Reagent Co., Ltd., Shanghai, China), acetylacetone (99%, Sinopharm Chemical Reagent Co., Ltd., Shanghai, China), and distilled water were used as the solvents. According to the stoichiometric ratio, two precursor solutions of BNBT and BMN were prepared by mixing the above raw materials and then mixed separately according to the ratio. To compensate for volatilization during annealing, an excess of 10 mol% sodium and bismuth was used. The preparation steps of the BNBT precursor solution were as follows. First, the distilled water and glacial acetic acid were heated to 80 °C; then, sodium acetate trihydrate, bismuth nitrate, and barium acetate were added and fully dissolved. After natural cooling to room temperature, acetylacetone was added and stirred for 20 min. Finally, tetrabutyl titanate was added and stirred for 20 min. The volume ratio of glacial acetic acid, distilled water, and acetylacetone is 4:1:1. BMN precursor solution was prepared by the same method and then mixed with BNBT precursor solution in proportion. Formamide, lactic acid, and polyvinyl pyrrolidone in a ratio of 1:1:1 were added to the mixed solution to increase viscosity and avoid cracks after film formation. The final concentration of the solution is 0.5 mol/L. Then, the resulting BNBT-xBMN precursor solutions were spin-coated on the Pt/Si layer at 3000 rpms for 30 s. After each coating, the wet layer was dried (150 °C, 3 min), pyrolyzed (410 °C, 10 min), and crystallized (700 °C, 3 min) by using a high-temperature tubular furnace. The above steps are repeated until the film reaches the desired thickness. Finally, the sample was annealed in a tube furnace at 700 °C to crystallize.

The crystal structure and microstructure of the films were analyzed by X-ray diffractometer (XRD Bruker D8 Advance diffractometer, Germany) and scanning electron microscope (FE-SEM, Zeiss GeminiSEM 300, Germany), respectively. The polarization electric field (*P-E*) hysteresis loops and leakage current density-electric field (*J-E*) curves were performed through ferroelectric test system (Radiant Technologies, Inc., Albuquerque, NM, USA). The energy storage performance of the film was calculated by *P-E* loops. The dielectric properties were investigated by Agilent E4980A LCR analyzer. The evolution of microstructure, domain structure, and the piezoelectric response of the films were measured by piezoelectric force microscopy (PFM, Bruker, Icon, Germany). The photovoltaic performance of the films was measured by using electrometers (Keithley, 6517B, USA) and a solar simulator (NBeT, Solar500, China).

3. Results and Discussion

The XRD pattern of BNBT-xBMN grown on the Pt/Si substrate measured at room temperature is shown in Figure 1a; all the samples show a typical perovskite structure. The diffraction peak of the second phase can be observed near 36°. JADE software (MDI jade 6) analysis indicates that this phase is most likely NbO, whose ICSD number is 61634. Compared with the x = 0 sample, the (200) peak of the samples doped with BMN is obviously shifted, which is due to the lattice distortion caused by the introduction of atoms with different ionic radii at both A and B positions. Figure 1b shows the surface microstructure of the sample with x = 0.050, and the grains are evenly distributed and closely arranged. The smaller grains increase the number of grain boundaries, which will increase the channel for carrier transport, helping to improve photovoltaic performance. The illustration in Figure 1b shows the morphology of the sample cross-section. The boundary between the dielectric layer and the substrate is obvious; the dielectric layer is uniform and continuous, and the thickness is about 200 nm. The other components also exhibit dense microstructure and smaller grains, as shown in Figure S1.

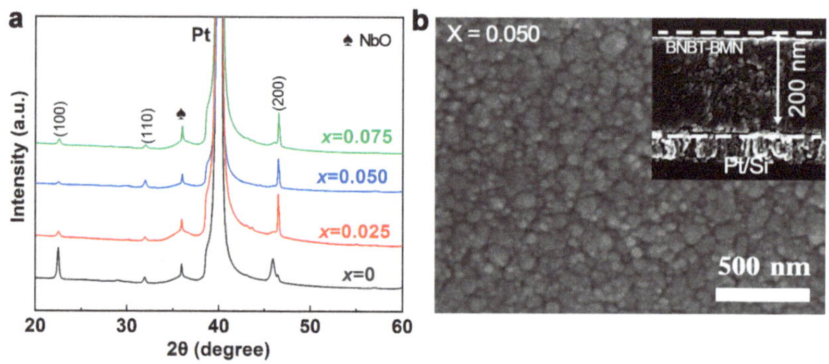

Figure 1. (**a**) XRD patterns of the BNBT-xBMN thin films with different compositions, and (**b**) SEM image of the sample with x = 0.050.

Figure 2a illustrates *P-E* loops measured for the BNBT-xBMN thin films under a critical electric field at a frequency of 10 Hz. Obviously, the x = 0 sample is broken down at a lower electric field and exhibits a large P_r and coercive field (E_c), resulting in a fatter ring of the *P-E* loops. With the increase in BMN doping, the breakdown electric field of the film sample obviously increases, the P_r decreases, and the shape of the curve gradually becomes thin. But, the introduction of BMN also caused a decrease in the P_{max}. The W_{rec} and η corresponding to Figure 2a were calculated, and their change curves were shown in Figure 2b. It can be seen that with the increase in BMN doping, W_{rec} increases first and then decreases, while η keeps increasing. Finally, it showed the best energy storage performance in the x = 0.050 sample. The *P-E* loops of x = 0.050 sample measured under different electric fields are shown in Figure 2c. Its maximum electric field is up to 2500 kV/cm, and due to relaxation properties, the curve always remains thin as the electric field changes. By calculation (Figure 2d), W_{rec} keeps increasing as the electric field increases, and η is the opposite. The W_{rec} and η of the x = 0.050 sample changed from 6.1 J/cm^3 and 88.7% at 625 kV/cm to 73.1 J/cm^3 and 86.2 % at 2500 kV/cm.

The temperature-dependent ε_r and tanδ of the BNBT-xBMN thin films at 10 kHz are depicted in Figure 3. It can be seen that the ε_r of the x = 0 samples increases sharply with the increase in temperature in the lower temperature region and shows an obvious inflection point near 150 °C, and the amplitude of change slows down. With the increase in BMN content, the magnitude of ε_r changes gradually decreases, and the inflection point around 150 °C becomes increasingly blurred. This suggests that the introduction of BMN leads to a decrease in polar coupling and an increase in relaxation behavior [25]. The dielectric loss also shows a similar phenomenon with the change in temperature. With the increase in BMN, the change of tanδ slows down, and the dielectric loss decreases overall.

Figure 4a,d and Figure S2a,d display atomic force microscope (AFM) images of BNBT-xBMN films. The arithmetic average roughness (R_a) of each component film was calculated using NanoScope Analysis software (NanoScope Analysis 2.0), resulting in values of 4.29 nm, 4.48 nm, 4.11 nm, and 4.46 nm. Notably, the BMN-doped film with x = 0.050 exhibited the smallest R_a value, suggesting an optimal solid solution ratio between BNBT and BMN. This particular film sample demonstrated superior grain growth and surface flatness. On the other hand, Figure 4b,e and Figure S2b,e show out-of-plane pressure power microscopy (PFM) images of BNBT-xBMN thin films, with different colors indicating various phase directions [26,27]. The x = 0 sample (Figure 4b) exhibited a distinct electric domain structure, indicative of a long-range ordered ferroelectric state. As the BMN content increased, the domain size reduced, and the number of domains multiplied, leading to the formation of polar nanoregions (PNRs). At x = 0.050 (Figure 4e), no continuous large-sized ferroelectric domains were observed, emphasizing the presence of PNRs as a key structural characteristic of relaxation ferroelectrics. The evolution of the domain structure clearly illustrates that the addition of BMN induces a transition from a ferroelectric to a relaxation state. The addition of BMN causes local lattice distortion and fluctuation in charge distribution, leading to the development of a random electric field internally. This electric field can either pin or distort domain walls, facilitating the formation of PNRs. With an increase in BNZ content, the local electric field strength rises, resulting in the creation of more PNRs and ultimately enhancing the relaxation of the system. To better understand the behavior of ferroelectric domains and PNRs, the phase and amplitude of the piezoelectric response were measured in selected areas of all samples as a function of applied DC voltage, as illustrated in Figure 4c,f and Figure S2c,f. The phase-voltage loop shows that at x = 0, the loop appears nearly square with evident electrical hysteresis. As BMN content increases, the square shape gradually diminishes, transforming into an approximate single line shape at x = 0.050, indicating the gradual formation of the PNRs region and a decrease in hysteresis. This suggests that PNRs are more easily influenced by external electric fields compared to traditional ferroelectric domains. Furthermore, the amplitude–voltage diagram at x = 0 displays a classic butterfly curve, demonstrating clear ferroelectric behavior. As BMN content increases, the butterfly curve becomes narrower, indicating heightened relaxation behavior [28]. Overall, the introduction of BMN leads to the emergence of PNRs, triggering the transition from a ferroelectric state to a relaxed state.

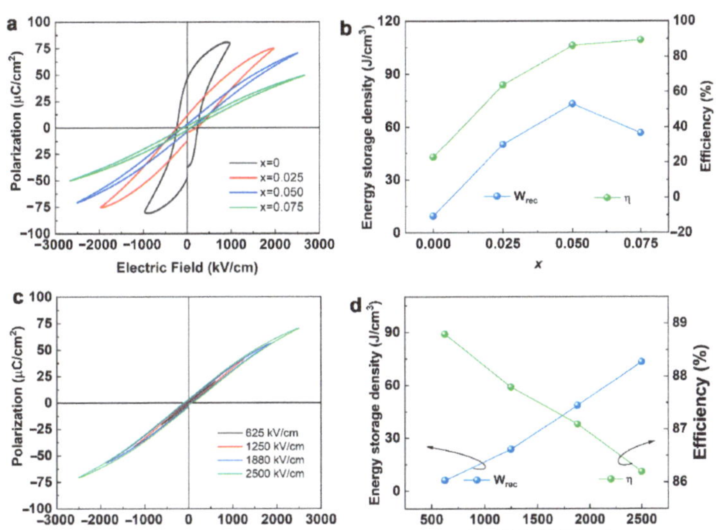

Figure 2. (**a**) *P-E* loops of different components at breakdown electric field and (**c**) *P-E* loops of the x = 0.050 simple at different electric fields tested at 10 Hz frequency and room temperature, (**b**) and (**d**) are the corresponding curves of W_{rec} and η, respectively.

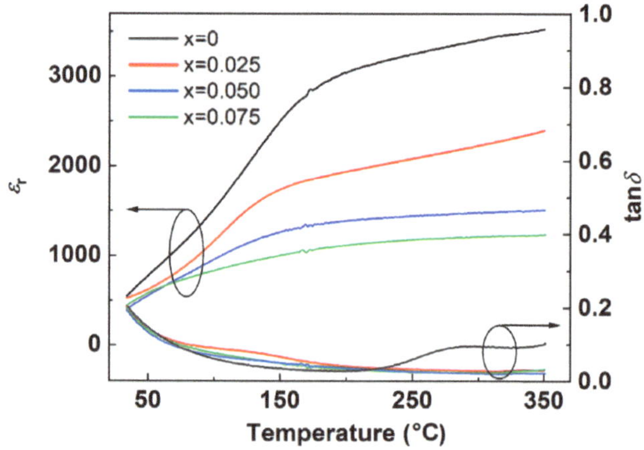

Figure 3. Temperature-dependent ε_r and tanδ of the BNBT-xBMN thin films at 10 kHz.

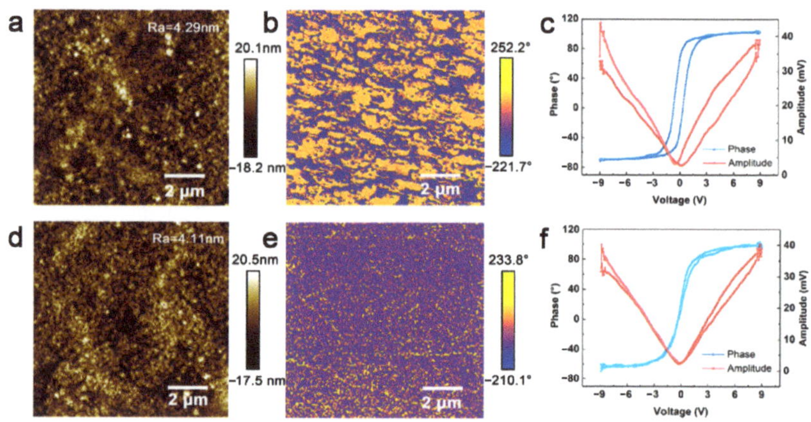

Figure 4. AFM image, out-of-plane phase PFM image, and voltage-dependent piezoresponse phase and amplitude of the BNBT-xBMN thin films: (**a**–**c**) x = 0; (**d**–**f**) x = 0.050.

In addition to the relaxation characteristics, the BDS is another important factor affecting the energy storage performance of the dielectric [29]. The leakage current density of a material is closely related to the strength of the BDS, the current densities for BNBT-xBMN thick films are plotted against the applied electric field (*J-E*) and shown in Figure 5a. Clearly, with the increase in BMN, the leakage current density gradually decreases. When x = 0.075, the leakage current density is the smallest, which corresponds to the changing law of the BDS in Figure 2a. In order to further explore its internal mechanism, the logarithmic plots of *J* as a function of *E* for BBT-xBMN films were drawn, and linear fitting was carried out, as shown in Figure 5b. It can be seen that the slope of the x = 0 sample after fitting is always the same, while the thin film after BMN doping has two different fitting results in the lower (below 100 kV/cm) and higher (over 100 kV/cm) electric fields, respectively. At lower electric fields, the slope is approximately 1. At higher electric fields, the slope decreases from 2.16 to 1.39 and gradually approaches 1. Obviously, the curves in the figure can be explained by the space-charge-limited current (SCLC) model [30]. When the slope is near 1, it is Ohmic conduction, while when the slope is near 2, the modified Child's law conduction is followed [31]. It can be concluded that when the conductivity mechanism in the film is modified by Child's law conduction, the leakage current density is large, and the film is easy to break down. With the increase in BMN content, the internal conductance model changes to Ohmic conduction, and the conductance mechanism dominated by Ohmic conduction helps to reduce the leakage current density of the film, thereby increasing the breakdown electric field strength.

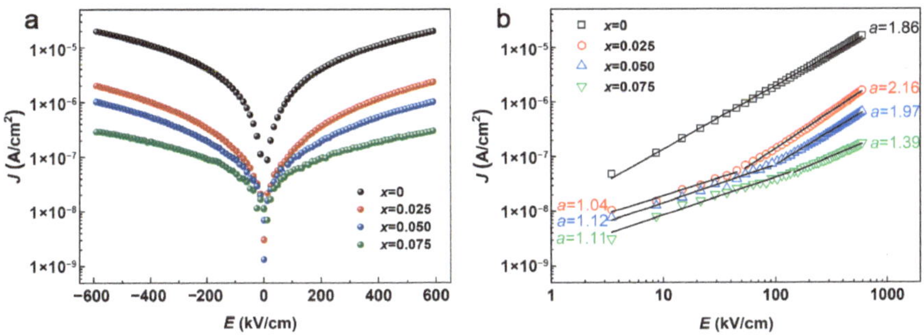

Figure 5. (**a**) *J-E* curves of the BNBT-xBMN thin films and (**b**) the fitting of unipolar *J-E* curves.

Temperature reliability is an important index to measure the practical application characteristics of electronic devices [28,32]. Figure 6a plots the unipolar *P-E* loops of the x = 0.050 film measured at various temperatures under 1000 kV/cm at 10 Hz. As the temperature was elevated from room temperature (~25 °C) to 200 °C, P_{max} increased gradually, but *P-E* loops widened slightly. This is due to temperature-induced leakage loss or ionic conduction. As can be seen from Figure 6b by calculating the corresponding energy storage performance, there is a slight decrease in W_{rec} from 23.6 J/cm^3 to 22.4 J/cm^3 and a significant reduction in η from 87.4% to 81.8%. Although η attenuates in the high-temperature environment, the relatively stable W_{rec} can still ensure the normal operation of the device at a high temperature. This proves that the x = 0.050 sample has excellent temperature reliability.

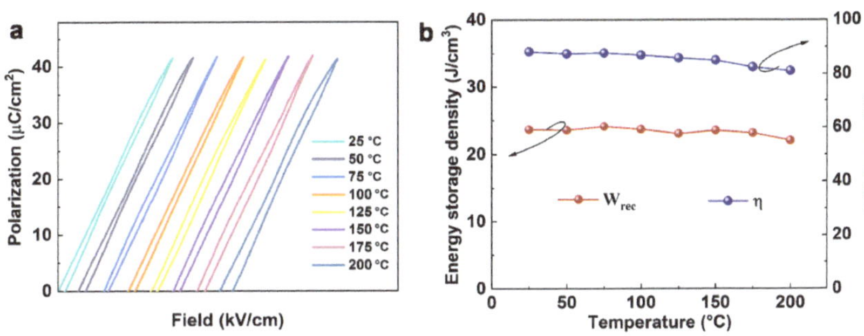

Figure 6. (**a**) *P-E* loops of x = 0.050 simple under 1000 kV/cm subjected to different temperatures and (**b**) the corresponding calculated W_{rec} and η.

To investigate the photovoltaic (PV) properties of the x = 0.050 sample, the electric current density–voltage (*J-V*) characteristics were tested in the unpolarized state and polarized state (±80 V). As can be seen from Figure 7a, the as-grown sample did not show obvious photovoltaic properties under illumination. After poling with a voltage of +80 V, a photocurrent density of 1.46 μA/cm^2 was obtained. After poling with a voltage of −80 V, the direction of the photocurrent reverses, exhibiting the switching effect of ferroelectric photovoltaic. Figure 7b shows that the x = 0.050 sample has extremely high current repeatability and stability under light switching. In order to explore the sensibility of photoresponse in the x = 0.050 sample, the short-circuit current (J_{sc}) under different light intensities was tested. As shown in Figure 7c, the film has obvious switching effects under different light intensities and has good stability. Under weak light intensity (20 mW/cm^2), a photocurrent density of 0.39 μA/cm^2 was still obtained. The temporal stability and repeatability of J_{sc} are also important indicators for measuring photodetection devices. As shown in Figure 7d, the stability and repeatability of J_{sc} within 2000 s under 100 mW/cm^2 illumination conditions were tested. It can be seen that during the entire test time range, the value of the photocurrent density only decreases slightly, indicating that the photodetector has very stable photoelectric output characteristics.

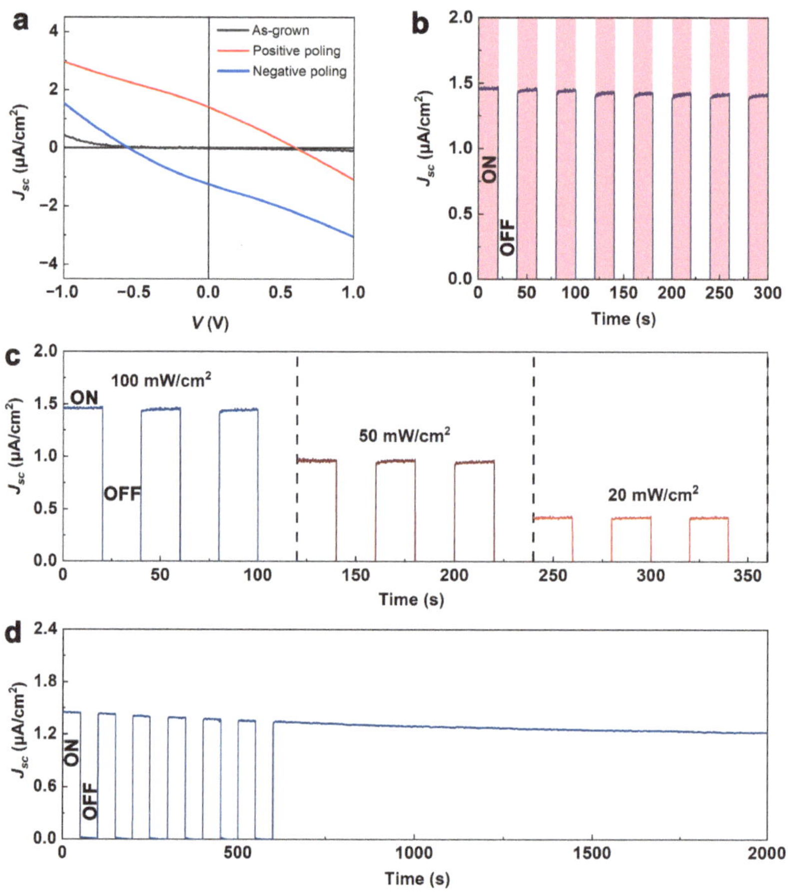

Figure 7. (**a**) *J-V* characteristics under different polarization states and (**b**) time-dependent J_{sc}; (**c**) J_{sc} at different light intensities; (**d**) stability diagram of J_{sc} under sunlight.

4. Conclusions

In summary, novel, lead-free (1-x)$Bi_{0.45}Na_{0.45}Ba_{0.1}TiO_3$-xBi($Mg_{1/3}Nb_{2/3}$)$O_3$ thin films were successfully deposited on a Pt/Si substrate by the sol-gel method. The excellent energy storage density of 73.1 J/cm^3 and an efficiency of 86.2% were obtained at x = 0.05, especially since the ultra-high energy storage efficiency exceeds that of most of the thin film energy storage capacitors. The results show that the introduction of BMN reduces the domain size and changes the conduction mode, which improves the relaxation characteristics and breakdown strength. In addition, BNBT-0.05BMN exhibits excellent thermal stability over a wide temperature range of 25 °C to 200 °C. The film is also suitable for photovoltaic detection. In addition, the film shows excellent potential for photodetection. This study demonstrates the potential application of BNBT-xBMN films in the field of hybrid electronic devices and provides ideas for the development of advanced multifunctional materials.

Supplementary Materials: The following supporting information can be downloaded at: https://www.mdpi.com/article/10.3390/ceramics7030068/s1, Figure S1: SEM images and particle size distribution statistics of x = 0, x = 0.025, and x = 0.075; Figure S2: AFM image, out-of-plane phase PFM image, and voltage-dependent piezoresponse phase and amplitude of the BNBT-xBMN thin films: (a–c) x = 0.025; (d–f) x = 0.075.

Author Contributions: Conceptualization, J.W. and Y.L. (Yong Li); methodology, J.W. and X.G.; data curation, J.W., Y.H. and L.N.; writing—original draft preparation, J.W. and X.L.; writing—review and editing, J.W. and T.Z.; project administration, Y.L. (Yong Li) and Y.L. (Yunying Liu); funding acquisition, Y.L. (Yong Li) and N.S. All authors contributed to the article. All authors have read and agreed to the published version of the manuscript.

Funding: This research received no external funding.

Institutional Review Board Statement: Not applicable.

Informed Consent Statement: Not applicable.

Data Availability Statement: The data presented in this study are available upon request from the corresponding author.

Acknowledgments: This work was supported by the Major Science and Technology Program of Ordos City (2021EEDSCXQDFZ014), Basic Research Funds for Universities Directly under Inner Mongolia (2023RCTD008, 2024QNJS002), Youth Science and Technology Talents Project of Inner Mongolia (NJYT22061), Scientific and Technological Development Foundation of the Central Guidance Local (2021ZY0008), Scientific research project of universities in Inner Mongolia (NJZZ23054), "Light of the West" Talent Training Program of Chinese Academy of Sciences, the Program for "Grassland Talents" of Inner Mongolia, Planning Project of Science and Technology of Ordos City (2022YY043), Talent Development Fund of Inner Mongolia, Natural Science Foundation of Inner Mongolia Autonomous Region (2024MS05016), the Fundamental Research Funds for Inner Mongolia University of Science & Technology(2024QNJS002), and the Scientific Research Project for China Northern Rare Earth (Group) High-tech Co., Ltd.

Conflicts of Interest: The authors declare no conflicts of interest.

References

1. Dang, Z.; Yuan, J.; Yao, S.; Liao, R. Flexible nanodielectric materials with high permittivity for power energy storage. *Adv. Mater.* **2013**, *25*, 6334–6365. [CrossRef] [PubMed]
2. Diao, C.; Li, H.; Yang, Y.; Hao, H.; Yao, Z.; Liu, H. Significantly improved energy storage properties of sol-gel derived Mn-modified $SrTiO_3$ thin films. *Ceram. Int.* **2019**, *45*, 11784–11791. [CrossRef]
3. Shangguan, D.; Duan, Y.; Wang, B.; Wang, C.; Li, J.; Bai, Y.; Zhang, F.; Li, Y.; Wu, Y.; Wang, Z.J. Enhanced energy-storage performances of (1-x)$PbZrO_3$-x$PbSnO_3$ antiferroelectric thin films under low electric fields. *J. Alloys Compd.* **2021**, *870*, 159440. [CrossRef]
4. Pan, Z.; Ding, J.; Hou, X.; Shi, S.; Yao, L.; Liu, J.; Li, P.; Chen, J.; Zhai, J.; Pan, H. Substantially improved energy storage capability of ferroelectric thin films for application in high-temperature capacitors. *J. Mater. Chem. A* **2021**, *9*, 9281–9290. [CrossRef]
5. Wang, G.; Lu, Z.; Li, Y.; Li, L.; Ji, H.; Feteira, A.; Zhou, D.; Wang, D.; Zhang, S.; Reaney, I.M. Electroceramics for high-energy density capacitors: Current status and future perspectives. *Chem. Rev.* **2021**, *121*, 6124–6172. [CrossRef]
6. Zhu, H.; Liu, M.; Zhang, Y.; Yu, Z.; Ouyang, J.; Pan, W. Increasing energy storage capabilities of space-charge dominated ferroelectric thin films using interlayer coupling. *Acta Mater.* **2017**, *122*, 252–258. [CrossRef]
7. Zhang, X.; Shen, Y.; Xu, B.; Zhang, Q.; Gu, L.; Jiang, J.; Ma, J.; Lin, Y.; Nan, C.W. Giant energy density and improved discharge efficiency of solution-processed polymer nanocomposites for dielectric energy storage. *Adv. Mater.* **2016**, *28*, 2055–2061. [CrossRef] [PubMed]
8. Yang, Z.; Du, H.; Jin, L.; Poelman, D. High-performance lead-free bulk ceramics for electrical energy storage applications: Design strategies and challenges. *J. Mater. Chem. A* **2021**, *9*, 18026–18085. [CrossRef]
9. Shen, B.; Li, Y.; Sun, N.; Zhao, Y.; Hao, X. Enhanced energy-storage performance of an all-inorganic flexible bilayer-like antiferroelectric thin film via using electric field engineering. *Nanoscale* **2020**, *12*, 8958–8968. [CrossRef]
10. Yan, F.; Yang, H.; Lin, Y.; Wang, T. Dielectric and Ferroelectric Properties of $SrTiO_3$–$Bi_{0.5}Na_{0.5}TiO_3$–$BaAl_{0.5}Nb_{0.5}O_3$ Lead-Free Ceramics for High-Energy-Storage Applications. *Inorg. Chem.* **2017**, *56*, 13510–13516. [CrossRef]
11. Xie, A.; Fu, J.; Zuo, R.; Jiang, X.; Li, T.; Fu, Z.; Yin, Y.; Li, X.; Zhang, S. Supercritical Relaxor Nanograined Ferroelectrics for Ultrahigh-Energy-Storage Capacitors. *Adv. Mater.* **2022**, *34*, 2204356. [CrossRef]
12. Diao, C.; Liu, H.; Lou, G.; Zheng, H.; Yao, Z.; Hao, H.; Cao, M. Structure and electric properties of sandwich-structured $SrTiO_3$/$BiFeO_3$ thin films for energy storage applications. *J. Alloys Compd.* **2019**, *781*, 378–384. [CrossRef]

13. Wang, Y.; Cui, J.; Yuan, Q.; Niu, Y.; Bai, Y.; Wang, H. Significantly enhanced breakdown strength and energy density in sandwich-structured barium titanate/poly(vinylidene fluoride) nanocomposites. *Adv. Mater.* **2015**, *27*, 6658–6663. [CrossRef] [PubMed]
14. Chen, T.; Wang, J.B.; Zhong, X.L.; Zeng, Y.K.; Wang, F.; Zhou, Y.C. Improved energy storage properties of $PbZrO_3$ thin films by inserting $0.88BaTiO_3$–$0.12Bi(Mg_{1/2},Ti_{1/2})O_3$ layer. *Appl. Surf. Sci.* **2013**, *285*, 744–747. [CrossRef]
15. Dong, H.; Luo, B.; Jin, K. Tunable dielectric and energy storage properties in nonstoichiometric $NaNbO_3$ thin films. *Ceram. Int.* **2022**, *48*, 16215–16220. [CrossRef]
16. Luo, B.; Dong, H.; Wang, D.; Jin, K. Large recoverable energy density with excellent thermal stability in Mn-modified $NaNbO_3$-$CaZrO_3$ lead-free thin films. *J. Am. Chem. Soc.* **2018**, *101*, 3460–3467. [CrossRef]
17. Pan, H.; Zhang, Q.; Wang, M.; Lan, S.; Meng, F.; Ma, J.; Gu, L.; Shen, Y.; Yu, P.; Lin, Y.H.; et al. Enhancements of dielectric and energy storage performances in lead-free films with sandwich architecture. *J. Am. Chem. Soc.* **2018**, *102*, 936–943. [CrossRef]
18. Ahn, Y.; Son, J.Y. Multiferroic and energy-storage characteristics of polycrystalline Ca-doped $BiFeO_3$ thin films on Si substrates. *J. Korean Ceram. Soc.* **2022**, *60*, 301–309. [CrossRef]
19. Thatikonda, S.K.; Huang, W.; Du, X.; Yao, C.; Ke, Y.; Wu, J.; Qin, N.; Bao, D. Sm-doping induced large enhancement of antiferroelectric and energy storage performances of (111) oriented $PbZrO_3$ thin films. *Ceram. Int.* **2019**, *45*, 23586–23591. [CrossRef]
20. Supriya, S. Crystal structure engineered non-toxic $Bi_{0.5}Na_{0.5}TiO_3$ based thin films-fabrication process, enhanced electrical performance, challenges and recent reports. *J. Inorg. Organomet. Polym. Mater.* **2023**, *33*, 3013–3026. [CrossRef]
21. Fan, Z.; Li, L.; Mei, X.; Zhao, F.; Li, H.; Zhuo, X.; Zhang, X.; Lu, Y.; Zhang, L.; Liu, M. Multilayer ceramic film capacitors for high-performance energy storage: Progress and outlook. *J. Mater. Chem. A* **2021**, *9*, 9462–9480. [CrossRef]
22. Wu, S.; Xu, L.; Zhu, K.; Song, B.; Yan, H.; Shen, B.; Zhai, J. Improved piezoelectricity and energy storage performance simultaneously achieved in [001]-preferentially oriented $Bi_{0.5}Na_{0.5}TiO_3$–$BaTiO_3$–$BiMnO_3$ thin films grown on Nb-doped $SrTiO_3$ single-crystalline substrates. *J. Eur. Ceram. Soc.* **2021**, *41*, 2539–2547. [CrossRef]
23. Xie, Y.; Hao, H.; Xie, J.; He, Z.; Zhang, S.; Li, Z.; Cao, M.; Yao, Z.; Liu, H. The energy-storage performance and dielectric properties of (0.94-x)BNT-0.06BT-xST thin films prepared by sol–gel method. *J. Alloys Compd.* **2021**, *860*, 158164. [CrossRef]
24. Chen, P.; Wu, S.; Li, P.; Zhai, J.; Shen, B. High recoverable energy storage density in (1-x)$Bi_{0.5}(Na_{0.8}K_{0.2})_{0.5}TiO_3$-x$SrZrO_3$ thin films prepared by a sol-gel method. *J. Eur. Ceram. Soc.* **2018**, *38*, 4640–4645. [CrossRef]
25. Fan, Y.; Zhou, Z.; Chen, Y.; Huang, W.; Dong, X. A novel lead-free and high-performance barium strontium titanate-based thin film capacitor with ultrahigh energy storage density and giant power density. *J. Mater. Chem. C* **2020**, *8*, 50–57. [CrossRef]
26. Zhuang, W.; Shi, C.; Zhang, Y.; Zhao, C.; Lin, T.; Wu, X.; Lin, C.; Gao, M. Energy storage behavior in flexible antiferroelectric $(Pb_{0.98},La_{0.02})(Zr_{0.95},Ti_{0.05})O_3$ thin film capacitors prepared via a direct epitaxial lift-off method. *Thin Solid Films* **2022**, *751*, 139206. [CrossRef]
27. Wu, J.; Mahajan, A.; Riekehr, L.; Zhang, H.; Yang, B.; Meng, N.; Zhang, Z.; Yan, H. Perovskite $Srx(Bi_{1-x}Na_{0.97-x}Li_{0.03})_{0.5}TiO_3$ ceramics with polar nano regions for high power energy storage. *Nano Energy* **2018**, *50*, 723–732. [CrossRef]
28. Ang, C.; Yu, Z. High remnant polarization in $(Sr_{0.7}Bi_{0.2})TiO_3$–$(Na_{0.5}Bi_{0.5})TiO_3$ solid solutions. *Appl. Phys. Lett.* **2009**, *95*, 232908. [CrossRef]
29. Diao, C.; Liu, H.; Li, Z.; Yao, Z.; Hao, H.; Cao, M. Simultaneously achieved high energy storage density and efficiency in sol-gel-derived amorphous Mn-doped $SrTiO_3$ thin films. *J. Alloys Compd.* **2020**, *845*, 155636. [CrossRef]
30. Gao, W.; Yao, M.; Yao, X. Achieving ultrahigh breakdown strength and energy storage performance through periodic interface modification in $SrTiO_3$ thin film. *ACS Appl. Mater. Interfaces* **2018**, *10*, 28745–28753. [CrossRef]
31. Wang, J.; Li, Y.; Sun, N.; Du, J.; Zhang, Q.; Hao, X. $Bi(Mg_{0.5}Ti_{0.5})O_3$ addition induced high recoverable energy-storage density and excellent electrical properties in lead-free $Na_{0.5}Bi_{0.5}TiO_3$-based thick films. *J. Eur. Ceram. Soc.* **2019**, *39*, 255–263. [CrossRef]
32. Pan, H.; Ma, J.; Ma, J.; Zhang, Q.; Liu, X.; Guan, B.; Gu, L.; Zhang, X.; Zhang, Y.-J.; Li, L.; et al. Giant energy density and high efficiency achieved in bismuth ferrite-based film capacitors via domain engineering. *Nat. Commun.* **2018**, *9*, 1813. [CrossRef] [PubMed]

ceramics MDPI

Article

Exploring Enhanced Structural and Dielectric Properties in Ag-Doped $Sr(NiNb)_{0.5}O_3$ Perovskite Ceramic for Advanced Energy Storage

Faouzia Tayari [1], Majdi Benamara [1,2], Madan Lal [3,4], Manel Essid [5], Priyanka Thakur [3], Deepak Kumar [4], S. Soreto Teixeira [1], M. P. F. Graça [1,*] and Kais Iben Nassar [1,6,*]

1 i3N-Physics Department, University of Aveiro, 3810-193 Aveiro, Portugal; faouziatayari12@gmail.com (F.T.); majdibenamara1@gmail.com (M.B.); silvia.soreto@ua.pt (S.S.T.)
2 Laboratory for Building Energy Materials and Components, Swiss Federal Laboratories for Materials Science and Technology (Empa), Überlandstrasse 129, 8600 Dübendorf, Switzerland
3 Department of Physics, Akal College of Basic Sciences, Eternal University, Sirmour HP-173101, India; madan.physics26@gmail.com (M.L.); thakurpriyanka672@gmail.com (P.T.)
4 Department Physics, Graphic Era (Deemed to be University), Dehradun UK-248002, India; deepakphy@gmail.com
5 Chemistry Department, College of Science, King Khaled University (KKU), P.O. Box 9004, Abha 61413, Saudi Arabia; maiseed@kku.edu.sa
6 CICECO—Aveiro Institute of Materials, Department of Chemistry, University of Aveiro, Campus Universitário de Santiago, 3810-193 Aveiro, Portugal
* Correspondence: mpfg@ua.pt (M.P.F.G.); kaisibnnassar12@gmail.com (K.I.N.)

Abstract: The ceramic $Sr(NiNb)_{0.5}O_3$, incorporating silver doping in the A site, was synthesized using a sol–gel route and subjected to comprehensive analysis through various experimental techniques. X-ray diffraction data analysis indicates a rhombohedral crystal structure. Scanning electron microscopy (SEM) examination reveals densely packed grains with minimal surface porosity. A thorough investigation of electrical properties, encompassing dielectric constant, loss tangent, electrical impedance, modulus, conductivity, etc., was conducted across a wide frequency range (10^3–10^6 Hz) and temperature range (260–340 K). This analysis provided valuable insights into structure–property relationships and conduction mechanisms. The discussion highlights the significance of interface effects, space charge polarization, and Maxwell–Wagner dielectric relaxation in achieving the material's high dielectric constant at low frequencies and elevated temperatures. Examination of temperature dependence through Nyquist plots elucidates the contributions of grain behavior to the material's resistive and capacitive properties. The dielectric permittivity, dissipation of energy, and electrical characteristics like impedance, modulus and conductivity are notably influenced by the frequency of the applied electric field and temperature. Overall, the material exhibits promising potential for industrial applications such as energy storage, given its intriguing properties.

Keywords: sol–gel; perovskite; ceramic; activation energy; impedance spectroscopy; relaxation

Citation: Tayari, F.; Benamara, M.; Lal, M.; Essid, M.; Thakur, P.; Kumar, D.; Teixeira, S.S.; Graça, M.P.F.; Nassar, K.I. Exploring Enhanced Structural and Dielectric Properties in Ag-Doped $Sr(NiNb)_{0.5}O_3$ Perovskite Ceramic for Advanced Energy Storage. *Ceramics* **2024**, *7*, 958–974. https://doi.org/10.3390/ceramics7030062

Academic Editors: Dawei Wang and Fayaz Hussain

Received: 28 May 2024
Revised: 28 June 2024
Accepted: 2 July 2024
Published: 10 July 2024

1. Introduction

Electrical equipment plays a pivotal role in the storage of electrical energy. These materials find applications in various devices, such as capacitors, resonators, and switches, which are essential components in virtually all electrical devices [1,2]. The critical factors for designing compact and efficient devices are a high dielectric constant and a minimal dielectric loss (tan δ) [3]. Achieving these properties is a challenging endeavor, particularly in ensuring the consistency of performance across a wide range of operational conditions [4]. Among the materials garnering global recognition for their significance in advancing green technologies are perovskite nanoparticles [5]. Researchers worldwide have embraced them as a key breakthrough in scientific studies. Furthermore, numerous other perovskite materials have been introduced and investigated to enhance their fundamental properties or even introduce novel ones. Perovskite materials have found applications in diverse fields, serving as electrolytes or electrode materials in solid oxide fuel cells (SOFCs) [5,6], which are highly efficient devices for clean energy conversion. They also play a role in oxygen sensors and exhibit promise as catalysts for various chemical reactions [7]. Their unique electronic properties render them attractive for catalytic processes in the chemical industry [8–12]. In recent years, perovskite oxides have garnered significant attention due to their extraordinary properties and their wide range of potential applications in the field of materials science and technology [13,14]. One such notable perovskite is Strontium Niobate ($SrNbO_3$), renowned for its intriguing electrical conductivity characteristics and distinctive behavior in impedance spectroscopy.

Furthermore, $SrNbO_3$ is a complex oxide characterized by its perovskite structure, showcasing intriguing electrical properties. The electrical conductivity of this material is intricately linked to its stoichiometry, crystal structure, and defect concentration. It is recognized as a mixed ionic and electronic conductor (MIEC) with the capacity to transport both oxygen ions and electrons. This dual conduction mechanism allows for precise tuning of its electrical conductivity through various methods, including doping with different elements, adjusting the oxygen partial pressure, or modifying the crystal structure. Such versatility positions $SrNbO_3$ as a highly promising candidate for a multitude of applications, particularly in solid oxide fuel cells (SOFCs), oxygen sensors, and other electrochemical devices [15–22]. Impedance spectroscopy emerges as a potent technique for delving into the electrical properties of materials. When applied to $SrNbO_3$, impedance spectroscopy grants valuable insights into its electrical and ionic conductivity, elucidates charge transfer processes, and unveils the nature of defects within the material. By scrutinizing impedance spectra, researchers can deepen their comprehension of how the material behaves under varying operational conditions. Such knowledge is pivotal for optimizing the performance of devices based on $SrNbO_3$, such as solid oxide fuel cells, and for customizing the material to suit specific industrial applications. In this study, $(Sr_{0.75}Ag_{0.25})(NiNb)_{0.5}O_3$ was meticulously prepared, and a comprehensive analysis of its structure, dielectric properties, and electrical behavior was undertaken. This investigation serves to enhance our understanding of the behavior of perovskite nanoparticles and paves the way for future applications in a multitude of relevant fields.

2. Materials and Methods

The $(Sr_{0.75}Ag_{0.25})(NiNb)_{0.5}O_3$ perovskite nanoparticles were synthesized using the sol–gel method, as depicted in Figure 1. Precisely measured quantities of high-purity strontium nitrate ($Sr(NO_3)_2$) (Sigma-Aldrich, 99.9%), silver nitrate ($AgNO_3$) (Merck, 99.8%), nickel nitrate ($Ni(NO_3)_2$) (Alfa Aesar, 99.9%), and niobium nitrate ($NbO(NO_3)_3$) (Sigma-Aldrich, 99.99%) were combined in the specified molar ratios to match the molecular formula of the prepared sample: 0.75 mol of $Sr(NO_3)_2$, 0.25 mol of $AgNO_3$, 0.5 mol of $Ni(NO_3)_2$, and 0.5 mol of $NbO(NO_3)_3$. The raw materials were sourced as follows: strontium nitrate ($Sr(NO_3)_2$) from Sigma-Aldrich, St. Louis, MO, USA; silver nitrate ($AgNO_3$) from Merck, Darmstadt, Germany; nickel nitrate ($Ni(NO_3)_2$) from Alfa Aesar, Haverhill, MA, USA; and niobium nitrate ($NbO(NO_3)_3$) from Sigma-Aldrich, St. Louis, MO, USA. These nitrates were dissolved in distilled water under thermal stirring at 80 °C, with the subsequent addition of citric acid, serving as a complexation agent for the various metal cations. The pH of the solution was carefully adjusted to approximately 7 by introducing ammonia. Following this step, ethylene glycol was introduced as a polymerization agent. After about 5 h, the formation of a viscous liquid gel became evident. This gel was then subjected to drying in an oven at 300 °C for 4 h. The resulting precursor was subsequently processed through several cycles of grinding, pelleting, and sintering. Finally, the well-defined structure of the sample was achieved through heat treatment at 1100 °C for duration of 24 h. X-ray diffraction (XRD) patterns were recorded utilizing the "Panalytical X'Pert Pro System, a two-circle automatic diffractometer operating at a copper wavelength (λ = 1.5406 Å). A

Nickel filter was employed to eliminate the K_β ray. The measurements were conducted in Bragg–Brentano geometry, with a divergent beam, employing a step size of 0.017° and a counting time of 18 s per step within an angular range spanning from 10 to 70 degrees ($10 \leq 2\theta \leq 80°$). Structural analysis was performed using the Rietveld method with the FullProf software [23]. The sample's morphology was examined in its pellet form using scanning electron microscopy (SEM) through a Philips XL 30 microscope. This microscope was equipped with an electron gun operating at an accelerating voltage of 15 kV. For electrical characterizations, a sample in the shape of a disk, with a diameter of 11.3 mm and a thickness of approximately 1.58 mm, was utilized. Measurements were conducted using an N4L-NumetriQ analyzer (model PSM1735) across a range of different temperatures and frequencies, spanning from 100 Hz to 1 MHz.

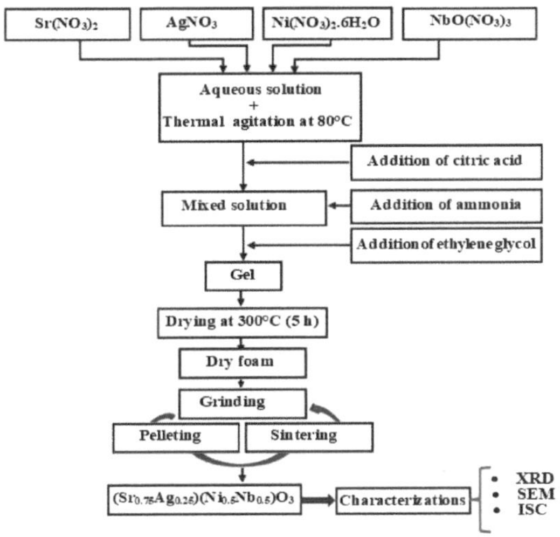

Figure 1. Synthesis process of $(Sr_{0.75}Ag_{0.25})(NiNb)_{0.5}O_3$ perovskite through the sol–gel method. SEM, XRD, and ISC stand for scanning electron microscopy, X-ray diffraction, and impedance spectroscopy complex, respectively.

3. Results and Discussion

3.1. Phase Structure and Microstructure

The X-ray diffraction (XRD) pattern for $(Sr_{0.75}Ag_{0.25})(NiNb)_{0.5}O_3$ perovskite is shown in Figure 2, revealing a distinct crystalline structure characterized primarily by the perovskite phase. For the Rietveld refinement analysis, the crystallographic data were processed using the International Centre for Diffraction Data (ICDD) database card number ICDD 01-073-8912. This card number corresponds to a relevant structure with compositional elements similar to those in our $(Sr_{0.75}Ag_{0.25})(NiNb)_{0.5}O_3$ perovskite, ensuring accurate refinement of the material's crystal structure parameters. This choice facilitated precise determination and validation of the perovskite phase observed in the XRD pattern. The refinement process confirmed the structural integrity and composition of our synthesized material, supporting its characterization for further study in advanced energy storage applications. The diffraction peaks of this perovskite were successfully identified and matched with the R3́C trigonal symmetry. The refined cell parameters for this crystal structure were determined to be a = b = 5.5859 (1) Å, c = 13.4263 (2) Å, with $\alpha = 90°$, $\beta = 90°$, and $\gamma = 120°$. The calculated unit cell volume was found to be V = 351.25 (1) Å3. The reliability factors obtained from the Rietveld refinement process are as follows: the profile factor Rp (%) = 5.91, the weighted profile factor Rwp (%) = 5.48, the structure factor RF (%) = 3.42, and the goodness of fit χ^2 (%) = 1.44 [24]. Ag^+ ions, when doped into the $Sr(NiMn)_{0.5}O_3$ lattice, introduce additional charge carriers. This increase in carrier concentration is primarily due to the substitution of Sr^{2+} ions by Ag^+ ions, which can create oxygen vacancies and enhance the mobility of charge carriers. These values align well with the existing literature, suggesting that the refined crystallographic parameters are accurate. Additionally, the average grain size of the material was estimated using the XRD peaks and the Scherer formula [25]. These values align well with the existing literature, suggesting that the refined crystallographic parameters are accurate. Additionally, the average grain size of the material was estimated using the XRD peaks and the Scherer formula [26].

$$D_{XRD} = \frac{K \cdot \lambda}{\beta \cdot \cos\theta} \quad (1)$$

where D is the average grain size, K is a dimensionless shape factor, typically assumed to be around 0.89, λ is the X-ray wavelength, β is the corrected full-width half-maxima (FWHM) of the XRD peaks and θ is the Bragg angle. With the given values, including the X-ray wavelength (λ), the corrected FWHM (β), and the Bragg angle (θ), the estimated average grain size of 38 nm was determined. This grain size estimation is useful in characterizing the crystal structure and properties of the material.

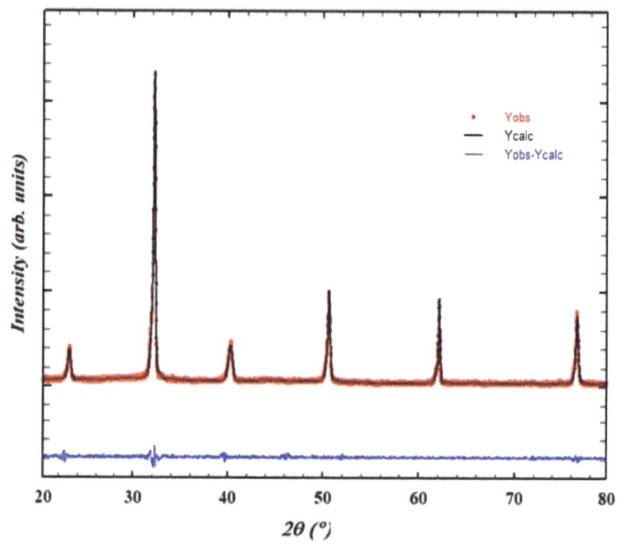

Figure 2. XRD pattern of of $(Sr_{0.75}Ag_{0.25})(NiNb)_{0.5}O_3$ ceramic.

In Figure 3a, the scanning electron microscopy (SEM) image reveals a uniform morphology with no discernible chemical contrast among the crystalline grains. This observation serves to confirm the absence of any secondary phases outside the perovskite phase in the prepared compound. Notably, this micrograph also indicates that the sample comprises small particles, with an average size of D_{MEB} (Diameter by Mean Equivalent Breadth) measuring approximately 175 nanometers. This average size is notably larger than the value determined from the X-ray diffraction (XRD) pattern. It is important to note that each particle analyzed through SEM is composed of a multitude of much smaller grains. As X-rays used in XRD have a considerably higher resolution compared to SEM, the grain sizes estimated from XRD appear significantly smaller than those obtained from SEM. Furthermore, in Figure 3b, the energy dispersive X-ray (EDX) spectroscopy of the sample is presented. The spectrum clearly indicates the presence of all the elements involved in the preparation process with no discernible impurities. This observation underscores the purity of the prepared material. Based on the EDX quantification results, the atomic percentages of the elements in the $(Sr_{0.75}Ag_{0.25})(NiNb)_{0.5}O_3$ perovskite nanoparticles are as follows:

strontium (Sr) constitutes 38.0%, silver (Ag) is 25.1%, nickel (Ni) accounts for 18.5%, and niobium (Nb) makes up 18.4%, with oxygen (O) adjusted for balance. These results confirm the intended composition and doping levels of the synthesized material.

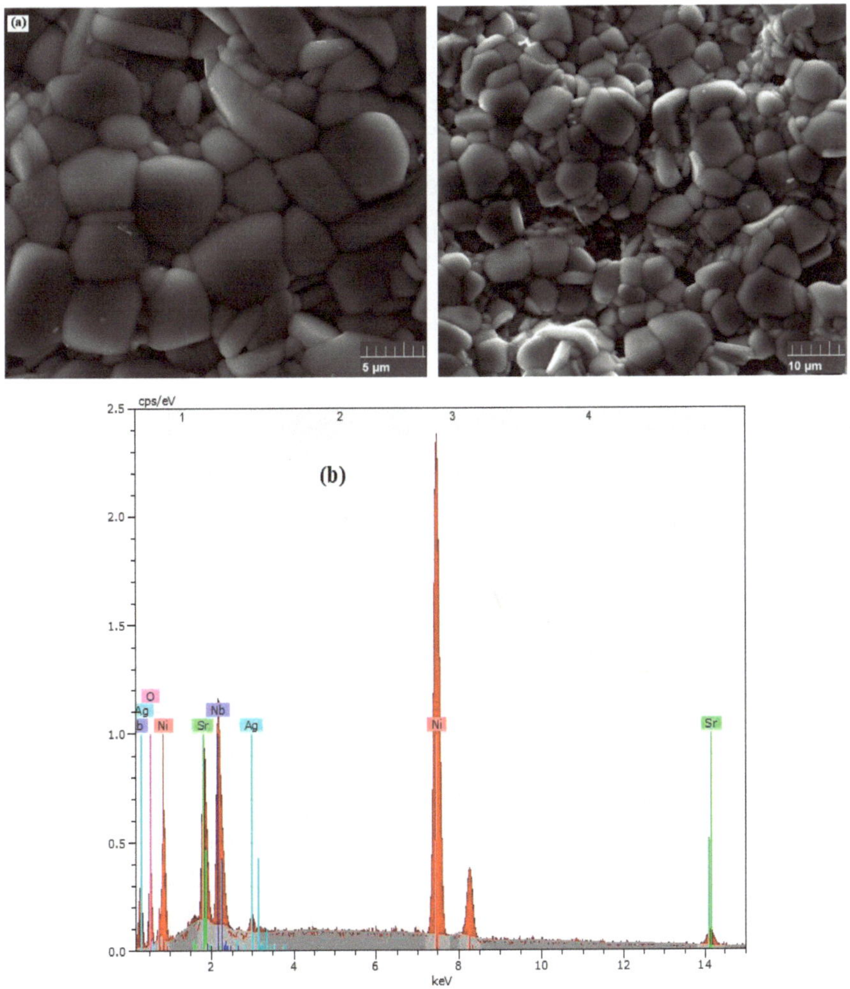

Figure 3. (**a**) SEM micrograph and (**b**) EDX compositions of the sample.

3.2. Impact of Frequency on Dielectric Parameters

Using geometric and physical parameters such as area (S), thickness (e), and capacitance (C) of the sample in a straightforward expression $\varepsilon = Ce/\varepsilon_0 S$, one can calculate the dielectric constant ε or permittivity of the sample. Figure 4 illustrates the effect of frequency on dielectric parameters, including (a) permittivity (ε) and (b) tangent loss (tan δ), within the frequency at selected temperatures. Initially, the value of ε smoothly decreases with the rise in frequency, eventually stabilizing at higher frequencies, indicating saturation. This trend persists across all temperatures. In the low-frequency region, various types of polarization typically occur, resulting in higher relative permittivity [27]. A decrease in dielectric constant with increasing frequency is expected in nearly all insulators due to relaxation processes within the material. As frequency increases, the total polarization of the material decreases because the mechanisms contributing to polarization become less effective, consequently reducing permittivity. The temperature and frequency dependence of dielectric loss or dissipation factor exhibit a similar behavior to the dielectric constant. This behavior aligns with the Maxwell and Wagner two-layer model, which is based on Koop's phenomenological theory. According to this model, the dielectric structure of the material consists of two layers: conducting grains and insulating grain boundaries [28]. At lower frequencies, electrons as charge carriers are more active at grain boundaries, whereas at higher frequencies, electrons primarily interact with conducting grains [29]. At low frequencies, electrons require more energy for motion due to the higher resistance of grain boundaries, resulting in a larger dissipation factor (tan δ) at high temperatures. Conversely, at high frequencies, electrons require less energy for motion due to lower resistance, leading to a smaller dissipation factor within the specified temperature range. The loss factor is significant for device quality as it describes the energy dissipation in insulators. A low tangent loss is desirable to ensure minimal power loss for an effective dielectric material [30].

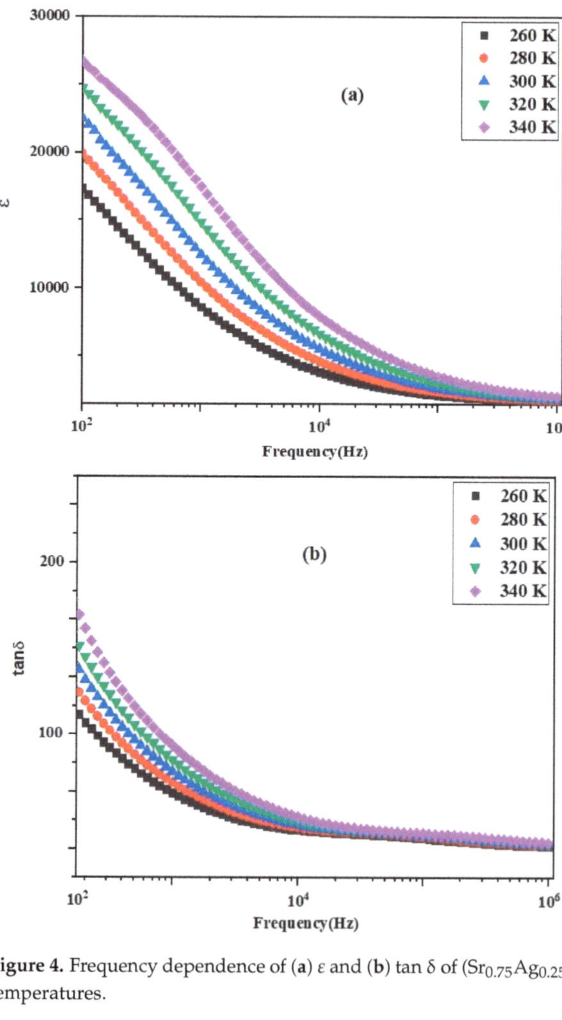

Figure 4. Frequency dependence of (**a**) ε and (**b**) tan δ of $(Sr_{0.75}Ag_{0.25})(NiNb)_{0.5}O_3$ ceramic for various temperatures.

3.3. Variation of Dielectric Properties with Temperature

Figure 5a illustrates the impact of temperature and frequency on the insulating (permittivity) properties of the material under investigation. It is observed that the dielectric parameters of the material remain unaffected by frequency and temperature, indicating the significant role of electronic and/or ionic polarization [31]. The dielectric permittivity (dielectric constant) experiences a notable increase with temperature elevation. This characteristic variation in the dielectric parameters can be attributed to the scattering of temperature-dependent charge carriers and/or defects/impurities present in the sample [32]. In Figure 5b, the temperature dependence of tan δ at a selected frequency of the AC electric field is elucidated. The trend in the variation of tan δ with temperature mirrors that of ε. Tan δ increases with temperature escalation and exhibits a sharp rise at higher temperatures. This abrupt increase is heavily influenced by electrical conductivity [33].

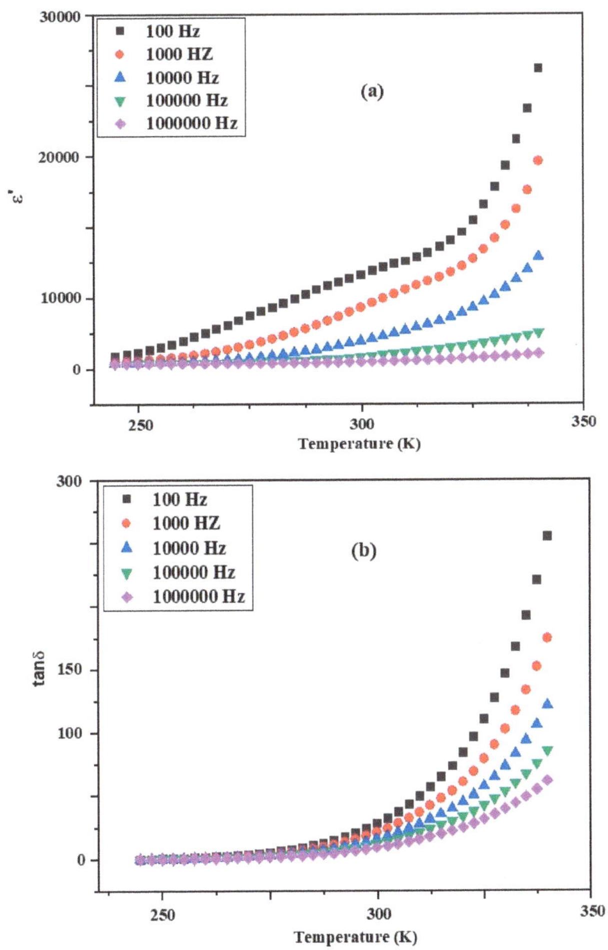

Figure 5. Temperature dependence of (**a**) ε and (**b**) tan δ of $(Sr_{0.75}Ag_{0.25})(NiNb)_{0.5}O_3$ ceramic for various frequencies.

3.4. Impedance Spectroscopy

The electrical characteristics of electro-ceramics and ionic conductors, encompassing impedance across grain boundaries and electrodes, can undergo assessment utilizing a non-invasive complex impedance approach. This method stands as a reliable means of characterizing these materials, providing insights into both the tangible (resistive) and abstract (reactive) elements of complex impedance. Spectroscopy enables the determination of impedance and elastic modulus, along with the extraction of various associated parameters. By subjecting the sample to an alternating current (AC) electric field, a sinusoidal disturbance is initiated, facilitating the examination of these characteristics across different frequencies and temperatures [34–37]. Typically, the frequency-dependent behavior of dielectric properties in materials is elucidated through complex capacitive and impedance parameters, which are articulated as follows: complex dielectric constant, complex impedance, electric modulus and dielectric loss.

$$\varepsilon^* = \varepsilon' - j\,\varepsilon' \quad (2)$$

$$Z^* = Z' - jZ'' = Rs + j\,(\omega L_s - 1/\omega C_s) \quad (3)$$

$$M^* = M' + j\,M'' = j\,\omega C_0\,Z^* \quad (4)$$

$$\tan\delta = \varepsilon'/\varepsilon'' \quad (5)$$

where R_s is the resistance, L_s is the inductance, C_s is the capacitance, ω is the angular frequency ($\omega = 2\pi f$, where f is the frequency), and j is the imaginary unit. The equations provided offer ample opportunity for creating visual representations to gauge the impacts of the parameters within the electrode/ceramic/electrode system. Figure 6a,b illustrate how Z' and Z'' vary with frequency and temperature. The influence of the applied AC electric field frequency on the sample is typically associated with various complex physical properties, including dielectric constant, impedance, electric modulus, and dielectric loss. Apart from the equations mentioned earlier, additional equations can also be employed to determine the real and imaginary components of impedance [38].

$$Z' = \frac{R}{1+(\omega\tau)^2} \quad (6)$$

$$Z'' = \frac{\omega R\tau}{1+(\omega\tau)^2} \quad (7)$$

The resistance measured as R corresponds with the angular frequency of the electric field denoted by ω, and the relaxation time, symbolized by τ, equals RC, where C signifies capacitance. The relaxation time for a bulk sample can be estimated using the formula $\tau = R_g C_g$. As depicted in Figure 6a, the decrease in the value of Z' with increasing temperature at low frequencies indicates the semiconductor characteristics of the material. Consequently, the negative temperature coefficient of resistance (NTCR) or semiconductor behavior of the material is observed at higher temperatures [39]. Therefore, the relaxation properties of the sample are established [32]. The Z' value smoothly rises with increasing temperature and frequency, suggesting an increase in AC conductivity at low frequency (100 kHz) with temperature. In the figure, it is observed that at higher frequencies, the Z' value merges into a line at higher temperatures, indicating the existence of a mechanism strongly related to temperature and the release of space charge [40]. As observed in Figure 6b, there is no peak, indicating less or no dissipation of current in the sample at low temperatures. The presence of a peak at higher temperatures suggests the existence of dielectric relaxation in the material. The frequency at which the imaginary value Z'' reaches its highest limit (Z_{max}'') shifts to the higher frequency region with increasing temperature. The broadening of peaks and the decrease in the highest limit of the imaginary impedance component (with respect to temperature) indicate that the dielectric relaxation mechanism is temperature dependent. Therefore, these dielectric processes related to relaxation occur due to lattice deformation and vacancies present in the material at higher temperatures [41,42].

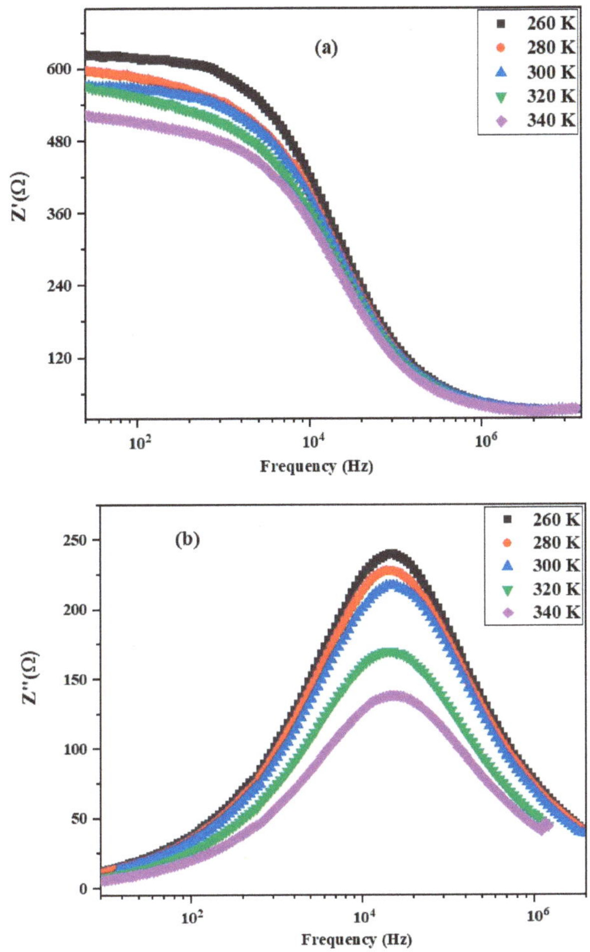

Figure 6. (**a**) Variation of real part of impedance (Z′) with frequency measured at different temperatures for $(Sr_{0.75}Ag_{0.25})(NiNb)_{0.5}O_3$. (**b**) Variation of imaginary part of impedance (Z″) with frequency measured at different temperatures for $(Sr_{0.75}Ag_{0.25})(NiNb)_{0.5}O_3$.

3.5. Nyquist Diagram

Figure 7 displays the Nyquist diagram, showcasing complex impedance plots (Z′ vs. Z″), within the frequency range of 1 kHz to MHz, depicting temperature-dependent semicircular arcs. By comparing experimental data (real and imaginary components of Z) with the components of a circuit comprising resistors and capacitors, fitting parameters are obtained. In this circuit, the capacitor represents polarization with energy storage, while the resistor signifies a conductive path. The decrease in impedance with rising temperature indicates dependence on the sample's conductivity. According to Debye theory, a perfect semicircle with its center precisely on the *x*-axis (real impedance component) signifies Debye-type dielectric relaxation, resulting in a single semicircle [43]. Conversely, when the semicircle's center is below the axis (typically observed at high temperatures), the relaxation process is non-Debye type, resulting in two circles or semicircles. The center of the first semicircle falling on the Z' axis indicates grain resistance or effect, while the second one is caused by grain boundary resistance. However, in this study, only a single semicircle (with its center below the x-axis) is observed even at higher frequencies and temperatures, attributed to the conduction of the material due to grain effect [44]. The Debye-type dielectric relaxation primarily accounts for the formation of such single semicircles, whose magnitude can be estimated using an electric circuit with a parallel combination of a capacitor C_g and resistance R_g. The relaxation time (τ) related to the circle can be estimated using the formula $\tau = R_g C_g$. As depicted in the figure, grain boundaries do not contribute to impedance or electrical parameters, particularly at the stated frequency and temperature ranges, indicating that only grains are responsible for transport properties or electrical conduction in the material. The value of grain or bulk resistance can be determined using the semicircle's diameter. The close agreement between experimental and model parameters obtained from the circuit using the software Zveiw underscores the accuracy and validity of the experimental data and the proposed theoretical model. The values of the parameters R_g and C_g are summarized in Table 1.

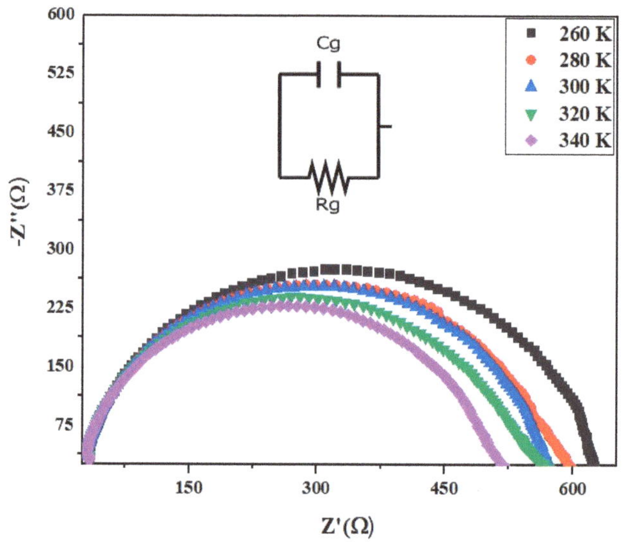

Figure 7. Nyquist plots for $(Sr_{0.75}Ag_{0.25})(NiNb)_{0.5}O_3$ ceramic at different temperatures.

Table 1. Electrical parameters of equivalent circuit deduced from Nyquist plots at different temperatures for $(Sr_{0.75}Ag_{0.25})(NiNb)_{0.5}O_3$ perovskite.

T (K)	260	280	300	320	340
R_g (KΩ)	36.25	28.57	22.78	18.75	13.54
C_g (10^{-9}F)	11.25	8.27	4.64	7.81	3.27

3.6. Modulus Analysis

To determine the resistive and capacitive characteristics of dielectrics under AC electrical fields, electric modulus spectroscopy is employed. Analyzing the frequency dependence of electrical modulus aids in understanding relaxation processes in materials. Through this technique, various electrical properties and processes of materials can be studied, including polarization, the role of grain boundaries in conduction, AC/DC electrical conductivity, dielectric relaxation, and more. Additionally, the influence of temperature and frequency on these parameters can be investigated using this technique. The equation for the modulus technique's general relation has been derived and presented as follows [45].

$$M' = A\,[\frac{(\omega RC)^2}{1+(\omega RC)^2}] \;=\; A\,[\frac{\tau^2\omega^2}{1+\tau^2\omega^2}] \tag{8}$$

$$M'' = A\,[\frac{(\omega RC)}{1+(\omega RC)^2}] \;=\; A\,[\frac{\omega\tau}{1+\tau^2\omega^2}] \tag{9}$$

In the equation, A represents the ratio of capacitance (C) in vacuum to dielectric (C_o/C), R stands for resistance, and ω denotes the angular frequency of the applied AC field. By leveraging the effects of grains and grain boundary in electric modulus technique, we can effectively investigate the heterogeneous nature of polycrystalline sample. This insight is not achievable through complex impedance graphs. Furthermore, this technique is instrumental in estimating the electrode contribution to the total resistance. It is noteworthy that impedance techniques are unable to discern electrode contributions in the resistive/capacitive characteristics of the material. Figure 8a,b illustrate the effects of electric field frequency and temperatures on modulus components, namely M′ and M″. A continuous decrease in the value of M′ is observed in the low-frequency region, followed by a gradual increase in its value with increasing frequency, eventually reaching a saturated asymptotic value at higher frequencies across all temperatures. To elucidate the transport properties and conduction mechanism in the material, the short-range mobility of charge carriers is taken into account. This implies that under the influence of a DC electric field, there is a lack of restoring force that affects the flow of charge carriers [46]. As depicted in Figure 8b, initially, the value of M″ increases with frequency up to its maximum (peak) value (M''_{max}), then decreases. Additionally, it is observed that with increasing temperature, the position of M″ max shifts to the higher frequency side. The shift of the peak position to the higher frequency side at higher temperatures indicates that thermal energy activates charge carriers. Moreover, as the broadening of the asymmetric peak increases with rising temperature, the relaxation time of charge carriers decreases. This phenomenon is observed only in the case of a non-Debye type of relaxation mechanism. The dielectric relaxation can be modeled using the Arrhenius relation, and the activation energy (Ea) estimated from Ln (F_{max}) vs. (1000/T) plot is determined as 0.266 eV (Figure 9).

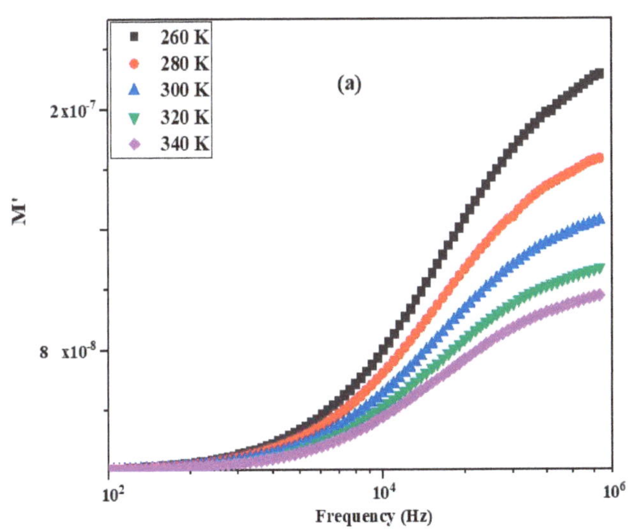

Figure 8. *Cont.*

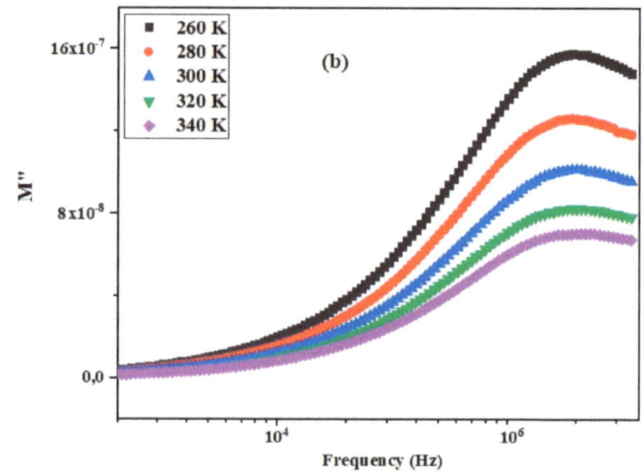

Figure 8. (**a**) Variation of reel part of electrical modulus (M′) with frequency at different temperatures for $(Sr_{0.75}Ag_{0.25})(NiNb)_{0.5}O_3$ ceramic. (**b**) Variation of imaginary part of electrical modulus (M″) with frequency at different temperatures for $(Sr_{0.75}Ag_{0.25})(NiNb)_{0.5}O_3$ ceramic.

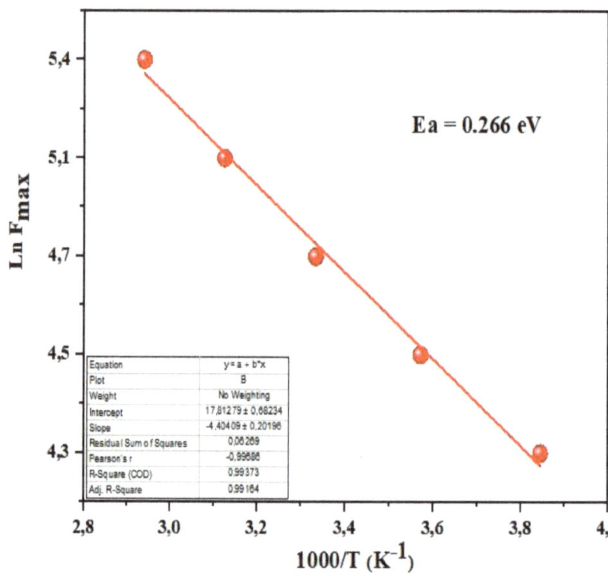

Figure 9. Arrhenius plot of $Ln(F_{max})$ vs. (1000/T) for $(Sr_{0.75}Ag_{0.25})(NiNb)_{0.5}O_3$ perovskite.

3.7. Electrical Conductivity Study

In many instances, the electrical conductivity (σ) in manganite systems such as $(Ln^{3+}{}_{1-x} M^{2+}{}_{x})(Mn^{3+}{}_{1-x}Mn^{4+}{}_{x})O^{-2}{}_{3}$, where Ln represents elements like La or Sr, is primarily ascribed to the process of electron hopping that takes place between adjacent Mn^{3+} and Mn^{4+} ions. This electron hopping is facilitated by the presence of the oxygen anion O^{2-} [40], and it serves as a fundamental mechanism influencing the electrical properties of these materials. To investigate the hopping conduction mechanism in the synthesized $(Sr^{3+}{}_{0.75}Ag^{+2}{}_{0.25})_{0.5}(Ni^{+3}Nb^{+4})_{0.5}O_3$ sample, the changes in its conductivity are depicted in

Figure 10 with respect to both frequency and temperature. The values of σ (conductivity) were determined using the following relationship [47]:

$$\sigma = G \frac{t}{A} \quad (10)$$

where G represents the electrical conductance, t is the thickness of the pellet, and A is the cross-sectional area of the pellet. Notably, conductivity values remain relatively stable in the low-frequency range, up to a frequency value ($f \leq 10^5$ Hz), identified as the hopping frequency. This constancy is attributed to limited electron hopping between Mn^{3+}–Mn^{4+} ions, primarily influenced by the heightened activity of weakly conductive grain boundaries within this frequency spectrum. The low-frequency conductivity corresponds to the dc conductivity (σ_{dc}). Subsequently, a notable exponential increase in conductivity occurs beyond frequency. This phenomenon is attributed to enhanced activity of conductive grains in this frequency range, intensifying electron hopping between Mn^{3+}–Mn^{4+} ions and thereby facilitating the conduction process in the sample. This high-frequency conductivity aligns with the ac conductivity (σ_{AC}). To modelize σ-values, the Jonscher power law [48,49] was employed as the mathematical relation.

$$\sigma_{Ac}(\omega) = \sigma_{dc} + A\omega^{s} \quad (11)$$

where σ_{dc} represents the dc conductivity, and A and s denote the pre-exponential and exponent factors, respectively.

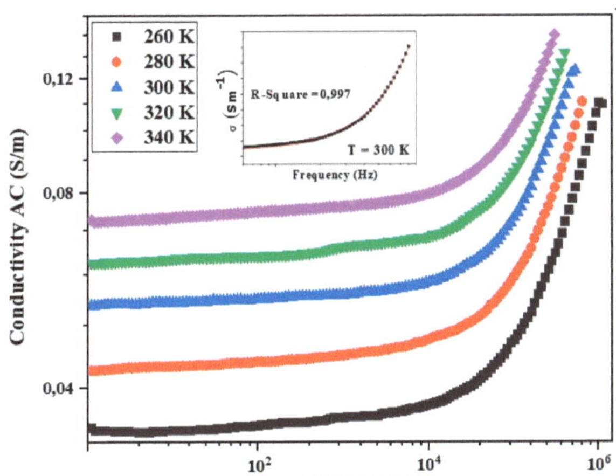

Figure 10. Variation of AC conductivity of $(Sr_{0.75}Ag_{0.25})(NiNb)_{0.5}O_3$ ceramic with frequency at different temperatures.

Furthermore, our analysis leads to the conclusion that the sample demonstrates semiconductor behavior without undergoing a transition into a metal-semiconductor phase. Notably, all s-values are below unity, indicating, according to the Funke criterion [50], that electron hopping between Mn^{3+}–Mn^{4+} ions occurs between neighboring sites. In the high-temperature range, the dc conductivity behavior aligns well with the Arrhenius relation [51], expressed by the following mathematical formula:

$$\sigma_{dc} = \sigma_0 \exp\left(\frac{-E_a}{K_B T}\right) \quad (12)$$

where E_a represents the activation energy, σ_0 is the pre-exponential term, and K_B is Boltzmann's constant. The activation energy (E_a) for the sample is calculated using Equation (12), and the graph of Ln (σ_{dc}) vs. (1000/T) is depicted in Figure 11. The expected E_a value, derived from the slope of the plot, is determined to be 0.268 eV. The activation energy derived from electrical conductivity closely aligns with the activation energy obtained from complex spectroscopic impedance. This close correspondence affirms that the conduction processes and relaxation phenomena within the prepared materials involve the same type of charge carrier. The electrical properties of $Sr_{0.75}Ag_{0.25}Ni_{0.5}Mn_{0.5}O_3$, a material doped with silver, were investigated using impedance spectroscopy across various temperatures and frequencies. The study revealed that silver doping significantly enhances the material's electrical conductivity compared to its undoped counterpart, $Sr(NiMn)_{0.5}O_3$. Specifically, the electrical conductivity of $Sr_{0.75}Ag_{0.25}Ni_{0.5}Mn_{0.5}O_3$ was found to range from 0.9×10^{-2} S/m to 1.25×10^{-2} S/m, indicating a notable increase attributed to the doping process. Moreover, the activation energy required for electrical conduction was observed to decrease upon silver doping. The undoped $Sr(NiMn)_{0.5}O_3$ exhibited an activation energy of 0.287 eV, whereas the doped $Sr_{0.75}Ag_{0.25}Ni_{0.5}Mn_{0.5}O_3$ showed a reduced activation energy of 0.268 eV. This reduction in activation energy signifies that the introduction of silver lowers the energy barrier for charge carrier movement within the material. Consequently, the doped material demonstrates improved electrical conductivity due to enhanced mobility of charge carriers at a given temperature.

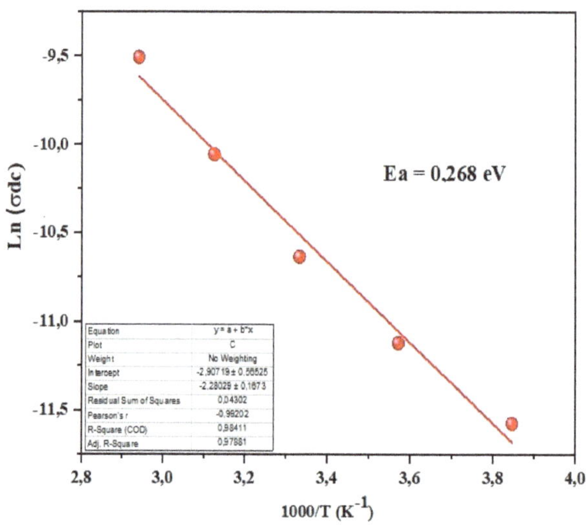

Figure 11. Variation of Ln(σ_{dc}) as a function of (1000/T) for $(Sr_{0.75}Ag_{0.25})(NiNb)_{0.5}O_3$ perovskite.

4. Conclusions

The polycrystalline $(Sr_{0.75}Ag_{0.25})(NiNb)_{0.5}O_3$ ceramic has been synthesized using the sol–gel technique. Initial structural analysis based on X-ray diffraction data indicates a single phase of the trigonal system. The dielectric parameters like permittivity and dissipation factor are influenced by the frequency of the applied AC electric field and temperature. These electrical parameters are obtained through dielectric and impedance measurements. The high dielectric constant of the material is attributed to the Maxwell–Wagner model and space charge polarization. Despite the dissipation factor being temperature-dependent, the compound exhibits low energy loss even at higher temperatures, which further decreases with increasing frequency. Due to its low energy dissipation factor, the material possesses a high-quality factor, making it suitable for use as a component in microwave devices. Complex-impedance plots offer valid explanations and applications of the conduction phenomenon in the material. The frequency dependence of the AC conductivity conforms to Jonscher's universal power law.

Author Contributions: Conceptualization, F.T. and M.B.; methodology, K.I.N.; software, M.E.; validation, M.L. and F.T.; formal analysis, K.I.N.; investigation, P.T.; resources, S.S.T.; data curation, S.S.T.; writing—original draft preparation, M.P.F.G.; writing—review and editing, F.T.; visualization, M.E.; supervision, M.P.F.G.; project administration, D.K.; funding acquisition, D.K. All authors have read and agreed to the published version of the manuscript.

Funding: The authors extend their appreciation to the Deanship of Research and Graduate Studies at King Khalid University for funding this work through Large Research Project under grant number RGP2/201/45. The authors are grateful to i3N (LA/P/0037/2020, UID-B/50025/2020, and UID-P/50025/2020) financed by COMPETE 2020 Program and national funds through the FCT/MEC and FEDER under the PT2020 Partnership Agreement. This work is funded by national funds (OE) through FCT—Fundação para a Ciência e a Tecnologia, I.P., in the scope of the framework contract foreseen in the numbers 4, 5, and 6 of article 23 of the Decree-Law 57/2016, of August 29, changed by Law 57/2017, of July 19.

Data Availability Statement: The data presented in this study are available from the corresponding author upon reasonable re-quest.

Acknowledgments: The authors extend their appreciation to the Deanship of Research and Graduate Studies at King Khalid University for funding this work through Large Research Project under grant number RGP2/201/45.

Conflicts of Interest: The authors declare no conflicts of interest.

References

1. Davies, P.K.; Roth, R.S. *National Institute of Standards and Technology Special Publication 804*; NIST: Gaithersburg, MD, USA, 1991.
2. Lines, M.E.; Glass, A.M. *Principles and Applications of Ferroelectrics and Related Materials*; Oxford University Press: New York, NY, USA, 1997.
3. Tang, C.C.; Roberts, M.A.; Azough, F.; Leach, C.; Freer, R. Synchrotron x-ray diffraction study of $Ba_{4.5}Nd_9Ti_{18}O_{54}$ microwave dielectric ceramics at 10–295 K. *J. Mater. Res.* **2022**, *17*, 675–682. [CrossRef]
4. Colla, E.L.; Reaney, I.M.; Setter, N.J. Effect of structural changes in complex perovskites on the temperature coefficient of the relative permittivity. *Appl. Phys.* **1993**, *74*, 3414–3425. [CrossRef]
5. Hanif, M.B.; Rauf, S.; Motola, M.; Babar, Z.U.; Li, C.J.; Li, C.X. Recent progress of perovskite-based electrolyte materials for solid oxide fuel cells and performance optimizing strategies for energy storage applications. *Mater. Res. Bull.* **2022**, *146*, 111612. [CrossRef]
6. Hanif, M.B.; Motola, M.; Rauf, S.; Li, C.J.; Li, C.X. Recent advancements, doping strategies and the future perspective of perovskite-based solid oxide fuel cells for energy conversion. *Chem. Eng. J.* **2022**, *428*, 132603. [CrossRef]
7. Kubicek, M.; Bork, A.H.; Rupp, J.L.M. Perovskite oxides—A review on a versatile material class for solar-to-fuel conversion processes. *J. Mater. Chem. A* **2017**, *5*, 11983–12000. [CrossRef]
8. Ofoegbuna, T.; Darapaneni, P.; Sahu, S.; Plaisance, C.; Dorman, J.A. Stabilizing the B-site oxidation state in ABO_3 perovskite nanoparticles. *Nanoscale* **2019**, *11*, 14303–14311. [CrossRef]
9. Oka, D.; Hirose, Y.; Nakao, S.; Fukumura, T.; Hasegawa, T. Intrinsic high electrical conductivity of stoichiometric $SrNbO_3$ epitaxial thin films. *Phys. Rev. B* **2015**, *92*, 205102. [CrossRef]
10. Hossain, A.; Bandyopadhyay, P.; Roy, S. An overview of double perovskites $A_2B'B''O_6$ with small ions at A site: Synthesis, structure and magnetic properties. *J. Alloys Compd.* **2018**, *740*, 414–427. [CrossRef]
11. Velinov, N.; Brashkova, N.; Kozhukharov, V. Synthesis, structure and conductivity of layered perovskites. *Ceram.-Silik.* **2005**, *49*, 29–33.
12. Arjun, N.; Pan, G.-T.; Yang, T.C.K. The exploration of Lanthanum based perovskites and their complementary electrolytes for the supercapacitor applications. *Results Phys.* **2017**, *7*, 920–926. [CrossRef]
13. Tran, M.-N.T.; Quach, H.Y.; Nguyen, Q.V.; Nguyen, T.D.; On, D.T. Synthesis of perovskite-based nanocomposites for $deNO_x$ catalytic activity. *Can. J. Chem.* **2016**, *94*, 215–220. [CrossRef]
14. Li, X.; Zhu, S.; Jia, Q.; Zhao, H.; Cao, Y.; Ma, Y.; Hu, S.; Cao, X. Fast synthesis of MoO_{3-x} and its catalytic effect on the thermal decomposition of ammonium perchlorate based molecular perovskite (DAP-4). *Can. J. Chem.* **2021**, *99*, 795–800. [CrossRef]
15. Gómez, L.; Galeano, V.; Parra, R.; Michel, C.R.; Paucar, C.; Morán, O. Carbon dioxide gas sensing properties of ordered oxygen deficient perovskite $LnBaCo_2O_{5+\delta}$ (Ln = La, Eu). *Sens. Actuators B Chem.* **2015**, *221*, 1455–1460. [CrossRef]
16. Polini, R.; Pamio, A.; Traversa, E. Effect of synthetic route on sintering behaviour, phase purity and conductivity of Sr- and Mg-doped $LaGaO_3$ perovskites. *J. Eur. Ceram. Soc.* **2004**, *24*, 1365–1370. [CrossRef]
17. Fumo, D.A.; Jurado, J.R.; Segadães, A.M.; Frade, J.R. Combustion synthesis of iron-substituted strontium titanate perovskites. *Mater. Res. Bull.* **1997**, *32*, 1459–1470. [CrossRef]
18. Alhokbany, N.; Almotairi, S.; Ahmed, J.; Al-Saeedi, S.I.; Ahamad, T.; Alshehri, S.M. Investigation of structural and electrical properties of synthesized Sr-doped lanthanum cobaltite ($La_{1-x}Sr_xCoO_3$) perovskite oxide. *J. King Saud Univ.-Sci.* **2021**, *33*, 101419. [CrossRef]

19. Jayakumar, G.; Poomagal, D.S.; Irudayaraj, A.A.; Raj, A.D.; Thresa, S.K.; Akshadha, P. Study on structural, magnetic and electrical properties of perovskite lanthanum strontium manganite nanoparticles. *J. Mater. Sci. Mater. Electron.* **2020**, *31*, 20945–20953. [CrossRef]
20. Intatha, U.; Eitssayeam, S.; Wang, J.; Tunkasiri, T. Impedance study of giant dielectric permittivity in $BaFe_{0.5}Nb_{0.5}O_3$ perovskite ceramic. *Curr. Appl. Phys.* **2010**, *10*, 21–25. [CrossRef]
21. Kodera, M.; Moriya, Y.; Katayama, M.; Hisatomi, T.; Minegishi, T.; Domen, K. Investigation on nitridation processes of $Sr_2Nb_2O_7$ and $SrNbO_3$ to $SrNbO_2N$ for photoelectrochemical water splitting. *Sci. Rep.* **2018**, *8*, 15849. [CrossRef]
22. Bigi, C.; Orgiani, P.; Sławińska, J.; Fujii, J.; Irvine, J.T.; Picozzi, S.; Panaccione, G.; Vobornik, I.; Rossi, G.; Payne, D.; et al. Direct insight into the band structure of $SrNbO_3$. *Phys. Rev. Mater.* **2020**, *4*, 025006. [CrossRef]
23. Rietveld, H.M. A profile refinement method for nuclear and magnetic structures. *J. Appl. Crystallogr.* **1969**, *2*, 65. [CrossRef]
24. Voorhoeve, R.J.H. *Advanced Materials in Catalysis*; Academic Press: Cambridge, MA, USA, 1977; p. 129.
25. Khadhraoui, S.; Triki, A.; Hcini, S.; Zemni, S.; Oumezzine, M. Variable-range-hopping conduction and dielectric relaxation in $Pr_{0.6}Sr_{0.4}Mn_{0.6}Ti_{0.4}O_{3\pm\delta}$ perovskite. *J. Magn. Magn. Mater.* **2014**, *371*, 69. [CrossRef]
26. Hsini, M.; Hamdaoui, N.; Hcini, S.; Bouazizi, M.L.; Zemni, S.; Beji, L. Effect of iron doping at Mn-site on complex impedance spectroscopy properties of $Nd_{0.67}Ba_{0.33}MnO_3$ perovskite. *Phase Transit.* **2018**, *91*, 316. [CrossRef]
27. Benamara, M.; Nassar, K.I.; Soltani, S.; Kallekh, A.; Dhahri, R.; Dahman, H.; El Mir, L. Light-enhanced electrical behavior of a Au/Al-doped ZnO/p-Si/Al heterostructure: Insights from impedance and current–voltage analysis. *RSC Adv.* **2023**, *13*, 28632–28641. [CrossRef] [PubMed]
28. Rosić, M.; Logar, M.; Devečerski, A.; Prekajski, M.; Radosavljević-Mihajlović, A.; Kusigerski, V.; Spasojević, V.; Matović, B. Synthesis, structural and magnetic properties of nanostructured $Ca_{0.9}Gd_{0.1}MnO_3$ obtained by modified glycine nitrate procedure (MGNP). *Ceram. Int.* **2011**, *37*, 1313–1319. [CrossRef]
29. Nassar, K.I.; Tayari, F.; Benamara, M.; Teixeira, S.S.; Graça, M.P. Exploring bismuth-doped polycrystalline ceramic $Ba_{0.75}Bi_{0.25}Ni_{0.7}Mn_{0.3}O_3$: Synthesis, structure, and electrical properties for advanced electronic applications. *RSC Adv.* **2023**, *13*, 24023–24030. [CrossRef] [PubMed]
30. Parida, S.K.; Choudhary, R.N.P.; Achary, P.G.R. Study of structural and electrical properties of polycrystalline $Pb(Cd_{1/3}Ti_{1/3}W_{1/3})O_3$ tungsten perovskite. *Int. J. Microstruct. Mater. Prop.* **2020**, *15*, 107–121.
31. Tayari, F.; Iben Nassar, K.; Maalem, M.B.; Teixeira, S.S.; Graça, M.P.F. Structural, morphology, Raman spectroscopy, magnetic and electrical proprieties of $BaNi_{0.5}Mn_{0.25}Fe_{0.25}O_3$ ce-ramic for electronic applications. *Indian J. Phys.* **2023**, *97*, 3545–3555. [CrossRef]
32. Elliot, S.R. Ac conduction in amorphous chalcogenide and pnictide semiconductors. *Adv. Phys.* **1987**, *36*, 135–217. [CrossRef]
33. Gudmundsson, T.; Svavarsson, H.G.; Gudjonsson, S.; Gislason, H.P. Frequency-dependent conductivity in lithium-diffused and annealed GaAs. *J. Phys. B* **2003**, *340*, 324–328. [CrossRef]
34. Benamara, M.; Iben Nassar, K.; Rivero-Antúnez, P.; Essid, M.; Soreto Teixeira, S.; Zhao, S.; Serrà, A.; Esquivias, L. Study of Electrical and Dielectric Behaviors of Copper-Doped Zinc Oxide Ceramic Prepared by Spark Plasma Sintering for Electronic Device Applications. *Nanomaterials* **2024**, *14*, 402. [CrossRef] [PubMed]
35. Hcini, S.; Khadhraoui, S.; Triki, A.; Zemni, S.; Boudard, M.; Oumezzine, M. Impedance Spectroscopy Properties of $Pr_{0.67}A_{0.33}MnO_3$ (A = Ba or Sr) Perovskites. *J. Supercond. Nov. Magn.* **2014**, *27*, 195. [CrossRef]
36. Padmasree, K.P.; Kanchan, D.D.; Kulkami, A.R. Impedance and Modulus studies of the solid electrolyte system $20CdI_2–80[xAg_2O - y (0.7 V_2O_5 - 0.3 B_2O_3)]$, where $1\leq x/y \leq 3$. *Solid State Ion.* **2006**, *177*, 475. [CrossRef]
37. Mahato, D.K.; Dutta, A.; Sinha, T.P. Impedance spectroscopy analysis of double perovskite Ho_2NiTiO_6. *J. Mater. Sci.* **2010**, *45*, 6757–6762. [CrossRef]
38. Datta, R.; Pradhan, S.K.; Majumdar, S.; De, S.K. Dielectric and impedance spectroscopy of Nd_2CoIrO_6 double perovskite. *J. Phys. Condens. Matter* **2020**, *32*, 495702.
39. Von Hauff, E.; Klotz, D. Impedance spectroscopy for perovskite solar cells: Characterisation, analysis, and diagnosis. *J. Mater. Chem. C* **2022**, *10*, 742–761. [CrossRef]
40. Khadhraoui, S.; Triki, A.; Hcini, S.; Zemni, S.; Oumezzine, M. Structural and impedance spectroscopy properties of $Pr_{0.6}Sr_{0.4}Mn_{1-x}TixO_{3\pm\delta}$ perovskites. *J. Alloys Compd.* **2013**, *574*, 290–298. [CrossRef]
41. Zarazua, I.; Sidhik, S.; Lopez-Luke, T.; Esparza, D.; De la Rosa, E.; Reyes-Gomez, J.; Garcia-Belmonte, I.M.-S. Operating Mechanisms of Mesoscopic Perovskite Solar Cells through Impedance Spectroscopy and J–V Modeling. *J. Phys. Chem. Lett.* **2017**, *8*, 6073–6079. [CrossRef] [PubMed]
42. Cordoba-Torres, P.; Mesquita, T.J.; Devos, O.; Tribollet, B.; Roche, V.; Nogueira, R.P. On the intrinsic coupling between constant-phase element parameters α and Q in electrochemical impedance spectroscopy. *Electro. Acta* **2012**, *72*, 172–178. [CrossRef]
43. Hirschorn, B.; Orazema, M.E.; Tribollet, B.; Vivier, V.; Frateur, I.; Musiani, M. Determination of effective capacitance and film thickness from constant-phase-element parameters. *Electro. Acta* **2010**, *55*, 6218–6227. [CrossRef]
44. Wang, Z.; Yuan, C.; Zhu, B.; Feng, Q.; Liu, F.; Xu, J.; Zhou, C.; Chen, G. Complex impedance spectroscopy of perovskite microwave dielectric ceramics with high dielectric constant. *J. Am. Ceram. Soc.* **2019**, *102*, 1852–1865. [CrossRef]
45. Tlili, D.; Hamdaoui, N.; Hcini, S.; Bouazizi, M.L.; Zemni, S. Above room temperature complex impedance analysis of properties of $La_{0.33}Sr_{0.67}Mn_{0.33}Ti_{0.67}O_{3\pm\delta}$ perovskite. *Phase Transit.* **2017**, *90*, 644. [CrossRef]
46. Pascoe, A.R.; Duffy, N.W.; Scully, A.D.; Huang, F.; Cheng, Y.B. Insights into planar $CH_3NH_3PbI_3$ perovskite solar cells using impedance spectroscopy. *J. Phys. Chem. C.* **2015**, *119*, 4444–4453. [CrossRef]

47. Tayari, F.; Nassar, K.I.; Benamara, M.; Essid, M.; Teixeira, S.S.; Graça, M.P.F. Sol–gel synthesized $(Bi_{0.5}Ba_{0.5}Ag)_{0.5}(NiMn)_{0.5}O_3$ perovskite ceramic: An exploration of its structural characteristics, dielectric properties and electrical conductivity. *Ceram. Int.* **2024**, *50*, 11207–11215. [CrossRef]
48. Achary, P.G.R.; Choudhary, R.N.P.; Parida, S.K. Structure, electric and dielectric properties of $PbFe_{1/3}Ti_{1/3}W_{1/3}O_3$ single perovskite compound. *Process. Appl. Ceram.* **2020**, *14*, 146–153. [CrossRef]
49. Chebaane, M.; Talbi, N.; Dhahri, A.; Oumezzine, M.; Khirouni, K.J. Structural and impedance spectroscopy properties of $La_{0.8}Ba_{0.1}Ca_{0.1}Mn_{1-x}RuxO_3$ perovskites. *Magn. Magn. Mater.* **2017**, *426*, 646. [CrossRef]
50. Halizan, M.Z.M.; Mohamed, Z.; Yahya, A.K. Simultaneously improved dielectric, optical and conductivity properties of $SrLa_{1-x}NdxLiTeO_6$ double perovskites. *Mater. Res. Express* **2020**, *7*, 086301. [CrossRef]
51. Boudad, L.; Taibi, M.; Belayachi, W.; Sajieddine, M. High temperature dielectric investigation, optical and conduction properties of $GdFe_{0.5}Cr_{0.5}O_3$ perovskite. *J. Appl. Phys.* **2020**, *127*, 174103. [CrossRef]

ceramics

MDPI

Article

Energy Storage Performance of $Na_{0.5}Bi_{0.5}TiO_3$–$CaHfO_3$ Lead-Free Ceramics Regulated by Defect Engineering

Zhuo Li *, Jing Zhang, Zixuan Wang, Xiaotian Wei, Dingjie Long, Xin Zhao and Yanhui Niu *

School of Materials Science and Engineering, Chang'an University, Xi'an 710061, China
* Correspondence: lizhuo@chd.edu.cn (Z.L.); niuyh@chd.edu.cn (Y.N.)

Abstract: Over the past decades, $Na_{0.5}Bi_{0.5}TiO_3$ (NBT)-based ceramics have received increasing attention in energy storage applications due to their high power density and relatively large maximum polarization. However, their high remnant polarization (P_r) and low breakdown field strength are detrimental for their practical applications. In this paper, a new solid solution $(1-x)Na_{0.5}Bi_{0.5}TiO_3$–$xCaHfO_3$ (x = 0.04, 0.08, 0.12, 0.16) was constructed by introducing $CaHfO_3$ into NBT, and thus was prepared using a conventional solid-state reaction. With the addition of $CaHfO_3$, the disorder of the structure increased, A-site vacancies formed, and thus oxygen vacancies were suppressed due to the replacement of the Na^+ by Ca^{2+}, resulting in the enhanced relaxation behavior and the reduced P_r, the refined grain, and improved breakdown strength. Furthermore, an optimal recoverable energy storage density (W_{rec}) of 1.2 J/cm^3 was achieved in $0.92Na_{0.5}Bi_{0.5}TiO_3$–$0.08CaHfO_3$ ceramics under the breakdown strength of 140 kV/cm, which is mainly attributed to the resultant defect of Na^+ vacancy.

Keywords: NBT; lead-free ceramics; defect engineering; energy storage density

Citation: Li, Z.; Zhang, J.; Wang, Z.; Wei, X.; Long, D.; Zhao, X.; Niu, Y. Energy Storage Performance of $Na_{0.5}Bi_{0.5}TiO_3$–$CaHfO_3$ Lead-Free Ceramics Regulated by Defect Engineering. *Ceramics* **2024**, *7*, 1002–1013. https://doi.org/10.3390/ceramics7030065

Academic Editor: Fanglin Chen

Received: 18 June 2024
Revised: 18 July 2024
Accepted: 24 July 2024
Published: 28 July 2024

1. Introduction

Due to their high power density and quick charge/discharge rate, dielectric capacitors have drawn more attention in recent decades in the field of energy and electricity, and electronic circuit systems [1,2]. Among them, ceramic-based dielectric materials have high breakdown field strength (E_b), low dielectric loss ($\tan(\delta)$), excellent fatigue resistance, and superior thermal stability, making them suitable in a variety of industries, including in oil exploration, aerospace, and military applications [3–5]. Usually, dielectric ceramic materials with large maximum polarization (P_{max}), small remnant polarization (P_r) and high E_b are perfect for creating energy storage capacitors, and energy storage performance can be evacuated by using the polarization-electric field (P-E) loop according to the subsequent Equations (1)–(3).

$$W_{tot} = \int_0^{P_{max}} EdP \tag{1}$$

$$W_{rec} = \int_{P_r}^{P_{max}} EdP \tag{2}$$

$$\eta = \frac{W_{rec}}{W_{tot}} \times 100\% = \frac{W_{rec}}{W_{rec} + W_{loss}} \times 100\% \tag{3}$$

where E, W_{tot}, W_{rec}, W_{loss}, and η denote the applied electric field, total energy storage density, recoverable energy storage density, energy loss density, and energy storage efficiency, respectively.

Although Pb-based ceramics exhibit excellent energy storage performance due to their larger P_{max} and higher E_b, the resultant heavy metal Pb due to the volatilization during high-temperature preparation process and the subsequent scrap of products all pose risks to the ecological environment and people's health [6,7]. Therefore, designing and researching lead-free materials are considered a feasible solution to solve this problem. In numerous lead-free dielectric ceramics, $Na_{0.5}Bi_{0.5}TiO_3$ (NBT) based ceramics have attracted much attention on account of high Curie temperature (T_c) (~320 °C) and large saturation polarization strength (P_s, 45 μC/cm^2) [8]. However, the energy storage properties of NBT ceramics were dissatisfied because of the high Pr (~38 μC/cm^2). In addition, the elements Bi and Na are easily volatile during the process of high-temperature sintering, which may lead to the generation of cationic vacancies and oxygen vacancies, and thus the E_b reduced [9,10]. As a consequence, many researchers have focused on the settlement of these issues [11–21]. Thereinto, Yan et al. [12] prepared $0.75Bi_{(0.5+x)}Na_{(0.5-x)}TiO_3$–$0.25SrTiO_3$ ceramics by defect engineering and suppressed the formation of oxygen vacancies by adjusting the non-stoichiometric ratio of the A-site, as well as refined the grains size to increase E_b significantly. Eventually, the outstanding W_{rec} (~5.63 J/cm^3) and the relatively high η (~94%) were achieved under 569 kV/cm as x = 0.08. The $0.9(Na_{0.4}Bi_{0.4}Ba_{0.06}Sr_{0.14}Ti_{(1-x)}Ta_xO_3)$–$0.1NaNbO_3$ ceramics designed and prepared by Jiang et al. [13] effectively suppressed oxygen vacancies and increased the E_b through defect engineering and finally realized a relative high W_{rec} (~3.12 J/cm^3) and excellent η (~87.68%). By replacing Bi^{3+} and Na^+ at A-site with La^{3+} to increase the concentration of cation vacancies and decrease oxygen vacancies, the 0.85NBT–$0.15CaTiO_3$–0.75La ceramics showed the improved energy storage performance with W_{rec} of ~2.15 J/cm^3 and η of ~81% [16]. Apparently, the defect engineering modulation is an effective way to improve the energy storage performance of NBT-based ceramics.

In this study, by incorporating $CaHfO_3$ into NBT ceramics from a defect engineering perspective, a binary solid solution $Na_{0.5}Bi_{0.5}TiO_3$–$CaHfO_3$ was designed. On the one hand, the introduction of Ca^{2+} at A-site may induce the cation vacancies and thus significantly suppress oxygen vacancies, along with Hf^{4+} replacing Ti^{4+} in the B-site with a relatively strong bond strength of Hf–O, ultimately all improving the E_b. On the other hand, due to the incorporation of non-equivalent cations as Na^+, Ca^{2+}, and Bi^{3+}, the creation of a local random field may break the long-range order ferroelectric domain and facilitate the formation of polar nanoregions (PNRs), which contributed to the decrease of P_r [22–24]. On the basis of the above analysis, $(1-x)Na_{0.5}Bi_{0.5}TiO_3$–$xCaHfO_3$ (NBT–$xCaHfO_3$, x = 0.04, 0.08, 0.12, 0.16) solid solution was prepared using the conventional solid state method, and the phase structure, microstructure, dielectric properties and energy storage characteristics were investigated in detail. The ceramics designed and fabricated using this strategy are expected to be used in the field of pulsed power capacitors.

2. Experimental Procedures

$(1-x)Na_{0.5}Bi_{0.5}TiO_3$–$xCaHfO_3$ (NBT$-xCaHfO_3$, x = 0.04, 0.08, 0.12, 0.16) ceramics were prepared using a conventional solid phase method [25,26]. The starting materials including Na_2CO_3 (analytical reagent (AR), ≥ 99.8%), Bi_2O_3 (AR, ≥ 99.9%), TiO_2 (chemically pure, ≥ 98.0%), $CaCO_3$ (AR, ≥ 99.0%), and HfO_2 (AR, ≥ 98.0%), which were all from Sinopharm Chemical Reagent Co., Ltd., Shanghai, China, were weighed according to the stoichiometric ratio of their components. The weighed powders were ball milled for 8 h by using the alcohol as medium. After that, the mixture was dried in an oven before calcination in a muffle furnace at 850 °C for 2 h. Following that, the powder was ball milled for another 6 h, and then was dried and sieved. The sieved powder was granulated with 6% polyvinyl alcohol (PVA) and then pressed into disks with a diameter of 13 mm under a pressure of 300 MPa. Finally, the disks were heated to 550 °C at the speed of 3 °C/min for 3 h to remove the PVA, and then sintered at 1100–1140 °C for 2 h at the rate of 5 °C/min to obtain the final ceramic samples.

X-ray diffraction (XRD, D8 Advanced, Bruker AXS Ltd., Karlsruhe, Germany) was used to analyze the crystal structure of the ceramic samples with diffraction angle 2θ in the range of 20–80°, and the accurate crystal structure of the ceramic materials were obtained using Rietveld full diffraction patterns fitting. The microstructures of the ceramic samples were observed using field emission scanning electron microscopy (SEM, S-4800, Hitachi, Japan). As for electrical testing, the ceramic samples were polished into a thickness of

~1 mm, covered with silver paste on the top and bottom surfaces, and then were fired at 850 °C for 10 min to form the electrodes. The LCR impedance meter (E4294A, Agilent, Santa Clara, USA) was used to measure the dielectric properties in the temperature range of 25–400 °C with a frequency in the range of 1 kHz to 1 MHz. In addition, the P-E curves of ceramic samples under different electric field strength were tested using a ferroelectric analyzer (TF2000, AixACCT, Aachen, Germany).

3. Results and Discussions

The XRD patterns of the NBT–x$CaHfO_3$ (x = 0.04, 0.08, 0.12, 0.16) ceramics sintered at the optimum temperature are shown in Figure 1a. As can be seen from this Figure, the X-ray diffraction patterns of all ceramic component exhibited a single perovskite structure, and no second phases were detected, which indicated that $CaHfO_3$ was completely diffused into the NBT lattice and formed a perfect solid solution. Figure 1b,c present the magnified plots of the diffraction peaks near 40.2° and 46.8° to further investigate the structural evolution of the ceramics. It can be seen that a significant splitting presented in both peaks near 40.2° and 46.8°, confirming that all ceramics were in the coexistence of the rhombohedral phase (R3c) and the tetragonal phase (P4bm) at room temperature [27]. The two split diffraction peaks weakened and merged gradually when $CaHfO_3$ content increased, which demonstrated that the coexistence of R3c and P4bm phases had evolved into a pseudo-cubic phase [28,29]. In addition, as $CaHfO_3$ doping content increased, the diffraction peaks were first shifted to higher angles (in particular, x = 0.12 is the most prominent), confirming the decreased cell volumes owing to the smaller radius Ca^{2+} (1.34 Å) replacing the larger radius Na^+ (1.39 Å). However, due to the limited solid solubility of the A-site, when $CaHfO_3$ doping content exceeds the threshold, the Ti^{4+} (0.605 Å) with a smaller radius at the B-site could be replaced by Ca^{2+} (1.34 Å) [30], and at the same time, more Hf^{4+} (0.71 Å) with a larger radius enters the B-site to replace Ti^{4+}, resulting in lattice volume expansion and diffraction peaks shifting towards small angles. In addition, the bulk density of NBT–x$CaHfO_3$ (x = 0.04, 0.08, 0.12, 0.16) ceramics sintered at different temperatures was plotted in Figure 1d. The bulk density of NBT–x$CaHfO_3$ increased initially with increasing sintering temperatures and then decreased after reaching the maximum value. It should be noted that the sintering temperature of the maximum bulk density was 1120 °C for all the different compositions.

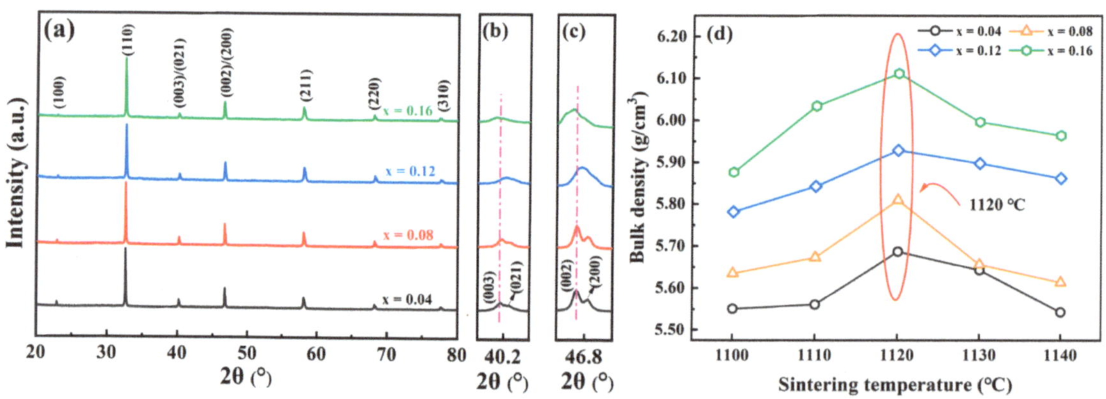

Figure 1. (**a**) XRD pattern of NBT–x$CaHfO_3$ (x = 0.04, 0.08, 0.12, 0.16) ceramics. The magnified view at 2θ (**b**) near 40.2° and (**c**) near 46.8°. (**d**) The bulk density of NBT–x$CaHfO_3$ ceramics sintered at different temperatures.

In order to further analyze the effect of $CaHfO_3$ doping on the phase structure and lattice parameters of the NBT–x$CaHfO_3$ ceramics, Figure 2 presents the Rietveld refinement of the full-scan XRD diffraction patterns based on the R3c and P4bm phase, and the corresponding phase proportion, lattice parameters, and reliability factors are shown in

Table 1. The reliability factors of weighted patterns (R_{wp}) and the goodness-of-fit indicator (χ^2) of all component ceramics are within the confidence interval, confirming that the results of XRD refinement are relatively reliable. Clearly, the fraction of the P4bm phase first increased and then decreased with the increase in $CaHfO_3$ content, as x = 0.08 reached the maximum. Usually, P4bm as the weakly polar phase could be converted into a polar ferroelectric state under an electric field and return to the original weakly polar state when the electric field was removed, hence the enhancement proportion of the P4bm phase could contribute to the reduction in P_r and thus promote the energy storage properties [31,32]. It can be assumed that the NBT–0.08$CaHfO_3$ ceramics may have optimal energy storage performance accompanied with relatively low P_r.

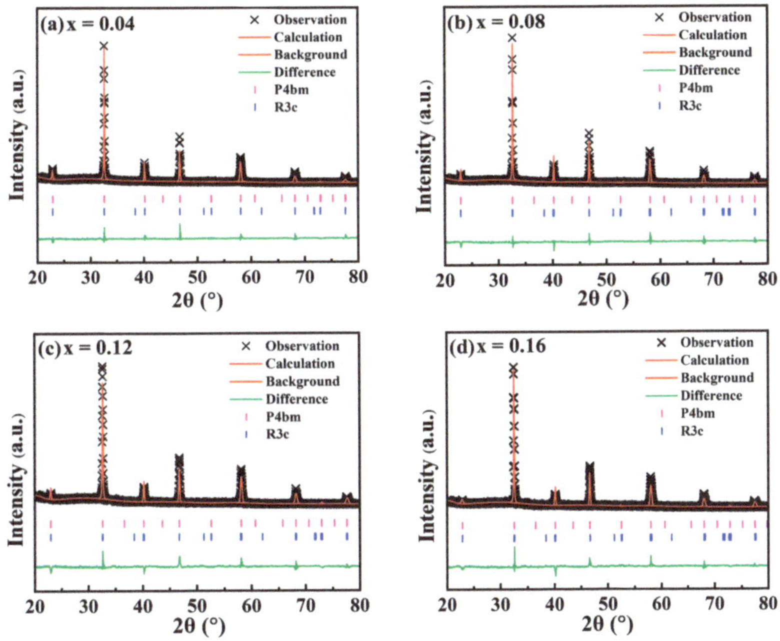

Figure 2. Rietveld Refined XRD Patterns of NBT–x$CaHfO_3$ ceramics: (**a**) x = 0.4, (**b**) x = 0.08, (**c**) x = 0.12, and (**d**) x = 0.16.

Table 1. The refinement results from full-diffraction pattern fitting of NBT–x$CaHfO_3$ ceramics.

x	Phase Structure	Relative Content (%)	Lattice Parameters			Reliability Factor	
			a(Å)	b(Å)	c(Å)	Rwp (%)	χ2
0.04	R3c P4bm	37.30% 62.70%	5.497 5.500	5.497 5.500	13.478 3.887	8.4%	1.76
0.08	R3c P4bm	7.47% 92.53%	5.497 5.503	5.497 5.503	13.494 3.889	8.7%	1.83
0.12	R3c P4bm	28.26% 71.74%	5.495 5.500	5.495 5.500	13.502 3.892	8.8%	2.26
0.16	R3c P4bm	22.85% 77.15%	5.494 5.507	5.494 5.507	13.520 3.895	9.4%	2.28

Figure 3 shows the natural surface micrographs of NBT–x$CaHfO_3$ ceramics. All samples present clear grain morphology and a relatively dense microstructure. In order to further investigate the effect of $CaHfO_3$ doping on the microscopic morphology of NBT ceramics, the average grain (AG) sizes of each component ceramics were analyzed, as shown in Figure 4. The average grain size of the ceramics decreased from 3.14 μm for x = 0.04 to 1.35 μm for x = 0.12, verifying that the addition of $CaHfO_3$ could refine grain size to a certain extent. Since Ca^{2+} tended to displace Na^+ at A-site due to the similar ionic radii ($R_{Ca^{2+}}$ = 1.34 Å, R_{Na^+} = 1.39 Å), and thus produced cation defects with positive charges, denoted as $Ca^{\bullet}_{Na}$ [32,33], and Na^+ vacancies (expressed as V'_{Na}) for charge compensation and for maintaining electrical neutrality. Consequently, the oxygen vacancy ($V_O^{\bullet\bullet}$), created by loss of oxygen from the crystal lattice at low oxygen partial pressure during sintering at high temperatures had to be inhabited [13,34,35], which further suppressed the conductivity and refined the grain size as well as beneficial to the enhancement of the E_b [36,37]. Furthermore, with the $CaHfO_3$ increasing, more V'_{Na} favor the mass transfer in ceramics during sintering [13,38], and as a consequence, the grain size increased up to 1.40 μm when x = 0.16. The equation for the defect is as follows (Equations (4) and (5)), in which the Kröger–Vink notation was employed.

$$CaO \overset{Na_2O}{\rightarrow} Ca^{\bullet}_{Na} + V'_{Na} + O_O \quad (4)$$

$$O_O \rightarrow V_O^{\bullet\bullet} + e' \quad (5)$$

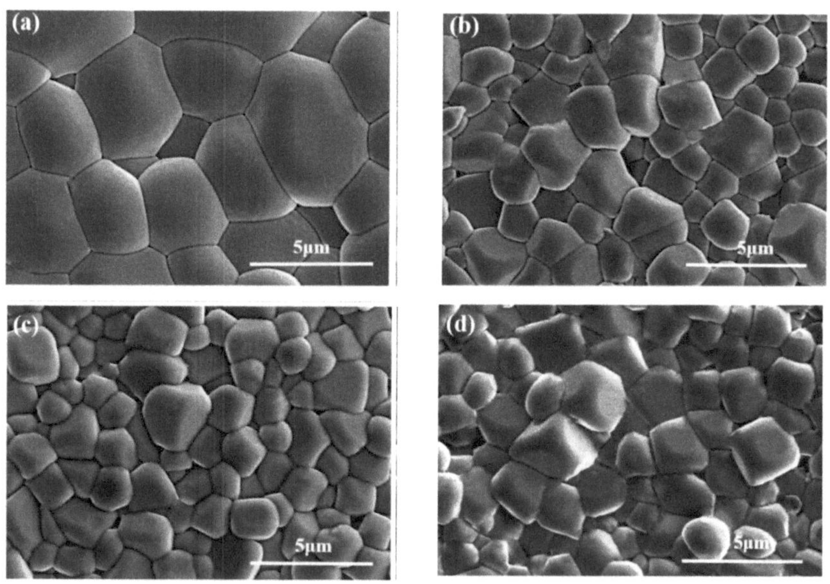

Figure 3. SEM images of NBT–x$CaHfO_3$ ceramics: (**a**) x = 0.4, (**b**) x = 0.08, (**c**) x = 0.12, and (**d**) x = 0.16.

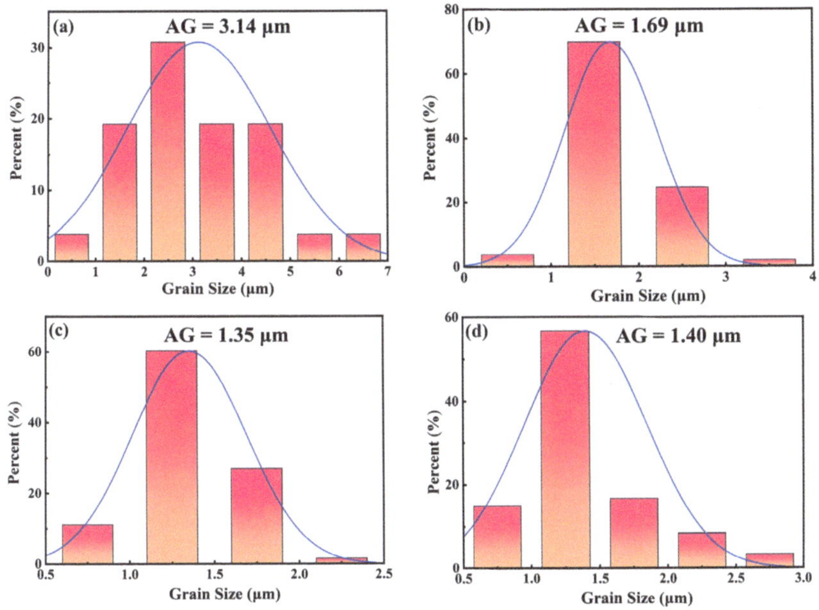

Figure 4. Grain size distribution diagram of NBT–x$CaHfO_3$ ceramics: (**a**) x = 0.4, (**b**) x = 0.08, (**c**) x = 0.12, (**d**) x = 0.16.

Figure 5 shows the variation curves of dielectric constant (ε_r) and dielectric loss (tan(δ)) under different frequency for NBT–x$CaHfO_3$ ceramics. The ε_r of all ceramics presents a decreasing trend to a different extent with the frequency increased, which may be due to the inability of defect dipoles to respond at high frequencies [39,40]. Besides that, when the $CaHfO_3$ content increased, ε_r at the frequency of 10 kHz decreased from 811 for the composition of x = 0.04 to 606 (x = 0.08) and then increased sharply to 1066 (x = 0.12), accomplished by the minimum dielectric loss (tan(δ) less than 0.022) emerging in the composition of x = 0.08 in a wide range of frequencies from 1k Hz to 1 MHz, implying that the appropriate amount of $CaHfO_3$ can reduce the concentration of oxygen vacancies and decrease the dielectric loss, which is conducive to the enhancement of the breakdown field strength. Figure 6 illustrates the temperature-dependent dielectric properties of NBT–x$CaHfO_3$ (x = 0.04, 0. 08, 0.12, and 0.16) ceramics at various frequencies. The dielectric anomaly peak (T_m) clearly observed refers to the ferro-paraelectric phase transition [41,42]. It is noteworthy that with the increase of $CaHfO_3$ content, the dielectric peak became flatter and broader, accompanied by the T_m shifting to the lower temperature, which may be caused by the diminished coupling effect of [TiO_6] octahedron and weakened structure stability, since A-site vacancies (V'_{Na}) formed as excessively Ca^{2+} doped [43,44]. In the meantime, as the frequency increased from 1 kHz to 1 MHz, ε_r decreased steadily and T_m shifted to the higher temperature, manifesting a relaxed behavior to a certain extent [45]. To further analyze the effect of $CaHfO_3$ content on the degree of relaxation, the diffuseness parameter γ between 1 (normal ferroelectric) and 2 (ideal relaxer) was calculated using the modified Curie–Weiss equation (Equation (6)) [46–48]:

$$\frac{1}{\varepsilon_r} - \frac{1}{\varepsilon_m} = \frac{(T - T_m)^{\gamma}}{C} \quad (6)$$

where C represents the Curie–Weiss constant and ε_m is the dielectric constant corresponding to T_m. The γ obtained by fitting the curves between $\ln(T-T_m)$ and $\ln((1/\varepsilon_r)-(1/\varepsilon_m))$ at 1 kHz, as shown in Figure 7, increased from 1.44 (x = 0.04) to 1.63 (x = 0.08) and then decreased to 1.57 (x = 0.12) and 1.54 (x = 0.16), hinting that the optimal relaxation characteristics can be obtained in the composition of x = 0.08. The optimizing relaxation performance may be related to the increased ionic disorder in the microstructure, where the non-equivalent substitution of Ca^{2+} for Na^{+} and Bi^{3+} broke the long-range ferroelectric domains and facilitated the formation of polar nanoregions [49–51].

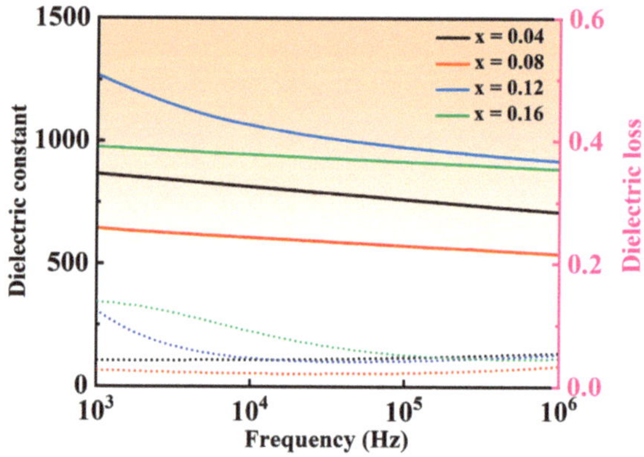

Figure 5. Frequency-dependent dielectric properties of NBT–x$CaHfO_3$ (x = 0.04, 0.08, 0.12, 0.16) ceramics at room temperature. Solid lines signify dielectric constant and dashed lines denote dielectric loss.

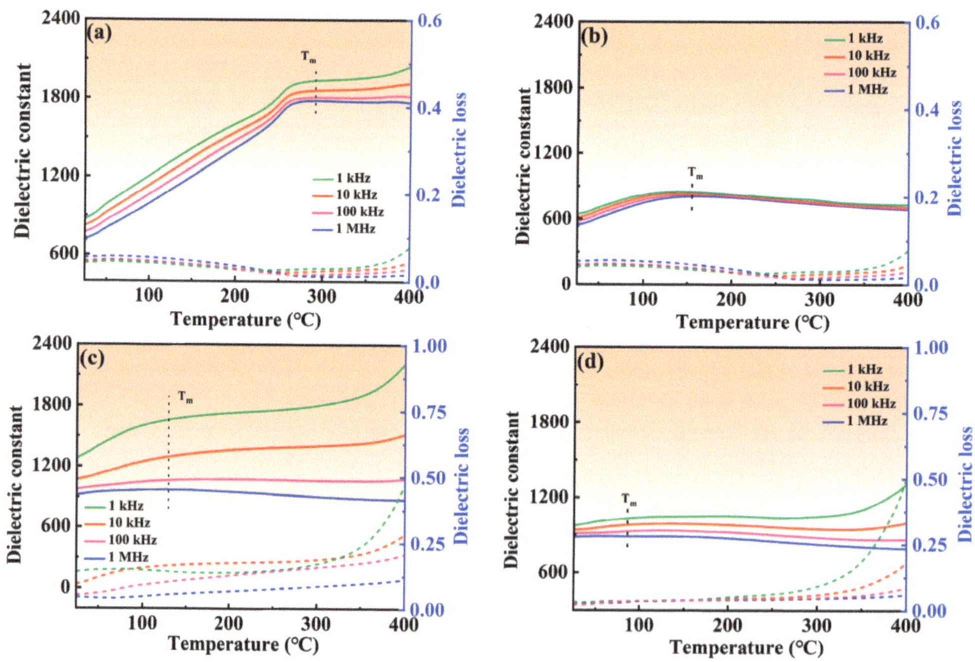

Figure 6. Temperature-dependent dielectric properties of NBT–x$CaHfO_3$ ceramics under various frequencies: (**a**) x = 0.04, (**b**) x = 0.08, (**c**) x = 0.12, and (**d**) x = 0.16. Solid lines signify dielectric constant and dashed lines denote dielectric loss.

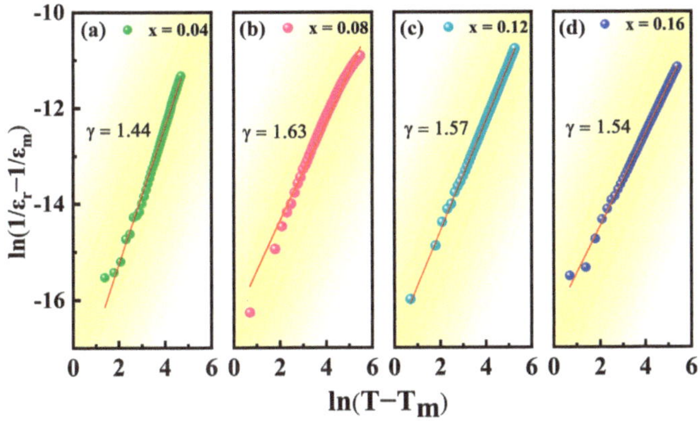

Figure 7. $\ln((1/\varepsilon_r)-(1/\varepsilon_m))$ as a function of $\ln(T-T_m)$ for NBT–xCaHfO$_3$ ceramics: (**a**) x = 0.04, (**b**) x = 0.08, (**c**) x = 0.12, and (**d**) x = 0.16. (solid lines—fitting to Equation (6)).

Figure 8 displays the P-E hysteresis loops of NBT–xCaHfO$_3$ ceramics under the electric fields of 120 kV/cm. Evidently, as CaHfO$_3$ content increases, the P-E loops become slimmer, especially for the NBT–0.08CaHfO$_3$ ceramics, which is in agreement with the result of diffuseness parameter γ. The slender P-E loops can be mainly ascribed to the broken long-range ferroelectric domain as the non-equivalent substitution of Na^+ and Bi^{3+} by Ca^{2+}. Thus, the formed polar microregions suppressed the early polarization saturation of the ceramics and reduced the P_r, which facilitated the achievement of excellent energy storage performance [24,52]. In addition, Figure 9 displays the P-E hysteresis loops of NBT–xCaHfO$_3$ ceramics under various electric fields. Apparently, P-E loops of all composition turn more slender with increasing applied electric filed, among which P_{max} and ΔP ($\Delta P = P_{max}-P_r$) of the composition with x = 0.08 increases significantly, while P_r increases slightly. Ultimately, under an electric fields of 140 kV/cm, P_{max} and ΔP reach their maximums, as displayed in the inset of Figure 9b.

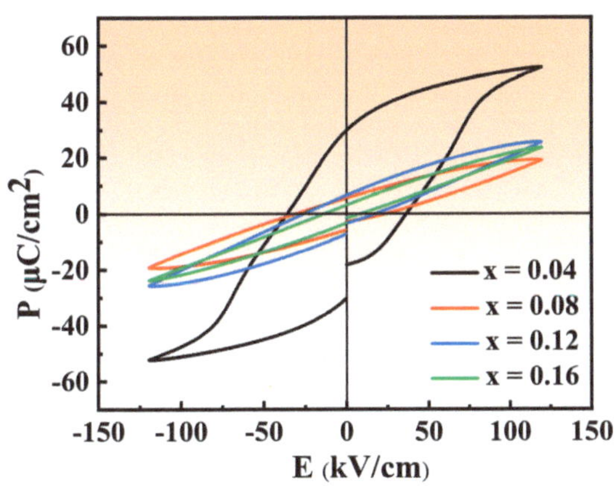

Figure 8. P-E curves of NBT–xCaHfO$_3$ ceramic samples under the electric field of 120 kV/cm.

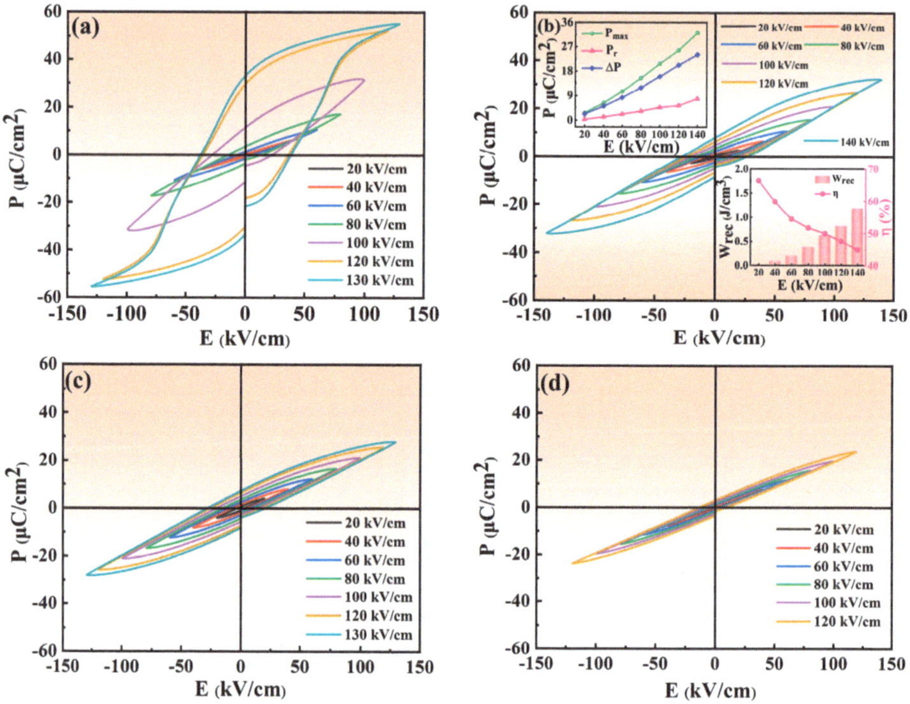

Figure 9. P-E curves of NBT–xCaHfO$_3$ ceramic samples under different electric field: (**a**) x = 0.04, (**b**) x = 0.08, (**c**) x = 0.12, and (**d**) x = 0.16.

The improvement of W_{rec} requires not only high P_{max} and low P_r, but also high applied field and, subsequently, the P-E hysteresis loops of NBT–xCaHfO$_3$ ceramics at 20 Hz under critical electric fields are displayed in Figure 10a. Compared to NBT–xCaHfO$_3$ ceramics with x = 0.04, 0.12, and 0.16, the NBT–0.08CaHfO$_3$ ceramics demonstrated the maximum E_b value of 140 kV/cm because of the dense microstructure and relatively small grain size (in Figure 3). The variation of W_{rec}, η and P_{max}, P_r, and ΔP for each ceramic under a critical electric field are shown in Figure 10b,c, and the relevant parameters are listed in Table 2. Overall, with the increase in the CaHfO$_3$ content, the P_{max} decreased from 55.39 μC/cm^2 (x = 0.04) to 23.66 μC/cm^2 (x = 0.16) together with P_r maintaining in a relatively low value. Finally, the ΔP attained its maximum for NBT–0.08CaHfO$_3$ ceramics, achieving the optimizing energy storage performance with W_{rec} of 1.2 J/cm^3 and η of 45%. Also, the energy storage performance and polarization properties of NBT–0.08CaHfO$_3$ ceramics under various electric fields from 20 kV/cm to 140 kV/cm are demonstrated in the inset of Figure 9b. Apparently, ΔP enhanced approximately linearly from 2.33 μC/cm^2 to 24.12 μC/cm^2, with the increase in applied electric field as a result of the significant increase of P_{max} along with a slight increase of P_r. The comparatively large ΔP and E_b facilitated excellent energy storage properties. Hence, the W_{rec} of 1.20 J/cm^3 was highlighted in 0.92NBT–0.08CaHfO$_3$ ceramics under the maximum electric field of 140 kV/cm, indicating certain potential in energy storage application.

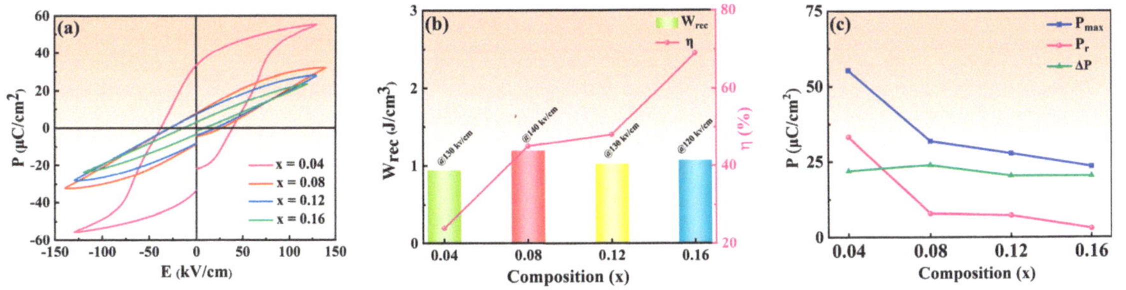

Figure 10. (**a**) P-E curves, (**b**) W_{rec} and η, (**c**) P_{max}, P_r, and ΔP of NBT–x$CaHfO_3$ ceramic samples under critical electric field.

Table 2. P_{max}, P_r, W_{rec}, and η of NBT–x$CaHfO_3$ ceramics under the critical electric field.

x	E_b (kV/cm)	P_{max} (μC/cm^2)	P_r (μC/cm^2)	W_{rec} (J/cm^3)	η (%)
0.04	130	55.39	33.30	0.95	24%
0.08	140	32.10	7.88	1.20	45%
0.12	130	27.93	7.31	1.03	48%
0.16	120	23.66	3.09	1.07	69%

4. Conclusions

NBT–x$CaHfO_3$ (x = 0.04, 0.08, 0.12, 0.16) ceramics were prepared using a conventional solid phase method. The introduction of Ca^{2+} substituting for Na^+ and Bi^{3+} at the A-site induced the cation vacancies and then suppressed the concentration of oxygen vacancies as well as refined the grain size, associated with Hf^{4+} replacing Ti^{4+} in the B-site with a relatively strong bond strength of Hf–O, eventually fulfilling the improvement of the E_b. Besides that, a local random field created by the incorporation of non-equivalent cations as Na^+, Ca^{2+}, Bi^{3+}, and cation vacancies, as well as bond strength of Hf–O, broke the long-range order ferroelectric domain and facilitated the formation of PNRs, resulting in an enhancement of relaxation properties and a decrease in P_r. Combined with enhanced relaxation behavior and increased E_b, an optimizing W_{rec} of 1.2 J/cm^3 was achieved in NBT–0.08$CaHfO_3$ ceramics under the E_b of 140 kV/cm. All of these demonstrated that the NBT–0.08$CaHfO_3$ ceramics have an attractive potential in the field of energy storage ceramics.

Author Contributions: Conceptualization, Z.L.; methodology, J.Z.; formal analysis, Z.L. and D.L.; investigation, X.W., D.L. and X.Z.; resources, Y.N.; data curation, X.Z. and Y.N.; writing—original draft, J.Z.; writing—review and editing, Z.L. and Z.W. All authors have read and agreed to the published version of the manuscript.

Funding: This research was funded by the Fundamental Research Funds for the Central Universities, CHD (No. 300102314904), the Natural Science Foundation of Shaanxi province, China (No. 2021JM-172) and the National Natural Science Foundations of China (Grant No. 52278427).

Data Availability Statement: The original contributions presented in the study are included in the article, further inquiries can be directed to the corresponding author.

Conflicts of Interest: The authors declare no conflicts of interest.

References

1. Yang, Z.; Du, H.; Jin, L.; Poelman, D. High-Performance Lead-Free Bulk Ceramics for Electrical Energy Storage Applications: Design Strategies and Challenges. *J. Mater. Chem. A* **2021**, *9*, 18026–18085. [CrossRef]
2. Palneedi, H.; Peddigari, M.; Hwang, G.; Jeong, D.; Ryu, J. High-Performance Dielectric Ceramic Films for Energy Storage Capacitors: Progress and Outlook. *Adv. Funct. Mater.* **2018**, *28*, 1803665. [CrossRef]
3. Yuan, Q.; Chen, M.; Zhan, S.; Li, Y.; Lin, Y.; Yang, H. Ceramic-Based Dielectrics for Electrostatic Energy Storage Applications: Fundamental Aspects, Recent Progress, and Remaining Challenges. *Chem. Eng. J.* **2022**, *446*, 136315.

4. Hu, Q.; Tian, Y.; Zhu, Q.; Bian, J.; Jin, L.; Du, H.; Alikin, D.O.; Shur, V.Y.; Feng, Y.; Xu, Z.; et al. Achieve Ultrahigh Energy Storage Performance in $BaTiO_3$–$Bi(Mg_{1/2}Ti_{1/2})O_3$ Relaxor Ferroelectric Ceramics via Nano-Scale Polarization Mismatch and Reconstruction. *Nano Energy* **2020**, *67*, 104264.
5. Muhammad, R.; Iqbal, Y.; Reaney, I.M. $BaTiO_3$–$Bi(Mg_{2/3}Nb_{1/3})O_3$ Ceramics for High-Temperature Capacitor Applications. *J. Am. Ceram. Soc.* **2016**, *99*, 2089–2095. [CrossRef]
6. Wang, H.; Liu, Y.; Yang, T.; Zhang, S. Ultrahigh Energy-Storage Density in Antiferroelectric Ceramics with Field-Induced Multiphase Transitions. *Adv. Funct. Mater.* **2019**, *29*, 1807321. [CrossRef]
7. Wang, H.; Hu, Q.; Liu, X.; Zheng, Q.; Jiang, N.; Yang, Y.; Kwok, K.W.; Xu, C.; Lin, D. A High-Tolerance BNT-Based Ceramic with Excellent Energy Storage Properties and Fatigue/Frequency/Thermal Stability. *Ceram. Int.* **2019**, *45*, 23233–23240.
8. Wang, G.; Lu, Z.; Li, Y.; Li, L.; Ji, H.; Feteira, A.; Zhou, D.; Wang, D.; Zhang, S.; Reaney, I.M. Electroceramics for High-Energy Density Capacitors: Current Status and Future Perspectives. *Chem. Rev.* **2021**, *121*, 6124–6172. [CrossRef] [PubMed]
9. Höfling, M.; Steiner, S.; Hoang, A.-P.; Seo, I.-T.; Frömling, T. Optimizing the Defect Chemistry of $Na_{1/2}Bi_{1/2}TiO_3$ -Based Materials: Paving the Way for Excellent High Temperature Capacitors. *J. Mater. Chem. C* **2018**, *6*, 4769–4776. [CrossRef]
10. Carter, J.; Aksel, E.; Iamsasri, T.; Forrester, J.S.; Chen, J.; Jones, J.L. Structure and Ferroelectricity of Nonstoichiometric $(Na_{0.5}Bi_{0.5})TiO_3$. *Appl. Phys. Lett.* **2014**, *104*, 112904. [CrossRef]
11. Yang, H.; Cai, Z.; Zhu, C.; Feng, P.; Wang, X. Ultra-High Energy Storage Performance in BNT-Based Ferroelectric Ceramics with Simultaneously Enhanced Polarization and Breakdown Strength. *ACS Sustain. Chem. Eng.* **2022**, *10*, 9176–9183. [CrossRef]
12. Yan, F.; Huang, K.; Jiang, T.; Zhou, X.; Shi, Y.; Ge, G.; Shen, B.; Zhai, J. Significantly Enhanced Energy Storage Density and Efficiency of BNT-Based Perovskite Ceramics via A-Site Defect Engineering. *Energy Storage Mat.* **2020**, *30*, 392–400. [CrossRef]
13. Jiang, Z.; Yang, H.; Cao, L.; Yang, Z.; Yuan, Y.; Li, E. Enhanced Breakdown Strength and Energy Storage Density of Lead-Free $Bi_{0.5}Na_{0.5}TiO_3$-Based Ceramic by Reducing the Oxygen Vacancy Concentration. *Chem. Eng. J.* **2021**, *414*, 128921. [CrossRef]
14. Zhang, L.; Jing, R.; Huang, Y.; Hu, Q.; Alikin, D.O.; Shur, V.Y.; Gao, J.; Wei, X.; Zhang, L.; Liu, G.; et al. Enhanced Antiferroelectric-like Relaxor Ferroelectric Characteristic Boosting Energy Storage Performance of $(Bi_{0.5}Na_{0.5})TiO_3$-Based Ceramics via Defect Engineering. *J. Mat.* **2022**, *8*, 527–536. [CrossRef]
15. Liu, G.; Dong, J.; Zhang, L.; Yu, L.; Wei, F.; Li, Y.; Gao, J.; Hu, J.; Yan, Y.; Li, Q.; et al. $Na_{0.25}Sr_{0.5}Bi_{0.25}TiO_3$ Relaxor Ferroelectric Ceramic with Greatly Enhanced Electric Storage Property by a B-Site Ion Doping. *Ceram. Int.* **2020**, *46*, 11680–11688. [CrossRef]
16. Yu, L.; Dong, J.; Tang, M.; Liu, Y.; Wu, F.; Yan, Y.; Liu, G.; Song, C. Enhanced Electrical Energy Storage Performance of Pb-Free A-Site La^{3+}-Doped $0.85Na_{0.5}Bi_{0.5}TiO_3$-$0.15CaTiO_3$ Ceramics. *Ceram. Int.* **2020**, *46*, 28173–28182. [CrossRef]
17. Li, T.; Chen, P.; Si, R.; Li, F.; Guo, Y.; Wang, C. High Energy Storage Density and Efficiency with Excellent Temperature and Frequency Stabilities under Low Operating Field Achieved in $Ag_{0.91}Sm_{0.03}NbO_3$-Modified $Na_{0.5}Bi_{0.5}TiO_3$-$BaTiO_3$ Ceramics. *J. Mater. Sci. Mater. Electron.* **2020**, *31*, 16928–16937. [CrossRef]
18. Liu, X.; Rao, R.; Shi, J.; He, J.; Zhao, Y.; Liu, J.; Du, H. Effect of Oxygen Vacancy and A-Site-Deficiency on the Dielectric Performance of BNT-BT-BST Relaxors. *J. Alloys Compd.* **2021**, *875*, 159999. [CrossRef]
19. Chen, Z.; Pu, Y.; Ning, Y.; Hui, Y.; Wu, C.; Zhang, L.; Zhang, X.; Wang, B. Prominent Energy Storage Density and Efficiency of $Na_{0.5}Bi_{0.5}TiO_3$ -based Ceramics via Multiscale Amelioration Strategy. *J. Am. Ceram. Soc.* **2024**, *107*, 2360–2370. [CrossRef]
20. Zhao, H.; Cao, W.; Liang, C.; Wang, C.; Wang, C. High Energy-Storage Performance in X9R-Type $Na_{0.5}Bi_{0.5}TiO_3$-Based Lead-Free Ceramics. *Ceram. Int.* **2023**, *49*, 15939–15946. [CrossRef]
21. Li, T.; Jiang, X.; Li, J.; Xie, A.; Fu, J.; Zuo, R. Ultrahigh Energy-Storage Performances in Lead-Free $Na_{0.5}Bi_{0.5}TiO_3$ -Based Relaxor Antiferroelectric Ceramics through a Synergistic Design Strategy. *ACS Appl. Mater. Interfaces* **2022**, *14*, 22263–22269. [CrossRef] [PubMed]
22. Roukos, R.; Zaiter, N.; Chaumont, D. Relaxor Behaviour and Phase Transition of Perovskite Ferroelectrics-Type Complex Oxides $(1-x)Na_{0.5}Bi_{0.5}TiO_3$–$xCaTiO_3$ System. *J. Adv. Ceram.* **2018**, *7*, 124–142. [CrossRef]
23. Luo, C.; Feng, Q.; Luo, N.; Yuan, C.; Zhou, C.; Wei, Y.; Fujita, T.; Xu, J.; Chen, G. Effect of Ca^{2+}/Hf^{4+} Modification at A/B Sites on Energy-Storage Density of $Bi_{0.47}Na_{0.47}Ba_{0.06}TiO_3$ Ceramics. *Chem. Eng. J.* **2021**, *420*, 129861. [CrossRef]
24. Wang, K.; Li, W.; Zhao, R.; Tang, X.; Zhang, S.; Zhang, Y.; Hu, J.; Shen, Z.; Jiang, Y.; Guo, X. High Energy Storage Density Obtained by $Bi(Ni_{0.5}Hf_{0.5})O_3$-Modified NBT-Based Ceramic at a Low Electric Field. *Ceram. Int.* **2024**, *50*, 5276–5284. [CrossRef]
25. Li, Z.; Zhang, J.; Wang, C.; Wang, Z.; Lei, N.; Zheng, L.; Long, D.; Wei, X.; Zhang, J.; Wang, Z.; et al. Phase Structure, Dielectric and Energy Storage Properties of $Na_{0.5}Bi_{0.5}TiO_3$-$BaTiO_3$ Ceramics with $Bi(Mg_{2/3}Nb_{1/3})O_3$ Modification. *Ceram. Int.* **2023**, *49*, 38735–38742. [CrossRef]
26. Zhang, L.; Pu, Y.; Chen, M.; Wei, T.; Keipper, W.; Shi, R.; Guo, X.; Li, R.; Peng, X. High Energy-Storage Density under Low Electric Fields and Improved Optical Transparency in Novel Sodium Bismuth Titanate-Based Lead-Free Ceramics. *J. Eur. Ceram. Soc.* **2020**, *40*, 71–77. [CrossRef]
27. Yang, F.; Bao, S.; Zhai, Y.; Zhang, Y.; Su, Z.; Liu, J.; Zhai, J.; Pan, Z. Enhanced Energy-Storage Performance and Thermal Stability in $Bi_{0.5}Na_{0.5}TiO_3$-Based Ceramics through Defect Engineering and Composition Design. *Mater. Today Chem.* **2021**, *22*, 100583. [CrossRef]
28. Qiao, X.; Wu, D.; Zhang, F.; Niu, M.; Chen, B.; Zhao, X.; Liang, P.; Wei, L.; Chao, X.; Yang, Z. Enhanced Energy Density and Thermal Stability in Relaxor Ferroelectric $Bi_{0.5}Na_{0.5}TiO_3$-$Sr_{0.7}Bi_{0.2}TiO_3$ Ceramics. *J. Eur. Ceram. Soc.* **2019**, *39*, 4778–4784. [CrossRef]

29. Yang, L.; Kong, X.; Cheng, Z.; Zhang, S. Ultra-High Energy Storage Performance with Mitigated Polarization Saturation in Lead-Free Relaxors. *J. Mater. Chem. A* **2019**, *7*, 8573–8580. [CrossRef]
30. Zhao, P.; Tang, B.; Si, F.; Yang, C.; Li, H.; Zhang, S. Novel Ca Doped $Sr_{0.7}Bi_{0.2}TiO_3$ Lead-Free Relaxor Ferroelectrics with High Energy Density and Efficiency. *J. Eur. Ceram. Soc.* **2020**, *40*, 1938–1946. [CrossRef]
31. Han, J.; Yin, J.; Wu, J. BNT-based Ferroelectric Ceramics: Electrical Properties Modification by Ta_2O_5 Oxide Addition. *J. Am. Ceram. Soc.* **2019**, *103*, 412–422. [CrossRef]
32. Fan, J.; He, G.; Cao, Z.; Cao, Y.; Long, Z.; Hu, Z. Ultrahigh Energy-Storage Density of a Lead-Free $0.85Bi_{0.5}Na_{0.5}TiO_3-0.15Ca(Nb_{0.5}Al_{0.5})O_3$ Ceramic under Low Electric Fields. *Inorg. Chem. Front.* **2023**, *10*, 1561–1573. [CrossRef]
33. Xie, A.; Fu, J.; Zuo, R.; Zhou, C.; Qiao, Z.; Li, T.; Zhang, S. $NaNbO_3$-$CaTiO_3$ Lead-Free Relaxor Antiferroelectric Ceramics Featuring Giant Energy Density, High Energy Efficiency and Power Density. *Chem. Eng. J.* **2022**, *429*, 132534. [CrossRef]
34. Prasertpalichat, S.; Cann, D.P. Hardening in Non-Stoichiometric $(1-x)Bi_{0.5}Na_{0.5}TiO_3-xBaTiO_3$ Lead-Free Piezoelectric Ceramics. *J. Mater. Sci.* **2016**, *51*, 476–486. [CrossRef]
35. Kumar, N.; Ansell, T.Y.; Cann, D.P. Role of Point Defects in Bipolar Fatigue Behavior of $Bi(Mg_{1/2}Ti_{1/2})O_3$ Modified $(Bi_{1/2}K_{1/2})TiO_3$-$(Bi_{1/2}Na_{1/2})TiO_3$ Relaxor Ceramics. *J. Appl. Phys.* **2014**, *115*, 154104. [CrossRef]
36. Li, M.; Pietrowski, M.J.; De Souza, R.A.; Zhang, H.; Reaney, I.M.; Cook, S.N.; Kilner, J.A.; Sinclair, D.C. A Family of Oxide Ion Conductors Based on the Ferroelectric Perovskite $Na_{0.5}Bi_{0.5}TiO_3$. *Nat. Mater.* **2014**, *13*, 31–35. [CrossRef]
37. Malik, R.A.; Hussain, A.; Maqbool, A.; Zaman, A.; Song, T.K.; Kim, W.-J.; Kim, M.-H. Giant Strain, Thermally-Stable High Energy Storage Properties and Structural Evolution of Bi-Based Lead-Free Piezoceramics. *J. Alloys Compd.* **2016**, *682*, 302–310. [CrossRef]
38. Zhu, X.; Gao, Y.; Shi, P.; Kang, R.; Kang, F.; Qiao, W.; Zhao, J.; Wang, Z.; Yuan, Y.; Lou, X. Ultrahigh Energy Storage Density in $(Bi_{0.5}Na_{0.5})_{0.65}Sr_{0.35}TiO_3$-Based Lead-Free Relaxor Ceramics with Excellent Temperature Stability. *Nano Energy* **2022**, *98*, 107276. [CrossRef]
39. Ding, J.; Pan, Z.; Chen, P.; Hu, D.; Yang, F.; Li, P.; Liu, J.; Zhai, J. Enhanced Energy Storage Capability of $(1-x)Na_{0.5}Bi_{0.5}TiO_3$-$xSr_{0.7}Bi_{0.2}TiO_3$ Free-Lead Relaxor Ferroelectric Thin Films. *Ceram. Int.* **2020**, *46*, 14816–14821. [CrossRef]
40. Zhang, L.; Hao, X.; Zhang, L. Enhanced Energy-Storage Performances of Bi_2O_3–Li_2O Added $(1-x)(Na_{0.5}Bi_{0.5})TiO_3-xBaTiO_3$ Thick Films. *Ceram. Int.* **2014**, *40*, 8847–8851. [CrossRef]
41. Wang, H.; Yuan, H.; Li, X.; Zeng, F.; Wu, K.; Zheng, Q.; Fan, G.; Lin, D. Enhanced Energy Density and Discharged Efficiency of Lead-Free Relaxor $(1-x)[(Bi_{0.5}Na_{0.5})_{0.94}Ba_{0.06}]_{0.98}La_{0.02}TiO_3$-$xKNb_{0.6}Ta_{0.4}O_3$ Ceramic Capacitors. *Chem. Eng. J.* **2020**, *394*, 124879. [CrossRef]
42. Huang, J.; Qi, H.; Gao, Y.; Xie, A.; Zhang, Y.; Li, Y.; Wang, S.; Zuo, R. Expanded Linear Polarization Response and Excellent Energy-Storage Properties in $(Bi_{0.5}Na_{0.5})TiO_3$-$KNbO_3$ Relaxor Antiferroelectrics with Medium Permittivity. *Chem. Eng. J.* **2020**, *398*, 125639. [CrossRef]
43. Li, D.; Lin, Y.; Zhang, M.; Yang, H. Achieved Ultrahigh Energy Storage Properties and Outstanding Charge–Discharge Performances in $(Na_{0.5}Bi_{0.5})_{0.7}Sr_{0.3}TiO_3$-Based Ceramics by Introducing a Linear Additive. *Chem. Eng. J.* **2020**, *392*, 123729. [CrossRef]
44. Li, X.; Cheng, Y.; Wang, F.; Xu, Q.; Chen, Y.; Xie, L.; Tan, Z.; Xing, J.; Zhu, J. Enhancement of Energy Storage and Hardness of $(Na_{0.5}Bi_{0.5})_{0.7}Sr_{0.3}TiO_3$-Based Relaxor Ferroelectrics via Introducing $Ba(Mg_{1/3}Nb_{2/3})O_3$. *Chem. Eng. J.* **2022**, *431*, 133441. [CrossRef]
45. Luo, C.; Wei, Y.; Feng, Q.; Wang, M.; Luo, N.; Yuan, C.; Zhou, C.; Fujita, T.; Xu, J. Significantly Enhanced Energy-Storage Properties of $Bi_{0.47}Na_{0.47}Ba_{0.06}TiO_3$-$CaHfO_3$ Ceramics by Introducing $Sr_{0.7}Bi_{0.2}TiO_3$ for Pulse Capacitor Application. *Chem. Eng. J.* **2022**, *429*, 132165. [CrossRef]
46. Yan, F.; Zhou, X.; He, X.; Bai, H.; Wu, S.; Shen, B.; Zhai, J. Superior Energy Storage Properties and Excellent Stability Achieved in Environment-Friendly Ferroelectrics via Composition Design Strategy. *Nano Energy* **2020**, *75*, 105012. [CrossRef]
47. Jiang, Z.; Yuan, Y.; Yang, H.; Li, E.; Zhang, S. Relaxor Regulation and Improvement of Breakdown Strength for $Bi_{0.5}Na_{0.5}TiO_3$-Based Ceramics by Co-Doping with Ca and Nb. *Ceram. Int.* **2022**, *48*, 9702–9709. [CrossRef]
48. Meng, K.; Li, W.; Tang, X.; Liu, Q.; Jiang, Y. The Defect Related Energy-Storage Properties of A-Site off-Stoichiometry Ferroelectric Ceramic. *Appl. Phys. A* **2021**, *127*, 337. [CrossRef]
49. Huang, Y.; Zhang, L.; Jing, R.; Hu, Q.; Alikin, D.O.; Shur, V.Y.; Islam, S.S.; Du, H.; Wei, X.; Feng, G.; et al. Thermal Stability of Dielectric and Energy Storage Performances of Ca-Substituted BNTZ Ferroelectric Ceramics. *Ceram. Int.* **2021**, *47*, 6298–6309. [CrossRef]
50. Krogstad, M.J.; Gehring, P.M.; Rosenkranz, S.; Osborn, R.; Ye, F.; Liu, Y.; Ruff, J.P.C.; Chen, W.; Wozniak, J.M.; Luo, H.; et al. The Relation of Local Order to Material Properties in Relaxor Ferroelectrics. *Nat. Mat.* **2018**, *17*, 718–724. [CrossRef]
51. Wang, Z.; Kang, R.; Zhang, L.; Mao, P.; Sun, Q.; Kang, F.; Wang, J. Remarkably Enhanced Energy-Storage Density and Excellent Thermal Stability under Low Electric Fields of $(Na_{0.5}Bi_{0.5})TiO_3$-Based Ceramics via Composition Optimization Strategy. *J. Eur. Ceram. Soc.* **2021**, *41*, 1917–1924. [CrossRef]
52. Li, X.; Dong, X.; Wang, F.; Tan, Z.; Zhang, Q.; Chen, H.; Xi, J.; Xing, J.; Zhou, H.; Zhu, J. Realizing Excellent Energy Storage Properties in $Na_{0.5}Bi_{0.5}TiO_3$-Based Lead-Free Relaxor Ferroelectrics. *J. Eur. Ceram. Soc.* **2022**, *42*, 2221–2229. [CrossRef]

ceramics

MDPI

Article

Origin of Temperature Coefficient of Resonance Frequency in Rutile $Ti_{1-x}Zr_xO_2$ Microwave Ceramics

Izaz Khan [1], Aneela Khan [1], Raz Muhammad [1,*], Minmin Mao [2], Dandan Han [3], Kaixin Song [2], Wen Lei [4] and Dawei Wang [5,*]

1 Department of Physics, Abdul Wali Khan University Mardan, Mardan 23200, Pakistan
2 College of Electronics and Information, Hangzhou Dianzi University, Hangzhou 310018, China; mmm@hdu.edu.cn (M.M.)
3 Key Laboratory for Special Functional Materials in Jilin Provincial Universities, Jilin Institute of Chemical Technology, Jilin 132022, China
4 School of Optical and Electronic Information, Huazhong University of Science and Technology, Wuhan 430074, China; wenlei@mail.hust.edu.cn
5 School of Instrumentation Science and Engineering, Harbin Institute of Technology, Harbin 150080, China
* Correspondence: raz@awkum.edu.pk (R.M.); wangdawei102@gmail.com (D.W.)

Abstract: In this study, we report the effect of Zr^{4+} doping on the optical energy gap and microwave dielectric properties of rutile TiO_2. Rietveld analysis explicitly confirmed that Zr^{4+} occupies the octahedral site, forming a single-phase tetragonal structure below the solubility limit ($x < 0.10$). Notably, at $x = 0.025$, a significant enhancement in $Q \times f_o$ was observed. This enhancement was attributed to the reduction in dielectric loss, associated with a decrease in oxygen vacancies and a lower concentration of Ti^{3+} paramagnetic centers. This conclusion was supported by Raman and electron paramagnetic resonance spectroscopy, respectively. The origin of high τ_f in rutile $Ti_{1-x}Zr_xO_2$ is explained on the basis of the octahedral distortion/tetragonality ratio, covalency, and bond strength.

Keywords: Zr-doped TiO_2; rutile; microwave dielectric properties

Citation: Khan, I.; Khan, A.; Muhammad, R.; Mao, M.; Han, D.; Song, K.; Lei, W.; Wang, D. Origin of Temperature Coefficient of Resonance Frequency in Rutile $Ti_{1-x}Zr_xO_2$ Microwave Ceramics. *Ceramics* **2024**, 7, 698–711. https://doi.org/10.3390/ceramics7020046

Academic Editors: John W. Halloran and Mari-Ann Einarsrud

Received: 15 March 2024
Revised: 2 May 2024
Accepted: 20 May 2024
Published: 23 May 2024

1. Introduction

TiO_2 is one of the most studied metal oxides, used in many conventional applications (white pigment in paintings, paper, plastics, etc.) and advanced applications (microwave (MW) dielectric resonators, photocatalysts, resistive random-access memory (RRAM), and high-energy density capacitors) [1–4]. Naturally, TiO_2 exists in different polymorphs i.e., brookite, rutile, and anatase. Among these phases, the thermodynamically stable phase is the rutile, as brookite and anatase phase can transform irreversibly to the rutile phase in the temperature range of 700–920 °C [5]. In several studies, colossal permittivity was found for (Mg^{2+} + Ta^{5+}) co-doped TiO_2 [6], ($Tb^{3+/4+}$ and Sb^{5+}) co-doped TiO_2 [7], (Zr and Ta) co-doped TiO_2 [8], (Lu^{3+}/Nb^{5+}) co-doped TiO_2 [9], and (Nb and Mn) co-doped TiO_2 [10]. Cohn et al. [2] reported MW properties of TiO_2 for dielectric resonators, which gained significant attention [11]. TiO_2 possesses a high relative permittivity (ε_r) of ~100 and a suitable quality factor ($Q \times f_o$) of ~10,000 GHz, but its poor temperature coefficient of resonant frequency (τ_f) of ~450 ppm/°C is not suitable for practical applications [12]. Several studies have proved that the dependency of the MW dielectric properties of rutile on the synthesis conditions, such as low oxygen partial pressure/high sintering temperature, results in a coring effect due to the reduction of Ti^{4+} to Ti^{3+} [13,14]. Noh et al. [15] sintered TiO_2 using spark plasma sintering and conventional sintering methods. The ε_r and $Q \times f_o$ of SPS-TiO_2 re-oxidized specimens were found to be 112.6 and 26,000, respectively. These properties were comparable to those samples sintered at 1300 °C using a solid-state route (ε_r = 101.3, $Q \times f_o$ = 41,600). Similarly, Weng et al. [16] sintered TiO_2 at 1040 °C/2 h and reported ε_r > 100, $Q \times f_o$ > 23,000 GHz, and τ_f = 200 ppm/°C. During sintering (in the air or under a low partial pressure of oxygen) at high temperatures, the generation of oxygen vacancies causes a coring effect, which significantly deteriorates the MW dielectric properties of TiO_2 [17]. Oxygen vacancies in dense rutile TiO_2 can be reduced by sintering it in an oxygen atmosphere or using a proper dopant. More recently, a tetravalent Hf^{4+} was doped in TiO_2 ($Hf_xTi_{1-x}O_2$), which exhibited ε_r = 17, $Q \times f_o$ = 84,020 GHz, and $\tau_f = -47$ ppm/°C (at x = 0.05) and ε_r = 37, $Q \times f_o$ = 53,600 GHz, and $\tau_f = -28$ ppm/°C (at x = 0.5) [18]. Mao et al. [19] studied the effect of Mn^{2+} and W^{6+} dopants in TiO_2, which improved the ε_r (~105) and $Q \times f_o$ (~39,000 GHz), but τ_f (~464.4 ppm/°C) further deteriorated. Similarly, other rutile TiO_2-based systems, such as $Zn_{0.15}Nb_{0.3}Ti_{0.55}O_2$ [20], Bi-doped TiO_2 [21], and $Ni_{0.5}Ti_{0.5}TaO_4$ [22] ceramics, exhibit a high τ_f. Similarly, Souza et al. [23] synthesized $TiO_2 - ZrO_2$ (x = 0, 0.1, and 0.2) and reported ε_r = 88, $Q \times f_o$ = 16,285 GHz for the pure TiO_2, ε_r = 85.4, $Q \times f_o$ = 9765 GHz for x = 0.1, and ε_r = 79.2, $Q \times f_o$ = 8883 GHz for x = 0.2; however, they did not report the temperature coefficient of the resonance frequency of these ceramics. Therefore, the purpose of the present work was to study the effect of the Zr^{4+} substation on the MW dielectric properties of rutile TiO_2, which exhibited a high τ_f value; however, the quality factor was improved due to a reduction in the coring effect. Furthermore, a structure–property relationship was developed for the unaffected τ_f in Zr-doped TiO_2.

2. Experimental Procedures

The $Ti_{1-x}Zr_xO_2$ (x = 0–0.1) ceramics were prepared through a conventional solid-state route using reagent-grade TiO_2 (99.9%) and ZrO_2 (99.9%) precursors. The precursors were dried at 700 °C for 6 h to remove moisture (if any). The precursors were weighed and then mixed/milled using a mortar and pestle for 1 h in acetone. The mixed powders were calcined at 1000 °C (x = 0) and 1200 °C (x > 0) for 4 h at a heating/cooling rate of 5 °C/min in air. The calcined powders were re-milled and then pressed uniaxially into 10 mm-in-diameter cylindrical pellets. The green pellets were sintered in the temperature range of 1300−1400 °C for 4 h at a heating/cooling rate of 5 °C/min in air.

The phase analysis of the samples was carried out using an X-ray diffractometer (Bruker D8 Advance, Germany) with Cu-Kα radiation. For Rietveld analysis, the GSAS + EXPGUI package was used [24,25]. For background fitting, the Shifted Chebyshev polynomial was used with 12 terms. To increase the level of accuracy, the damping factor was kept at 9. To study the vibrational modes, the Raman spectra of all the samples were collected at room temperature using a Micro Ramboss spectrometer with a laser source of 532 nm and a power of 80 mW, with a frequency of 150–800 cm^{-1}. Electron paramagnetic resonance (EPR) spectroscopy was carried out at room temperature using a Bruker BioSpin GMBH apparatus, Germany, with an X-band frequency of 9.843 GHz. The gyromagnetic factor (g) value was calculated by using the Equation (1),

$$g = \frac{h\upsilon_\circ}{\beta H} \tag{1}$$

where h is the Planck constant ($h = 6.626 \times 10^{-34}$ Js), $\upsilon_\circ$ is the MW frequency, β is the Bohr magneton ($\beta = 5.788 \times 10^{-15}$ eVT^{-1}), and H is the strength of the magnetic field. The microstructure and elemental distribution of the samples were examined via scanning electron microscopy with energy-dispersive X-ray analysis (SEM/EDX, Thermo Fisher Scientific, Waltham, MA, USA). The diffuse reflectance spectrum of the sample was obtained using a Jasco-V750 UV–vis spectrometer. The MW dielectric properties of the samples were measured using a network analyzer (Keysight, E5071C ENA, Santa Rosa, CA, USA) with the $TE_{01\delta}$ shielded cavity method [26]. The temperature coefficient of resonant frequency (τ_f) was calculated with the following formula:

$$\tau_f = \frac{f_1 - f_2}{f_1(T_1 - T_2)} \tag{2}$$

where f_1 and f_2 are the resonant frequencies at temperature T_1 and T_2, respectively.

3. Results and Discussions

Figure 1a shows the room-temperature XRD patterns of $Ti_{1-x}Zr_xO_2$ (x = 0–0.1) ceramics sintered at their optimum sintering temperatures. All diffraction peaks of the samples matched JCPDS #89−4920 for rutile TiO_2, having a tetragonal crystal structure with a space group *P4₂/mnm*. Figure 1b shows the Rietveld fitted and experimental data of $Ti_{1-x}Zr_xO_2$ (x = 0.025). Fitted plots for other samples are given in the Supplementary Information (Figure S1). The refined lattice parameters, cell volume, R factors (goodness of fitting or reliability factor), tetragonality (c/a), bond lengths (d), and theoretical density are given in Table 1. Due to the substitutionally doped Zr^{4+}, the lattice parameters and cell volume have been increased and the diffraction peaks were shifted to the lower 2θ values with an increase in x [27]. The shift in the diffraction peaks is attributed to the larger ionic radius of Zr^{4+} (r_{VI} = 0.72 Å) than Ti^{4+} (r_{VI} = 0.605 Å) [28,29]. A secondary phase (ZrO_2) was observed at x = 0.1, which suggests that the solubility limit of Zr^{4+} in TiO_2 is less than 0.10 [30,31]. The tetragonality ratio has been increased from 0.6445 to 0.6476 with the increase in the Zr concentration.

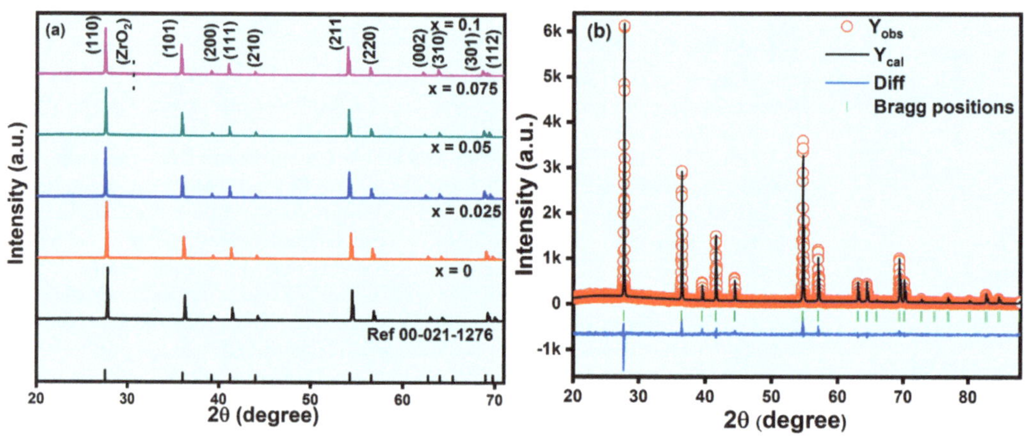

Figure 1. (**a**) XRD patterns of rutile $Ti_{1-x}Zr_xO_2$ (x = 0−0.1), and (**b**) fitted XRD patterns for x = 0.025.

Raman spectroscopy was carried out to confirm the rutile phase of $Ti_{1-x}Zr_xO_2$ (x = 0–0.1) ceramic samples. Rutile has 15 optical modes (i.e., $1A_{1g} + 1A_{2g} + 1A_{2u} + 1B_{1g} + 1B_{2g} + 2B_{1u} + 1E_g + 3E_u$) in which the E_g, B_{1g}, B_{2g}, and A_{1g} are Raman active [32,33]. Figure 2 shows the bands that are indicative of the rutile symmetry, which were observed at about 144 cm^{-1} (B_{1g}), 242 cm^{-1} (two phonon process), 441 cm^{-1} (E_g), and 606 cm^{-1} (A_{1g}) for all the samples. Furthermore, when the Zr concentrations increased, the E_g peak moved to a lower wavenumber, showing that the doping Zr ions enter the host lattice, which expands [33]. The 144 cm^{-1} (B_{1g}) peak is associated with the bending vibration of the O—Ti—O bond [27]. The peak at 242 cm^{-1} is the complex vibration peak due to the multi-phonon scattering (second-order effect), associated with the O—Ti—O bond's complex vibration on rutile TiO_2 [34]. Generally, the second-order effect was believed to be due to internal stress/strain or might be because of the Zr having a different ionic radius than that of the host [35]. The shift in the phonon peak towards the lower frequencies shows the reduction in the anharmonic behavior of the crystals, which shows an increase in crystal stability and enhancement of optical, dielectric, and thermal transport properties [36]. The E_g mode (441 cm^{-1}) represents the vibrational motion of oxygen atoms along the c-axis, whereas the A_{1g} mode (606 cm^{-1}) is related to the Ti−O stretching vibration [27]. When Ti^{4+} ions in the lattice were replaced by Zr^{4+}, the O^{2-} ions could be more closely bound, because the Zr^{4+} ions have a higher electro-positivity than Ti^{4+}. As a result of this, the vibrational motion of oxygen atoms along the c-axis is hindered, resulting in a decrease in the vibrational energy of oxygen atoms that corresponds to the E_g mode (441 cm^{-1}), as well as a decrease in frequency. On the other hand, the replacement of Zr^{4+} ions has no effect on the A_{1g} mode (606 cm^{-1}), matching the Ti−O stretching vibration energy. Therefore, the peak at 606 cm^{-1} is not shifted with an increase in the Zr^{4+} concentration [27].

Table 1. Rietveld refined lattice parameters, atomic positions, and refinement factors of $Ti_{1-x}Zr_xO_2$ (x = 0–0.1) ceramics.

x Values		0	0.025	0.05	0.075	0.1
a = b (Å)		4.5890 (7)	4.5923 (2)	4.6001 (5)	4.6061 (5)	4.6137 (9)
c (Å)		2.9579 (8)	2.9635 (7)	2.9732 (6)	2.9811 (1)	2.9882 (8)
α = β = γ (°)		90	90	90	90	90
c/*a*		0.6445	0.6453	0.6463	0.6471	0.6476
V_{cell} $(Å)^3$		62.29	62.45	62.92	63.26	63.61
R_{wp} (%)		11.36	14.12	12.06	11.98	12.97
R_p (%)		8.82	10.91	9.17	9.21	10.09
χ^2		1.49	2.14	1.66	1.71	1.96
Atomic positions (*x*, y, z)	Ti	(0, 0, 0)	(0, 0, 0)	(0, 0, 0)	(0, 0, 0)	(0, 0, 0)
	O	(0.3025, 0.3025, 0)	(0.3053, 0.3053, 0)	(0.3026, 0.3026, 0)	(0.3050, 0.3050, 0)	(0.2956, 0.2956, 0)
	Zr		(0, 0, 0)	(0, 0, 0)	(0, 0, 0)	(0, 0, 0)

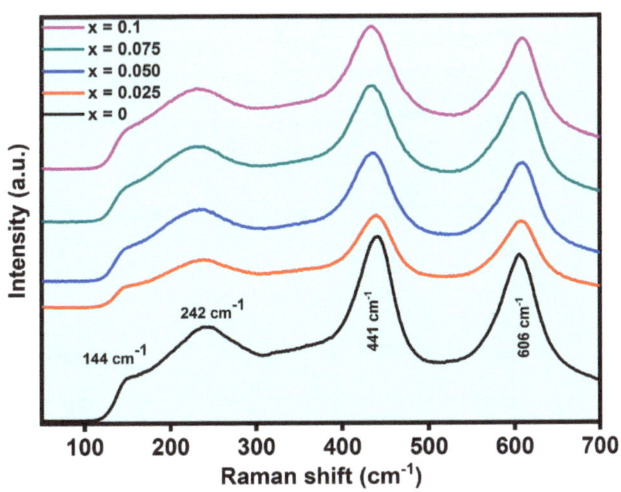

Figure 2. Raman spectra of rutile $Ti_{1-x}Zr_xO_2$ (x = 0–0.1).

TiO_2 belongs to a class of reducible oxides as it readily loses oxygen upon sintering at high temperatures in air with the formation of excess electrons and oxygen vacancies. The reduction in TiO_2 can be expressed using Kroger–Vink notations, in terms of either the formation of Ti^{4+} interstitials, oxygen vacancies, Ti^{3+} interstitials, or Ti^{3+} species and oxygen vacancies in octahedral sites given by the Equations (3)–(6), respectively [14].

$$Ti_{Ti}^{\times} + 2O_o^{\times} \Leftrightarrow Ti_i^{\bullet\bullet\bullet\bullet} + 4e' + O_2 \quad (3)$$

$$2O_o^{\times} \Leftrightarrow 2V_o^{\bullet\bullet} + 4e' + O_2 \quad (4)$$

$$Ti_{Ti}^{\times} + 2O_o^{\times} \Leftrightarrow Ti_i^{\bullet\bullet\bullet} + 3e' + O_2 \quad (5)$$

$$2Ti_{Ti}^{\times} + O_o^{\times} \Leftrightarrow 2Ti_{Ti}^{'} + V_o^{\bullet\bullet} + \frac{1}{2}O_2 \quad (6)$$

The excess electrons formed due to oxygen vacancies are usually stabilized at cations by forming Ti^{3+} as illustrated in Equation (5). This fact has been firmly confirmed, based on accurate theoretical calculations, that the electrons generated as a result of oxygen vacancies are preferentially stabilized on *d* orbitals of Ti [37] and were also shown by Chester [38] and Zimmermann [39]. For the detection of paramagnetic species, we have used EPR spectroscopy to investigate the bulk and surface defects in the rutile TiO_2. $Ti_{1-x}Zr_xO_2$ (x = 0–0.1) samples were examined via conventional continuous wave-EPR (CW-EPR) by homogeneously changing the magnetic field in a specific range until a resonance condition was achieved, which entails the transition between spin states. Figure 3 displays the room temperature EPR spectra of rutile $Ti_{1-x}Zr_xO_2$ (x = 0–0.1), where the signal belongs to the paramagnetic center Ti^{3+} formed due the generation of excess electrons due to oxygen vacancies, consistent with the previous result [40]. The line of the EPR signal for rutile TiO_2 recorded at room temperature tends to broaden and eventually disappear. This is because of the relaxation time of Ti^{3+}, which causes the weakening of the EPR signal's intensity at a temperature higher than 77 K [40]. With an increase in the Zr^{4+} concentration, the intensity of the signal decreases, which means the concentration of the paramagnetic center, implying that Ti^{3+} has been reduced in the sample. The corresponding g-value (1.97) belongs to the inner Ti^{3+} paramagnetic centers [41–43].

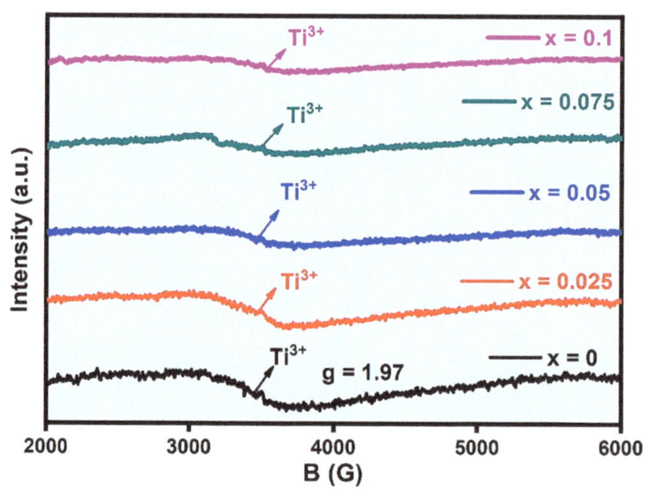

Figure 3. The room-temperature EPR spectra of rutile $Ti_{1-x}Zr_xO_2$ (x = 0–0.1).

The relative density of rutile $Ti_{1-x}Zr_xO_2$ (x = 0–0.1) with an increasing Zr concentration was measured. At an optimum sintering temperature, a relative density of ~93% was achieved. Figure 4a–e shows the SEM micrographs of sintered, thermally etched, and polished rutile $Ti_{1-x}Zr_xO_2$ (x = 0–0.1) samples, illustrating their morphologies and grain-size distributions. The SEM images show a well-developed grain morphology and an almost dense microstructure with some pores or voids. The grain size increased from 5 to 19 μm as the Zr concentration increased from 0 to 0.1 (Figure 4f). This is because Zr promotes the grain-growth [44]. The composition ($x = 0$, sintered at 1300 °C) exhibited a uniform distribution of grain size, while the doped samples sintered at high temperatures (1375−1400 °C) displayed a bimodal distribution, which is an indication of the Ostwald-ripening mechanism for the grain-growth [45]. During sintering, grain growth is dominated by the grain boundary and volume diffusion, which results in larger grains [11]. The pores/voids of the doped samples have been decreased as compared to the pure rutile TiO_2, but residual porosity has been observed in all the compositions. To further confirm the homogeneity in the samples, elemental mappings for $Ti_{1-x}Zr_xO_2$ ($x = 0.025$) are shown in Figure 4g–j, which are also supported by the EDX spectra, shown in Figure S2. At $x = 0.10$, the contrast of some small grains was different, consistent with the elemental mapping (Figure 5), which is in agreement with the XRD data (Figure 1), showing the formation of a secondary phase (ZrO_2) [30,31]. This confirmed the presence of Ti and O in the TiO_2 sample, while for the doped samples, Ti, O, and Zr were homogeneously distributed. However, for samples with $x = 0.10$, there are Zr-rich regions that show the formation of the secondary phase (Figure 5), consistent with the XRD and SEM.

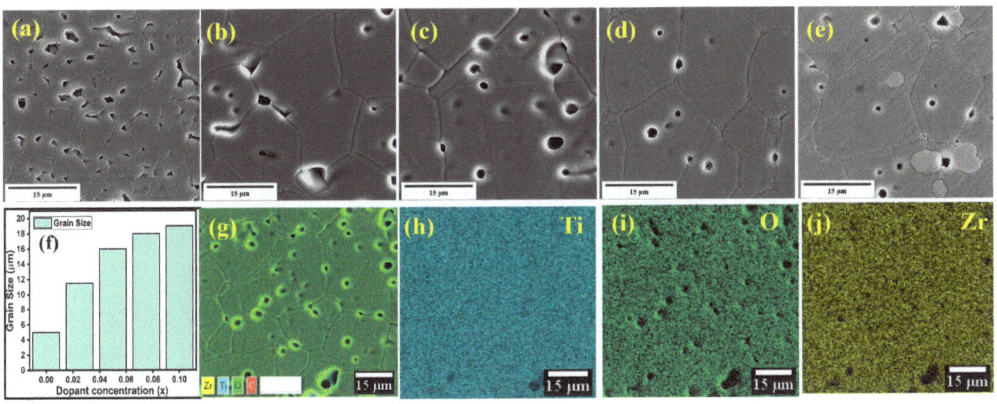

Figure 4. SEM micrographs of thermally etched and polished $Ti_{1-x}Zr_xO_2$ at (**a**) $x = 0$, (**b**) $x = 0.025$, (**c**) $x = 0.05$, (**d**) $x = 0.075$, (**e**) $x = 0.1$; (**f**) grain size distribution as a function of the Zr concentration; (**g**) elemental distribution and EDX mapping of (**h**) Ti, (**i**) O, and (**j**) Zr for a sample with $x = 0.025$.

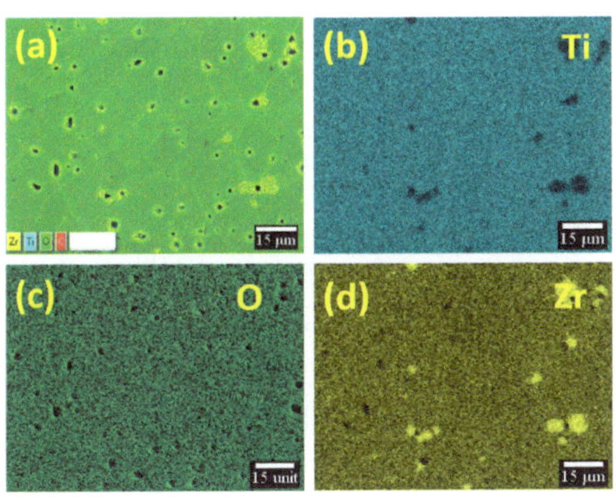

Figure 5. EDX mapping of (**a**) $Ti_{1-x}Zr_xO_2$ ($x = 0.1$), (**b**) Ti, (**c**) O, and (**d**) Zr.

Diffuse reflectance spectroscopy (DRS) studies were performed in the region 200–800 nm to evaluate the optical bandgap and the effect of Zr^{4+} doping on the bandgap of rutile TiO_2. Besides studying the possible transitions between conduction and the valence band, we can also find any other transition due to impurities with DRS as well. Figure 6a shows the DRS spectrum of $Ti_{1-x}Zr_xO_2$ (x = 0–0.1) samples, indicating a sharp decrease in reflectance at about 310 nm for the un-doped rutile sample, caused by strong absorption. After increasing the concentration of Zr^{4+}, the absorption edge shifted to a higher wavelength for $x \leq 0.075$, showing a decrease in the optical band gap energy. This absorption may be attributed to the charge transfer between O_{2p} and M_{3d} (M = Ti, Zr) [46]. The optical band gap (E_g) was calculated by solving the Kubelka-Munk equation (Equation (7)).

$$[F(R_\infty)hv] = \alpha(hv - E_g)^n \tag{7}$$

where $F(R_\infty)$ is the Kubelka–Munk function, hv is energy, E_g is band gap energy, and α is the proportionality constant. The power n describes the type of electronic transition and has a constant value, i.e., n = 0.5 for direct allowed and n = 2 for indirect allowed E_g. Furthermore, $n = 3/2$ and n = 3 are used for direct and indirect forbidden energy gaps, respectively [47,48]. The key parameters affecting the bandgap of materials are the crystallinity, processing technique, cell parameters, and average particle size [49]. The E_g values were calculated from the Tauc plot (Figure 6b). E_g for the x = 0 sample was estimated to be 2.92 eV, which decreased to 2.62 eV with the Zr concentration from x = 0 to 0.075 because Zr doping will modify the band structure and hence band gap. It has been speculated that the position of the conduction band has decreased while the position of the valence band has remained unchanged [46,50,51]. With a further increase in the Zr concentration (x = 0.1), the bandgap increased to 2.66 eV due to the presence of the ZrO_2 phase, which has a higher band gap value, i.e., 4.6−5.25 eV [52,53].

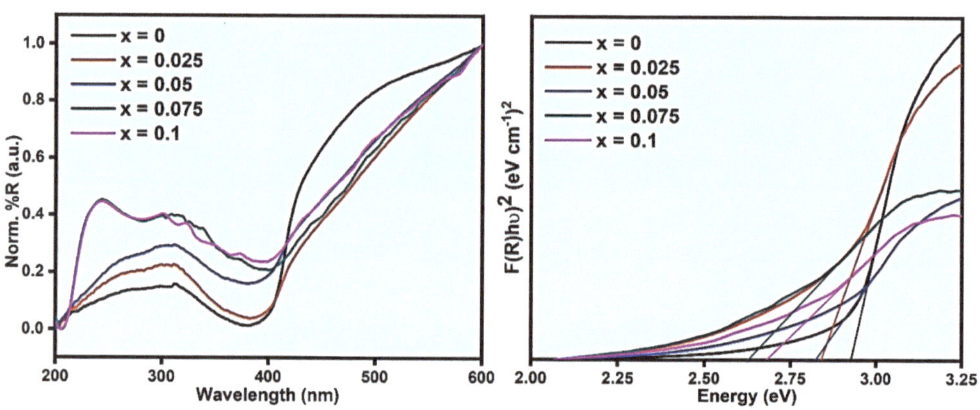

Figure 6. (**a**) Reflection spectra and (**b**) Tauc plots of $Ti_{1-x}Zr_xO_2$ (x = 0–0.1).

Table 2 shows the MW dielectric properties of the rutile $Ti_{1-x}Zr_xO_2$ (x = 0−0.1) ceramics. A high ε_r for rutile $Ti_{1-x}Zr_xO_2$ (96–103) was, observed which may be due to their characteristic extended apical bonds as compared to the equatorial bonds causing the rattling of cations in a rutile structure [20]. ε_r increased from 96 (x = 0) to 103 (x = 0.075) with an increase in the dopant (Zr) concentration and then decreased to 99 for sample x = 0.10. The enhancement of the ε_r could be related to the Zr substitution at the Ti lattice site [54]. This increase may also be attributed to the complex process of grain growth and densification of the ceramic [55,56]. Another crucial factor contributing to the enhancement of ε_r is the ionic polarizability. The ionic polarizability of Zr^{4+} (3.25 Å^3) is higher than that of Ti^{4+} (2.93 Å^3), which may be the phenomenon behind the increase in the ε_r [57]. Therefore, theoretical dielectric polarizability and ionic polarizabilities (α_{theo}) were calculated using Shannon's additivity rule [57] as follows:

$$\alpha_{theo} = (1-x)\alpha_{Ti^{4+}} + x\alpha_{Zr^{4+}} + 2\alpha_{O^{2-}} \quad (8)$$

where $\alpha_{Ti^{4+}}$, $\alpha_{Zr^{4+}}$, and $\alpha_{O^{2-}}$ are the ionic polarizabilities of Ti, Zr, and O, respectively. The increasing trend of the α_{theo} is in good agreement with the increasing trend of ε_r; therefore, it can be concluded that in rutile $Ti_{1-x}Zr_xO_2$, the change in ε_r arises from the ionic polarizability difference, as shown in Table 2. The non-linear behavior of ε_r for sample $x = 0.1$ is because of the presence of ZrO_2 as a secondary phase, which has an ε_r value of 15–40, which depends upon the frequency and crystal structure [58], leading to the decrease in the ε_r of the $x = 0.1$ sample. In addition, Ravichandran et al. [59] established a relationship between the ε_r and optical bandgap energy, i.e., the ε_r is inversely proportional to the E_g [60], which is consistent with our study. With a decrease in the E_g value, the ε_r values increased.

Table 2. MW dielectric properties and optical bandgap of $Ti_{1-x}Zr_xO_2$.

x	ST (°C)	Relative Density (%)	ε_r	α_{theo} (Å^3)	$Q \times f_o$ (GHz)	τ_f (ppm/°C)	Packing Fraction (%)	Bandgap Energy (eV)
0	1300	93	96	6.95	9500	417	70.6	2.92
0.025	1375	91	99	6.81	32,360	424	70.4	2.84
0.05	1375	89	101	6.96	29,710	432	70.0	2.81
0.075	1400	89	103	6.97	21,050	429	69.6	2.62
0.1	1400	--	99	6.98	18,800	453	69.3	2.68

Generally, the microstructures, densities, grain size, crystal lattice, packing fraction, secondary phases, and valency of Ti in the Ti-based ceramics affect the $Q \times f_o$ [61–63]. The $Q \times f_o$ initially increased from 9500 ($x = 0$) to 32,366 GHz (at $x = 0.025$) and then decreased to 18,798 GHz (at $x = 0.1$). The relative density of the samples showed a decreasing trend with an increase in the Zr concentration, which is consistent with the previous study [23]. At 1300 °C, TiO_2 has a density of ~93% but shows oxygen deficiency due to a low diffusion coefficient, making it unable to allow the oxygenation of the ceramic. However, oxygen diffused only to the surface of the ceramic because the core appears dark due to the presence of the reduced Ti^{3+} species, i.e., the bulk non-stoichiometry of oxygen is less, but it may be sufficiently large in the core to provoke the formation of condensation/shear structures of oxygen vacancies, just like that of Magnelli phases [17]. The reduction in TiO_2 increases the dielectric loss, leading to the deterioration of the $Q \times f_o$. In comparison, the doping of Zr^{4+} in TiO_2 has overcome the problem and enhanced the $Q \times f_o$ value. This phenomenon can be explained using the shift in the E_g band towards the lower wavenumber (Figure 2) with an increasing Zr concentration, which is attributed to the decrease in the oxygen vacancies in the ceramics [32]. Similarly, this fact is also confirmed using EPR spectroscopy, which showed a decrease in the concentration of the Ti^{3+} paramagnetic center. The $Q \times f_o$ of ceramics, having a rutile structure, also strongly depends on the c/a ratio [19,64], which agrees with the present findings (Table 1). The slight decrease in the $Q \times f_o$ from 32,366 ($x = 0.025$) to 18,798 GHz ($x = 0.1$) may be due to an increase in the number of voids (Figure 4). The $Q \times f_o$ and lattice vibrations of the rutile TiO_2 can also be influenced by the packing fraction (*PF*). The *PF* shows the density of atomic stacking and can be calculated using Equation (9) [65]:

$$\%PF = \left[\frac{volume\ of\ the\ packed\ ions}{volume\ of\ the\ unit\ cell}\right] \times Z \quad (9)$$

where Z shows the formula units per unit cell. As the $Q \times f_o$ and *PF* are directly proportional, the decrease in $Q \times f_o$ may also be attributed to the decrease in *PF* with an increase in the Zr concentration, as given in Table 2. At $x = 0.10$, the presence of the ZrO_2 phase may also be responsible for the decrease in the $Q \times f_o$ [23].

τ_f slightly increased with an increase in the Zr concentration. The mechanism responsible for the increase in the τ_f is mostly the increase in ε_r because the materials with high ε_r have high τ_f and vice versa [66]. This may also be attributed to the substitution of Zr^{4+}, a larger cation, for Ti^{4+}, a smaller cation, which distorts the octahedra, increasing the τ_f of the samples. Figure 7b shows the dependence of the τ_f on the unit cell volume of $Ti_{1-x}Zr_xO_2$ (x = 0–0.1) ceramics. τ_f is the combined effect of thermal expansion (α_L), as well as the temperature coefficient of permittivity (τ_ε), as shown in the Equation (10) [67]:

$$\tau_f = -\frac{1}{2}\tau_\varepsilon - \alpha_L \quad (10)$$

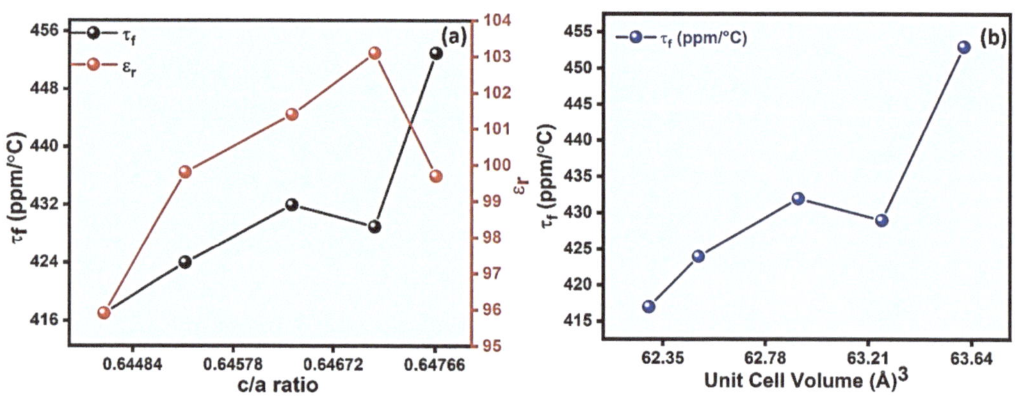

Figure 7. (**a**) Relationship between the c/a ratio and τ_f value and/or ε_r and (**b**) the variation in τ_f with the unit cell volume of $Ti_{1-x}Zr_xO_2$ (x = 0–0.1).

The α_L value for ceramics is constant and in the range of 10 ppm/°C; therefore, the magnitude and sign of the τ_f depend on τ_ε. According to the Clausius–Mossotti relation, τ_ε is given by Equations (11) and (12) [67].

$$\tau_\varepsilon = \frac{1}{\varepsilon}\left(\frac{\partial \varepsilon}{\partial T}\right) = \frac{(\varepsilon - 1)(\varepsilon + 2)}{\varepsilon}(A + B + C) \quad (11)$$

$$\left(A = -\frac{1}{3V}\left(\frac{\partial V}{\partial T}\right)_p, B = \frac{1}{3\alpha_m}\left(\frac{\partial \alpha_m}{\partial V}\right)_T\left(\frac{\partial V}{\partial T}\right)_p, C = \frac{1}{3\alpha_m}\left(\frac{\partial \alpha_m}{\partial T}\right)_V\right) \quad (12)$$

Here, α_m represents polarizability, while Havinga and Bosman [67,68] relate the terms A and B to the expansion of volume, but their effects cancel out each other due to their equal magnitude and opposite sign. As shown by Lee [67], the term C is the restoring force acting upon the ions and correlates with the potential well's shape, which shows its high dependence on the lattice parameters and structure. The τ_ε proportionally decreased with the increase in the volume of the unit cell, as with an increase in x, which in turn increased the τ_f.

The τ_f is also related to the structural characteristics of the oxygen octahedra, such as the bond strength (s), bond length, and degree of covalency of constituents [65,69]. These characteristics can be calculated based on the degree of covalency, which is related to covalency (f_c) and s using the following equations [70,71].

$$s = \left(\frac{R}{R_1}\right)^{-N} \quad (13)$$

$$f_c = a\left(\frac{R}{R_1}\right)^{-NM} \tag{14}$$

$$degree\ of\ covalency = \frac{a\left(\frac{R}{R_1}\right)^{-NM}}{\left(\frac{R}{R_1}\right)^{-N}} \times 100 = \frac{as^M}{s} \times 100 = \frac{f_c}{s} \times 100 \tag{15}$$

where R_1, N, a, and M are constants. The values of R_1 and N depend on the cation site, while a and M depend on the number of electrons. The values of these constants are already reported in Refs. [70–72]. R is the average bond length obtained from the Rietveld refinement results. The degree of the Ti/Zr−O bond covalency increases from 38.441 to 38.756% with an increase in the Zr concentration as given in Table 3. Therefore, an increase in τ_f values may be associated with the decrease in s, which is in good agreement with $Q \times f_o$, i.e., a decreasing trend from x = 0.025–0.1.

Table 3. Bond strength and covalency of $Ti_{1-x}Zr_xO_2$ (x = 0–0.1).

x	Bond Type	R (Å)	R_1	N	s	f_c	Covalency (%)	Degree of Covalency (%)
0	Ti−O	1.9601	1.806	5.2	0.6532	0.2511	38.441	38.441
0.025	Ti−O	1.9633	1.806	5.2	0.6477	0.2477	38.255	38.496
	Zr−O		1.950	6	0.9600	0.4596	47.874	
0.05	Ti−O	1.9664	1.806	5.2	0.6424	0.2446	38.077	38.554
	Zr−O		1.950	6	0.9509	0.4528	47.616	
0.075	Ti−O	1.9699	1.806	5.2	0.6365	0.2411	37.876	38.585
	Zr−O		1.950	6	0.9408	0.4453	47.327	
0.1	Ti−O	1.9710	1.806	5.2	0.6346	0.2400	37.814	38.756
	Zr−O		1.950	6	0.9377	0.4429	47.237	

The relationship among the relative permittivity, structure parameter, and τ_f value of rutile $Ti_{1-x}Zr_xO_2$ (x = 0–0.1) is also important. ε_r and τ_f, for rutile tetragonal ceramic, mainly dependent on the tetragonality (c/a ratio) [73] or octahedral distortion [74]. The distortion of each TiO_6 octahedra is affected by the c/a ratio and oxygen positional parameter (μ). The relationship between μ and c/a ratio is in the given (Equation (16)) [73].

$$\mu = \frac{2 - \left(4 - 2\left(1 - \left(\frac{d_e}{d_a}\right)^2\right)\left(\left(\frac{c}{a}\right)^2 + 2\right)\right)^{\frac{1}{2}}}{4\left(1 - \left(\frac{d_e}{d_a}\right)^2\right)} \tag{16}$$

where d_e and d_a are the lengths of four equatorial and two apical Ti—O bonds, respectively, and can be calculated from a, c, and μ [64]. In special cases, when $\mu = \mu^* = \frac{1}{4}\left[1 + \frac{1}{2}\left(\frac{c}{a}\right)^2\right]$, all the d_e and d_a bonds are equal. A perfect octahedron in a tetragonal structure would have an ideal $c/a \approx 0.586$ and $\mu_{ideal} = 1/2\ (c/a)_{ideal}$, giving all twelve O—O bonds equal lengths [64]. The degree of octahedral distortion can be indicated by the c/a ratio up to some extent [64]. Figure 7a shows the dependence of τ_f and ε_r on the c/a ratio of $Ti_{1-x}Zr_xO_2$ (x = 0–0.1) ceramics. It is observed that the c/a ratio of $Ti_{1-x}Zr_xO_2$ increased with an increase in the Zr concentration showing an increase in ε_r and τ_f. The c/a ratio for the $Ti_{1-x}Zr_xO_2$ (x = 0–0.1) ceramic was higher than the ideal c/a ratio (~0.586) and increased with an increase in x, which resulted in a high distortion of the TiO_6 octahedra. When the temperature increased, the rise in thermal energy was supposed to be absorbed completely in recovering the octahedral distortion, rather than in restoring the direct dependence of the polarizability on temperature [74], which means that the tetragonality ratio would change towards the ideal c/a ratio [64]. When the c/a ratio is closer to the $(c/a)_{ideal}$ (0.586), then there will be a slight change in c/a upon the increase in temperature from 25 to 85 °C. Therefore, an increase in the τ_f value for the $Ti_{1-x}Zr_xO_2$ solid solution is due to an increase in the c/a ratio.

4. Summary

In this study, $Ti_{1-x}Zr_xO_2$ ceramics were processed and their crystal structure, defect chemistry, microstructure, optical bandgap energy, and MW dielectric properties were investigated. Rutile TiO_2 crystallized into a tetragonal structure (*P4₂/mnm*). At $x = 0.1$, a secondary phase (ZrO_2) was observed, which shows that the solubility limit of Zr is below 0.10. The SEM micrographs showed an increase in the grain size from 5 μm to 19 μm with an increasing Zr concentration. The electron paramagnetic resonance spectroscopy showed the existence of Ti^{3+} species in the rutile sample, which reduces with the increase in the Zr^{4+} concentration. The Tauc plots showed a decreasing trend in the bandgap energies of $Ti_{1-x}Zr_xO_2$ (x = 0–0.075), i.e., 2.92−2.62 eV. The relative permittivity slightly increased, due to the extended apical bonds, as compared to the equatorial bonds causing the rattling of cations in the rutile structure. The quality factor initially increased with an increase in the Zr concentration (i.e., $x = 0.025$), which was explained based on a decrease in the coring effect and hence a decrease in oxygen vacancies. Above $x = 0.025$, the quality factor started decreasing due to a decrease in the packing fraction. Similarly, the temperature coefficient of the resonance frequency increased with an increase in x, due to an increase in the tetragonality (c/a ratio), which is indirectly associated with the distortion of the octahedra. Similarly, the degree of covalency also decreased, which is also a reason for the increase in τ_f.

Supplementary Materials: The following supporting information can be downloaded at: https://www.mdpi.com/article/10.3390/ceramics7020046/s1, Figure S1: Rietveld refined and experimental pattern of rutile $Ti_{1-x}Zr_xO_2$ (a) $x = 0$, (b) $x = 0.05$, (c) $x = 0.075$, (d) $x = 0.1$; Figure S2: Energy dispersive X-Ray analysis of $Ti_{1-x}Zr_xO_2$ (a) $x = 0$, (b) $x = 0.025$, (c) $x = 0.05$, (d) $x = 0.075$, (e) $x = 0.1$.

Author Contributions: Conceptualization, R.M.; methodology, I.K. and A.K.; formal analysis, D.H., W.L. and D.W.; investigation, I.K. and A.K; resources, R.M., D.H., K.S., W.L. and D.W.; writing—original draft preparation, I.K.; writing—review and editing, K.S., M.M., D.W., W.L., R.M. and I.K. All authors have read and agreed to the published version of the manuscript.

Funding: The authors acknowledge the financial support extended by the Directorate General of Science and Technology, Khyber Pakhtunkhwa, Pakistan through Project No. DGST/BSTIET/MS/2022/8773.

Institutional Review Board Statement: Not applicable.

Informed Consent Statement: Not applicable.

Data Availability Statement: The data presented in this study are available in this article and Supplementary Information.

Conflicts of Interest: The authors declare no conflicts of interest.

References

1. Balunarayanan, R.; Aiswarya, S.; Sreekala, C.; Menon, S.K. Nano Cylindrical Dielectric Resonator Antenna Using Titanium dioxide for Wi-Fi Applications. Proceddings of the 2021 Fourth International Conference on Electrical, Computer and Communication Technologies (ICECCT), Coimbatore, India, 15–17 September 2021; IEEE: Piscataway, NJ, USA, 2021; pp. 1–6.
2. Cohn, S.B. Microwave bandpass filters containing high-Q dielectric resonators. *IEEE Trans. Microw. Theory Tech.* **1968**, *16*, 218–227. [CrossRef]
3. Kanehara, K.; Hoshina, T.; Takeda, H.; Tsurumi, T. Terahertz permittivity of rutile TiO_2 single crystal measured by anisotropic far-infrared ellipsometry. *J. Ceram. Soc. Jpn.* **2015**, *123*, 303–306. [CrossRef]
4. Mett, R.R.; Sidabras, J.W.; Anderson, J.R.; Klug, C.S.; Hyde, J.S. Rutile dielectric loop-gap resonator for X-band EPR spectroscopy of small aqueous samples. *J. Magn. Reson.* **2019**, *307*, 106585. [CrossRef] [PubMed]
5. Hanaor, D.A.; Sorrell, C.C. Review of the anatase to rutile phase transformation. *J. Mater. Sci.* **2011**, *46*, 855–874. [CrossRef]

6. Thongyong, N.; Chanlek, N.; Srepusharawoot, P.; Thongbai, P. Origins of giant dielectric properties with low loss tangent in rutile $(Mg_{1/3}Ta_{2/3})_{0.01}Ti_{0.99}O_2$ ceramic. *Molecules* **2021**, *26*, 6952. [CrossRef]
7. Thanamoon, N.; Chanlek, N.; Moontragoon, P.; Srepusharawoot, P.; Thongbai, P. Microstructure, low loss tangent, and excellent temperature stability of Tb + Sb-doped TiO_2 with high dielectric permittivity. *Results Phys.* **2022**, *37*, 105536. [CrossRef]
8. Peng, P.; Chen, C.; Cui, B.; Li, J.; Xu, D.; Tang, B. Influence of the electric field on flash-sintered (Zr + Ta) co-doped TiO_2 colossal permittivity ceramics. *Ceram. Int.* **2022**, *48*, 6016–6023. [CrossRef]
9. Thanamoon, N.; Chanlek, N.; Srepusharawoot, P.; Swatsitang, E.; Thongbai, P. Microstructural Evolution and High-Performance Giant Dielectric Properties of Lu^{3+}/Nb^{5+} Co-Doped TiO_2 Ceramics. *Molecules* **2021**, *26*, 7041. [CrossRef]
10. Yang, C.; Wei, X.; Hao, J. Disappearance and recovery of colossal permittivity in (Nb + Mn) co-doped TiO_2. *Ceram. Int.* **2018**, *44*, 12395–12400. [CrossRef]
11. Hsu, T.-H.; Huang, C.-L. Microwave dielectric properties of ultra-low-temperature-sintered TiO_2 as a τ f compensator. *Appl. Phys. A* **2023**, *129*, 20. [CrossRef]
12. Zhang, J.; Yue, Z.; Zhou, Y.; Peng, B.; Zhang, X.; Li, L. Temperature-dependent dielectric properties, thermally-stimulated relaxations and defect-property correlations of TiO_2 ceramics for wireless passive temperature sensing. *J. Eur. Ceram. Soc.* **2016**, *36*, 1923–1930. [CrossRef]
13. Grabstanowicz, L.R.; Gao, S.; Li, T.; Rickard, R.M.; Rajh, T.; Liu, D.-J.; Xu, T. Facile oxidative conversion of TiH_2 to high-concentration Ti^{3+}-self-doped rutile TiO_2 with visible-light photoactivity. *Inorg. Chem.* **2013**, *52*, 3884–3890. [CrossRef] [PubMed]
14. Templeton, A.; Wang, X.; Penn, S.J.; Webb, S.J.; Cohen, L.F.; Alford, N.M. Microwave dielectric loss of titanium oxide. *J. Am. Ceram. Soc.* **2000**, *83*, 95–100. [CrossRef]
15. Noh, J.H.; Jung, H.S.; Lee, J.-K.; Kim, J.-R.; Hong, K.S. Microwave dielectric properties of nanocrystalline TiO_2 prepared using spark plasma sintering. *J. Eur. Ceram. Soc.* **2007**, *27*, 2937–2940. [CrossRef]
16. Weng, Z.; Wu, C.; Xiong, Z.; Feng, Y.; AminiRastabi, H.; Song, C.; Xue, H. Low temperature sintering and microwave dielectric properties of TiO_2 ceramics. *J. Eur. Ceram. Soc.* **2017**, *37*, 4667–4672. [CrossRef]
17. Pullar, R.C.; Penn, S.J.; Wang, X.; Reaney, I.M.; Alford, N.M. Dielectric loss caused by oxygen vacancies in titania ceramics. *J. Eur. Ceram. Soc.* **2009**, *29*, 419–424. [CrossRef]
18. Ding, Y.H.; Liu, L.; Yang, Z.J.; Li, L.; Chen, X.M. Structure and microwave dielectric characteristics of $Hf_{1-x}Ti_xO_2$ ceramics. *J. Am. Ceram. Soc.* **2022**, *105*, 1127–1135. [CrossRef]
19. Mao, S.; Yang, J.; Gong, M.; Ao, L.; Fang, Z.; Kashif, K.; Qu, C.; Zhang, X.; Yang, H.; Xiong, Z.; et al. The improved microwave dielectric characteristics of TiO_2 ceramics produced by Mn^{2+} and W^{6+} co-substitution. *J. Mater. Sci. Mater. Electron.* **2022**, *33*, 27041–27052. [CrossRef]
20. Kim, E.S.; Kang, D.H. Relationships between crystal structure and microwave dielectric properties of $(Zn_{1/3}B_{2/3}{}^{5+})_xTi_{1-x}O_2$ (B^{5+} = Nb, Ta) ceramics. *Ceram. Int.* **2008**, *34*, 883–888. [CrossRef]
21. Fukuda, K.; Awai, R.K. Microwave characteristics of TiO_2-Bi_2O_3 dielectric resonator. *Jpn. J. Appl. Phys.* **1993**, *32*, 4584. [CrossRef]
22. Wu, X.; Jing, Y.; Li, Y.; Su, H. Novel Tri-rutile $Ni_{0.5}Ti_{0.5}TaO_4$ Microwave Dielectric Ceramics: Crystal Structure Chemistry, Raman Vibration Mode, and Chemical Bond Characteristic In-Depth Studies. *J. Phys. Chem. C* **2022**, *126*, 14680–14692. [CrossRef]
23. Souza, J.V.C.; Castro, P.J.; Nono, M.d.C.d.A.; Mineiro, S.L. Microstructure, crystalline phase, and dielectric property analyses of TiO_2 composition with ZrO_2 addition. *Mater. Sci. Forum.* **2010**, *660*, 641–645. [CrossRef]
24. Larson, A.C.; Von Dreele, R.B. *General Structure Analysis System*; Report LAUR 86-748; Los Alamos National Laboratory: Los Alamos, NM, USA, 2004.
25. Toby, B.H. EXPGUI, a graphical user interface for GSAS. *J. Appl. Crystallogr.* **2001**, *34*, 210–213. [CrossRef]
26. Krupka, J. Frequency domain complex permittivity measurements at microwave frequencies. *Meas. Sci. Technol.* **2006**, *17*, R55. [CrossRef]
27. Wang, J.; Yu, Y.; Li, S.; Guo, L.; Wang, E.; Cao, Y. Doping behavior of Zr^{4+} ions in Zr^{4+}-doped TiO_2 nanoparticles. *J. Phys. Chem. C* **2013**, *117*, 27120–27126. [CrossRef]
28. Chang, S.-M.; Doong, R.-A. Characterization of Zr-doped TiO_2 nanocrystals prepared by a nonhydrolytic sol−gel method at high temperatures. *J. Phys. Chem. B.* **2006**, *110*, 20808–20814. [CrossRef] [PubMed]
29. Shannon, R.D. Revised effective ionic radii and systematic studies of interatomic distances in halides and chalcogenides. *Acta Crystallogr. Sect. A Cryst. Phys. Diffr. Theor. Gen. Crystallogr.* **1976**, *32*, 751–767. [CrossRef]
30. Gnatyuk, Y.; Smirnova, N.; Korduban, O.; Eremenko, A. Effect of zirconium incorporation on the stabilization of TiO_2 mesoporous structure. *Surf. Interface Anal.* **2010**, *42*, 1276–1280. [CrossRef]
31. Yu, J.C.; Lin, J.; Kwok, R.W. $Ti_{1-x}Zr_xO_2$ Solid Solutions for the Photocatalytic Degradation of Acetone in Air. *J. Phys. Chem. B.* **1998**, *102*, 5094–5098. [CrossRef]
32. Swamy, V.; Muddle, B.C.; Dai, Q. Size-dependent modifications of the Raman spectrum of rutile TiO_2. *Appl. Phys. Lett.* **2006**, *89*, 163118. [CrossRef]
33. Zhang, Y.; Harris, C.X.; Wallenmeyer, P.; Murowchick, J.; Chen, X. Asymmetric lattice vibrational characteristics of rutile TiO_2 as revealed by laser power dependent Raman spectroscopy. *J. Phys. Chem. C* **2013**, *117*, 24015–24022. [CrossRef]
34. Huang, W.; Zhang, Y.; Lu, J.; Chen, J.; Gao, L.; Omran, M.; Chen, G. Microwave drying method investigation for the process and kinetics of drying characteristics of high-grade rutile TiO_2. *Ceram. Int.* **2023**, *49*, 15618–15628. [CrossRef]

35. Wang, Z.; Chen, H.; Wang, T.; Xiao, Y.; Nian, W.; Fan, J. Enhanced relative permittivity in niobium and europium co-doped TiO_2 ceramics. *J. Eur. Ceram. Soc.* **2018**, *38*, 3847–3852. [CrossRef]
36. Yu, C. *Dielectric Materials for High Power Energy Storage*; Queen Mary University of London: London, UK, 2017.
37. Di Valentin, C.; Pacchioni, G.; Selloni, A. Reduced and n-type doped TiO_2: Nature of Ti^{3+} species. *J. Phys. Chem. C* **2009**, *113*, 20543–20552. [CrossRef]
38. Chester, P. Cross-Doping Agents for Rutile Masers. *J. Appl. Phys.* **1961**, *32*, 866–868. [CrossRef]
39. Kiwi, J.; Suss, J.; Szapiro, S. EPR spectra of niobium-doped TiO_2 and imiplications for water photocleavage processes. *Chem. Phy. Lett.* **1984**, *106*, 135–138. [CrossRef]
40. Chiesa, M.; Paganini, M.C.; Livraghi, S.; Giamello, E. Charge trapping in TiO_2 polymorphs as seen by Electron Paramagnetic Resonance spectroscopy. *Phys. Chem. Chem. Phys.* **2013**, *15*, 9435–9447. [CrossRef] [PubMed]
41. Fresno, F.; Hernández-Alonso, M.D.; Tudela, D.; Coronado, J.M.; Soria, J. Photocatalytic degradation of toluene over doped and coupled (Ti, M) O_2 (M = Sn or Zr) nanocrystalline oxides: Influence of the heteroatom distribution on deactivation. *Appl. Catal. B* **2008**, *84*, 598–606. [CrossRef]
42. Kumar, C.P.; Gopal, N.O.; Wang, T.C.; Wong, M.-S.; Ke, S.C. EPR investigation of TiO_2 nanoparticles with temperature-dependent properties. *J. Phys. Chem. B* **2006**, *110*, 5223–5229. [CrossRef]
43. Livraghi, S.; Maurelli, S.; Paganini, M.C.; Chiesa, M.; Giamello, E. Probing the local environment of Ti^{3+} ions in TiO_2 (rutile) by ^{17}O HYSCORE. *Angew. Chem. Int. Ed.* **2011**, *50*, 8038–8040. [CrossRef]
44. Zhong, M.; Li, J.; Shao, J.; Cao, Y.; Li, K.; Zhao, W. An investigation into the enhanced permittivity properties of Zr co-doped $(Ga_{0.5}Nb_{0.5})_{0.03}Ti_{0.97}O_2$ ceramics. *Ceram. Int.* **2019**, *45*, 14983–14990. [CrossRef]
45. Chao, S.; Petrovsky, V.; Dogan, F. Effects of sintering temperature on the microstructure and dielectric properties of titanium dioxide ceramics. *J. Mater. Sci.* **2010**, *45*, 6685–6693. [CrossRef]
46. Zhang, J.; Li, L.; Liu, D.; Zhang, J.; Hao, Y.; Zhang, W. Multi-layer and open three-dimensionally ordered macroporous TiO_2–ZrO_2 composite: Diversified design and the comparison of multiple mode photocatalytic performance. *Mater. Des.* **2015**, *86*, 818–828. [CrossRef]
47. Khan, R.U.; Khan, I.; Ali, B.; Muhammad, R.; Samad, A.; Shah, A.; Song, K.; Wang, D. Structural, dielectric, optical, and electrochemical performance of $Li_4Mo_5O_{17}$ for ULTCC applications. *Mater. Res. Bull.* **2023**, *160*, 112142. [CrossRef]
48. Lee, S.; Woodford, W.H.; Randall, C.A. Band gap energy of perovskite structured ABO_3 compounds. In Proceedings of the 2008 17th IEEE International Symposium on the Applications of Ferroelectrics, Santa Fe, NM, USA, 24–27 February 2008; Volume 1, p. 1.
49. Yadav, P.; Sinha, E. Structural, photophysical and microwave dielectric properties of α-$ZnMoO_4$ phosphor. *J. Alloys Compd.* **2019**, *795*, 446–452. [CrossRef]
50. Du, F.; Yu, S. Preparation and Characterization of Zr–N-Codoped TiO_2 Nano-Photocatalyst and Its Activity Enhanced-Mechanism. *J. Nanosci. Nanotechnol.* **2014**, *14*, 6965–6969. [CrossRef] [PubMed]
51. Duan, B.; Zhou, Y.; Huang, C.; Huang, Q.; Chen, Y.; Xu, H.; Shen, S. Impact of Zr-doped TiO_2 photocatalyst on formaldehyde degradation by Na addition. *Ind. Eng. Chem. Res.* **2018**, *57*, 14044–14051. [CrossRef]
52. Hadi, E.H.; Sabur, D.A.; Chiad, S.S.; Fadhil, N. Physical properties of nanostructured li-doped ZrO_2 thin films. *J. Green Eng.* **2020**, *10*, 8390–8400.
53. Tomar, L.J.; Chakrabarty, B. Synthesis, structural and optical properties of TiO_2-ZrO_2 nanocomposite by hydrothermal method. *Adv. Mater. Lett.* **2013**, *4*, 64–67. [CrossRef]
54. Juma, A.; Acik, I.O.; Oluwabi, A.; Mere, A.; Mikli, V.; Danilson, M.; Krunks, M. Zirconium doped TiO_2 thin films deposited by chemical spray pyrolysis. *Appl. Surf. Sci.* **2016**, *387*, 539–545. [CrossRef]
55. Padmamalini, N.; Ambujam, K. Structural and dielectric properties of ZrO_2–TiO_2–V_2O_5 nanocomposite prepared by CO-precipitation calcination method. *Mater. Sci. Semicond.* **2016**, *41*, 246–251. [CrossRef]
56. Zhao, E.; Hao, J.; Xue, X.; Si, M.; Guo, J.; Wang, H. Rutile TiO_2 microwave dielectric ceramics prepared via cold sintering assisted two step sintering. *J. Eur. Ceram. Soc.* **2021**, *41*, 3459–3465. [CrossRef]
57. Shannon, R.D. Dielectric polarizabilities of ions in oxides and fluorides. *J. Appl. Phys.* **1993**, *73*, 348–366. [CrossRef]
58. Oh, Y.; Bharambe, V.; Mummareddy, B.; Martin, J.; McKnight, J.; Abraham, M.A.; Walker, J.M.; Rogers, K.; Conner, B.; Cortes, P.; et al. Microwave dielectric properties of zirconia fabricated using NanoParticle Jetting™. *Addit. Manuf.* **2019**, *27*, 586–594. [CrossRef]
59. Ravichandran, R.; Wang, A.X.; Wager, J.F. Solid state dielectric screening versus band gap trends and implications. *Opt. Mater.* **2016**, *60*, 181–187. [CrossRef]
60. Hervé, P.; Vandamme, L. General relation between refractive index and energy gap in semiconductors. *Infrared Phy. Techn.* **1994**, *35*, 609–615. [CrossRef]
61. Fang, Z.-X.; Tang, B.; Li, E.; Zhang, S.-R. High-Q microwave dielectric properties in the $Na_{0.5}Sm_{0.5}TiO_3$+Cr_2O_3 ceramics by one synthetic process. *J. Alloys Compd.* **2017**, *705*, 456–461. [CrossRef]
62. Fang, Z.; Yang, H.; Yang, H.; Xiong, Z.; Zhang, X.; Zhao, P.; Tang, B. Ilmenite-type $MgTiO_3$ ceramics by complex $(Mn_{1/2}W_{1/2})^{4+}$ cation co-substitution producing improved microwave characteristics. *Ceram. Int.* **2021**, *47*, 21388–21397. [CrossRef]
63. Yu, S.; Tang, B.; Zhang, S.; Zhou, X. The effect of Mn addition on phase development, microstructure and microwave dielectric properties of $ZrTi_2O_6$–$ZnNb_2O_6$ ceramics. *Mater. Lett.* **2012**, *80*, 124–126. [CrossRef]

64. Pang, L.X.; Wang, H.; Zhou, D.; Yao, X. Sintering behavior, structures, and microwave dielectric properties of $(Li_xNb_{3x})Ti_{1-4x}O_2$. *J. Am. Ceram. Soc.* **2008**, *91*, 2947–2951. [CrossRef]
65. Kim, E.S.; Chun, B.S.; Freer, R.; Cernik, R.J. Effects of packing fraction and bond valence on microwave dielectric properties of A^{2+} B^{6+} O_4 (A^{2+}: Ca, Pb, Ba; B^{6+}: Mo, W) ceramics. *J. Eur. Ceram. Soc.* **2010**, *30*, 1731–1736. [CrossRef]
66. Reaney, I.M.; Iddles, D. Microwave dielectric ceramics for resonators and filters in mobile phone networks. *J. Am. Ceram. Soc.* **2006**, *89*, 2063–2072. [CrossRef]
67. Lee, H.-J.; Hong, K.-S.; Kim, S.-J.; Kim, I.-T. Dielectric Properties of MNb_2O_6 Compounds (Where M = Ca, Mn, Co, Ni, or Zn). *Mater. Res. Bull.* **1997**, *32*, 847–855. [CrossRef]
68. Bosman, A.J.; Havinga, E.E. Temperature dependence of dielectric constants of cubic ionic compounds. *Phys. Rev.* **1963**, *129*, 1593. [CrossRef]
69. Yang, H.; Zhang, S.; Yang, H.; Chen, Y.; Yuan, Y.; Li, E. Effects of ZrO2 substitution on crystal structure and microwave dielectric properties of $Zn_{0.15}Nb_{0.3}(Ti_{1-x}Zr_x)_{0.55}O_2$ ceramics. *Ceram. Int.* **2018**, *44*, 22710–22717. [CrossRef]
70. Brown, I.t.; Shannon, R. Empirical bond-strength–bond-length curves for oxides. *Acta Crystallogr. Sect. A Cryst. Phys. Diffr. Theor. Gen. Crystallogr.* **1973**, *29*, 266–282. [CrossRef]
71. Brown, I.t.; Wu, K.K. Empirical parameters for calculating cation–oxygen bond valences. *Acta Crystallogr. Sect. B Struct. Crystallogr. Cryst. Chem.* **1976**, *32*, 1957–1959. [CrossRef]
72. Ding, Y.H.; Wang, X.; Guo, R.Z.; Li, L.; Chen, X.M. Improvement of τf in $HfTiO_4$ microwave dielectric ceramics with Zr-and Sn-substitution. *J. Am. Ceram. Soc.* **2024**, *107*, 2337–2347. [CrossRef]
73. Choi, J.W.; Van Dover, R. Correlation between temperature coefficient of resonant frequency and tetragonality ratio. *J. Am. Ceram. Soc.* **2006**, *89*, 1144–1146. [CrossRef]
74. Kim, E.S.; Kang, D.H.; Kim, S.J. Effect of crystal structure on microwave dielectric properties of $(Ni_{1/3}B_{2/3})_{1-x}Ti_xO_2$ (B = Nb and Ta). *Jpn. J. Appl. Phys.* **2007**, *46*, 7101. [CrossRef]

ceramics

MDPI

Review

Lead-Free $NaNbO_3$-Based Ceramics for Electrostatic Energy Storage Capacitors

Sairatun Nesa Soheli [1], Zhilun Lu [2,*], Dongyang Sun [1] and Islam Shyha [1]

1 School of Computing, Engineering and Built Environment, Edinburgh Napier University, 10 Colinton Road, Edinburgh EH10 5DT, UK; sairatunnesa.soheli@napier.ac.uk (S.N.S.); d.sun@napier.ac.uk (D.S.); i.shyha@napier.ac.uk (I.S.)
2 School of Chemical and Process Engineering, University of Leeds, Leeds LS2 9JT, UK
* Correspondence: z.lu@leeds.ac.uk

Abstract: The burgeoning significance of antiferroelectric (AFE) materials, particularly as viable candidates for electrostatic energy storage capacitors in power electronics, has sparked substantial interest. Among these, lead-free sodium niobate ($NaNbO_3$) AFE materials are emerging as eco-friendly and promising alternatives to lead-based materials, which pose risks to human health and the environment, attributed to their superior recoverable energy density and dielectric breakdown strength. This review offers an insightful overview of the fundamental principles underlying antiferroelectricity and the applications of AFE materials. It underscores the recent advancements in lead-free $NaNbO_3$-based materials, focusing on their crystal structures, phase transitions, and innovative strategies devised to tailor their electrostatic energy storage performance. Finally, this review delineates the prevailing challenges and envisages future directions in the realm of $NaNbO_3$-based electrostatic energy storage capacitors, with the goal of fostering further advancements in this pivotal field.

Keywords: sodium niobate; antiferroelectric; lead-free material; energy storage capacitor

Citation: Soheli, S.N.; Lu, Z.; Sun, D.; Shyha, I. Lead-Free $NaNbO_3$-Based Ceramics for Electrostatic Energy Storage Capacitors. *Ceramics* **2024**, *7*, 712–734. https://doi.org/10.3390/ceramics7020047

Academic Editor: John W. Halloran

Received: 14 April 2024
Revised: 18 May 2024
Accepted: 21 May 2024
Published: 23 May 2024

1. Introduction

In recent years, energy storage has emerged as a topic of significant interest because of the rapid advancement of technologically complex electronic devices and electrical power systems [1–3]. Electrostatic capacitors, electrochemical capacitors, batteries, and solid oxide fuel cells (SOFCs) are most of the technologies that can store energy, as shown in Figure 1. Electrostatic capacitors, also known as dielectric capacitors, offer many advantages over electrochemical capacitors, batteries, and SOFCs, including swift charging–discharging rates, ultrahigh power density, and excellent thermal stability, though they have a lower energy density. Among capacitors, electrochemical capacitors (e.g., supercapacitors) generally have a higher energy density compared to dielectric capacitors, primarily due to the significantly increased surface area of the electrodes. However, they also have a lower operating voltage, typically less than 3 V, exhibit significant leakage current (in mA), and are more expensive, costing around 9500 USD/kWh [4–7]. Therefore, dielectric capacitors are more suitable for low-cost, high-voltage, and large-scale applications [8].

When it involves analysing power and energy densities for energy storage devices (ESDs), the Ragone plot is an indispensable instrument. According to the evidence shown in Figure 1, it is not possible for a single ESD to concurrently possess both a high energy density and a high power density. It is possible to determine the application of each ESD based on its own characteristic time, which is determined by the energy-to-power ratio or charge/discharge rate of the device, as shown by the straight dashed lines in Figure 1 [9]. It is essential to remain conscious of the charge or discharge time in which external factors, such as the resistance of the load, might influence ESD encounters. As energy is stored in a capacitor by the displacement of bound charged elements, capacitors have a higher charge or discharge rate and power density than batteries and SOFCs [10].

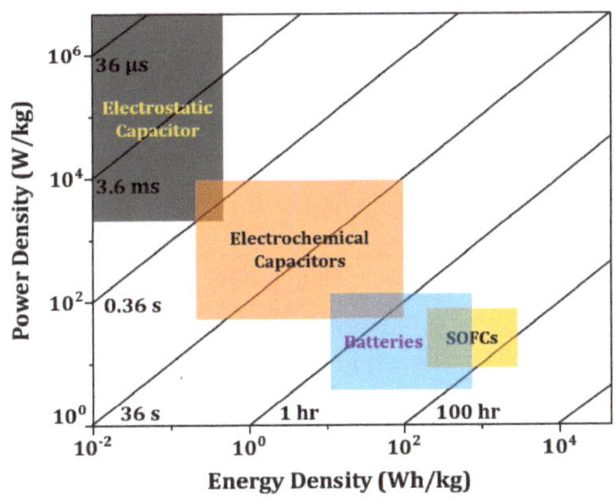

Figure 1. Ragone plot of various energy storage devices with characteristic times represented by straight dashed lines.

Dielectric capacitors are particularly useful for high-power/pulse-power systems due to their high charge or discharge rate, as indicated by their short characteristic time in Figure 1. This feature is also significant for effectively capturing energy from evanescent and intermittent renewable sources [11]. On the other hand, dielectric capacitors with enhanced energy densities offer significant benefits for a range of applications, including consumer electronics, commercial defibrillators, and pulsed power applications, due to their downsizing potential and easy integration [5,9,11–15]. Furthermore, increasing the energy density of dielectric capacitors to levels comparable with electrochemical capacitors or batteries [5] could significantly expand their use in energy storage, opening up a broader spectrum of applications.

Ceramic-based dielectric capacitors possess a rapid charge/discharge cycle and a high power density because of their ability to store energy via dipole moments as opposed to chemical reactions [10,16]. In addition, ceramics exhibit commendable mechanical properties and stability. These characteristics enable them to become fundamental elements of intermittent power systems. However, the application areas of ceramic capacitors are significantly limited because their integration and miniaturisation are impeded by their comparatively low recoverable energy storage density (W_{rec}) and energy storage efficiency (η). Therefore, the development of energy storage ceramics with superior efficacy is critical. Ceramic capacitors with high permittivity dielectrics are ideal for storing more energy due to their superior volumetric efficiency. These capacitors typically use ferroelectric (FE) materials, characterised by a spontaneous electric polarisation that can be modified with an external electric field [17]. In contrast, antiferroelectric (AFE) materials, while similar to FE materials, consist of adjacent dipoles in antiparallel orientations, leading to no net spontaneous polarisation [18,19]. AFE phases can undergo field-induced transitions to polar FE phases, leading to high polarisation at a high electric field. This opens up potential applications in energy storage, provided that the dielectric breakdown polarisation is high enough to induce the AFE-FE phase transition.

However, compared to FEs, AFEs are less commonly used due to the limited variety of AFE compounds and their less diverse functionalities [20,21]. Extensive research has been conducted on lead-based AFEs over the past decade, owing to their promising properties. Additionally, the phase stability of these materials can be easily adjusted through chemical substitution. The orthorhombic symmetry of the first lead-based AFE compound, $PbZrO_3$, was established by Sawaguchi et al. in 1951 [22]. Currently, over 40 types of AFE materials have been identified [23]. Among these, perovskite lead-based compounds are particularly notable, including $PbHfO_3$ [24], $(Pb,La)(Zr,Ti)O_3$ [25], $(Pb,La)(Zr,Sn,Ti)O_3$ [26,27], $Pb_{0.97}La_{0.02}(Zr_{0.5}Sn_{0.43}Ti_{0.07})O_3$ [28], $Pb(In_{0.5}Nb_{0.5})O_3$ [29], and $Pb_{0.97}La_{0.02}(Zr,Sn,Ti)O_3$ [30].

It is worth noting that lead oxides and lead-containing compounds typically comprise about 70% of the raw material mass in certain applications [23]. Lead, a heavy metal, is highly toxic and can cause symptoms like abdominal pain, neurasthenia, anaemia, and toxic encephalopathy if it accumulates in the body beyond safe levels [31,32]. Recognising its danger to human and environmental health, many countries have regulated lead in electrical items [33,34]. This has spurred global research into lead-free dielectric materials, with notable progress in lead-free FEs [35–37], in contrast to the modest advances in lead-free AFEs [18,23,38].

As of the current date, $NaNbO_3$ and $AgNbO_3$ are recognised as leading examples of lead-free AFE materials. Figure 2 depicts the scholarly articles on lead-free AFE materials for energy storage published in Scopus-indexed journals from 2015 to 2024. Comparing the synthesis and processing of $NaNbO_3$ to $AgNbO_3$, it is observed that $NaNbO_3$ is more cost-effective, owing to its reliance on less expensive raw materials and the elimination of the need for a protective environment during the sintering process [39–41]. Despite the advantage of not requiring a protective atmosphere like oxygen, which is necessary for $AgNbO_3$, the higher sintering temperatures and the volatilisation of sodium at elevated temperatures present challenges in achieving high-quality $NaNbO_3$-based ceramics. Therefore, the hydrothermal method has been recently employed for the synthesis of both $NaNbO_3$ and $AgNbO_3$, leading to the production of high-quality ceramics [42,43].

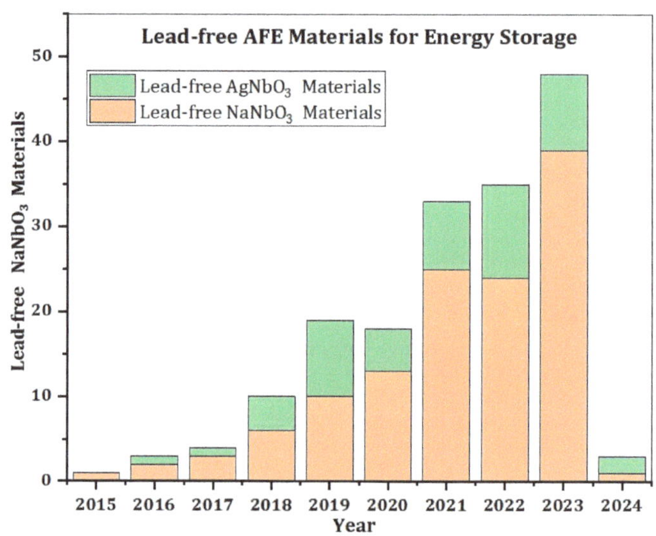

Figure 2. Publications on lead-free AFE materials for electrostatic energy storage (2015–2024).

The potential AFE properties of $NaNbO_3$ were initially identified in a publication in 1951 [44], coinciding with Kittel's proposal of the antiferroelectricity concept [45]. Recent advances have highlighted $NaNbO_3$ as a material of significant functional importance, particularly due to the discovery of unique AFE characteristics in some of its derived solid solutions [46,47]. These $NaNbO_3$-based solid solutions have drawn substantial interest for their ability to exhibit double loops and their relatively low density, estimated at 4.55 g/cm^3 [48]. Furthermore, the burgeoning scientific pursuit of innovative energy storage materials since 2015 has catalysed the development of compositionally altered $NaNbO_3$-based AFEs, characterised by their reversed field-induced phase transitions.

This review aims to provide a comprehensive summary of the latest advancements in lead-free $NaNbO_3$-based AFE ceramics for energy storage, a topic that has garnered considerable interest. Initially, it offers a succinct overview of the fundamental concepts and applications of AFE ceramics. It then delves into recent developments in lead-free

$NaNbO_3$-based AFE materials, covering aspects such as crystal structures, phase transitions, chemical modifications (doping, solid solutions), and strategies devised to improve their energy storage capabilities [23,38,49,50]. Finally, the review addresses the outstanding challenges and future outlook in the study of $NaNbO_3$-based AFE ceramics and capacitors.

2. Fundamentals of AFEs

2.1. Definition of AFEs

AFE materials are closely related to FE materials but exhibit distinct characteristics. In FEs, the application of an external electric field aligns the inherent polarisation, resulting in a single polarisation hysteresis loop with a non-zero remanent polarisation (P_r). This residual polarisation persists even after the external field is removed, indicative of the FE nature.

In contrast, AFEs are characterised by a double polarisation hysteresis loop during field-induced phase transitions from the AFE to the FE state. Structurally, AFEs lack a macroscopic spontaneous polarisation, often described as inhibited FEs [51]. This inhibition is due to the equivalent antiparallel alignment of neighbouring dipoles, effectively cancelling out the macroscopic polarisation. This unique alignment and structural characteristic define AFEs as distinct yet related to traditional FE materials.

2.2. Applications of AFEs

In the early stages of their discovery, AFE materials were primarily regarded for actuation applications [52]. However, mechanical fractures often occurred due to the significant and abrupt strain changes during phase transitions, compromising the reliability of these actuators. At present, AFE materials are increasingly recognised for their potential in constructing high-energy-storage capacitors. Their potential to exhibit a higher electrostatic energy density than FEs and linear dielectrics makes them particularly suitable for high-density power applications in electronics, including DC-link and snubber capacitors [23,53–57].

The unique advantage of AFE materials in energy storage stems from the minimal energy differential between their anisotropic AFE phase and the field-induced FE phase. Under high electric fields, AFEs can transition to a highly polarised FE phase and revert to their original AFE state once the field is eliminated [50,58–60]. This property is in stark contrast to FEs and other dielectrics, which generally show a decrease in dielectric permittivity with increasing DC bias. Conversely, AFEs demonstrate an increase in permittivity at or just before the critical field that triggers the AFE-to-FE phase transition [61].

Furthermore, AFE materials are gaining attention for emerging applications in ultrafast neuromorphic computing [62], tunnel junctions [63], hybrid charge trap memories [64], ferroelectric random-access memories [65], and solid-state ferroic refrigerators [66]. The distinctive AFE properties in HfO_2-based materials are attributed to field-induced transitions between the polar orthorhombic phase $(Pca2_1)$ [67] and the nonpolar tetragonal phase $(P4_2/nmc)$, differing from the antiparallel dipole configuration found in materials like $PbZrO_3$ and $NaNbO_3$ [68]. Comprehensive discussions and evaluations of AFE materials are elaborated in other studies [38,69].

3. Dielectric Capacitors for Energy Storage

Dielectric materials are used in the middle layer between two conducting electrodes to construct a parallel-plate capacitor. The following equation is used to calculate the capacitance (C), representing a capacitor's ability to store energy:

$$C = \frac{\varepsilon_0 \varepsilon_r A}{d} \tag{1}$$

where ε_0 is the permittivity of free space (8.85×10^{-12} F/m), ε_r is the relative permittivity, A is the overlapping area of two electrodes, and d is the thickness of the dielectric layer.

When an external voltage is supplied, a charging process occurs, leading to the accumulation of charges at the electrodes. These charges, having opposite signs but equal magnitude, create an internal electric field. This field is directed opposite to the surrounding electric field. As more charges accumulate, the magnitude of this internal electric field within the system increases proportionally. The charging process completes when the internal electric field, induced by the collected charges (Q), equals the external electric field. The C of the capacitor is represented by Q/V. The amount of stored energy can be determined using the following equation:

$$W = \int_0^{Q_{max}} Vdq \tag{2}$$

where Q_{max} is the maximum charge attained upon completion of the charging process, and dq denotes the incremental increase in charge.

Energy density (J) is a key metric for evaluating the energy storage performance of electrostatic capacitors. It quantifies the amount of energy stored relative to the capacitor's volume, providing a measure of how efficiently space is used for energy storage, as shown in Equation (3):

$$J = \frac{W}{Ad} = \frac{\int_0^{Q_{max}} Vdq}{Ad} = \int_0^{D_{max}} EdD \tag{3}$$

where D is the electrical displacement within the dielectric layer. Consequently, D_{max} denotes the maximum electric displacement in the material when subjected to the highest applied electric field or dielectric breakdown strength, denoted as E_{max}.

In dielectrics with high permittivity, we often find that the electric displacement (D) closely approximates the polarisation (P). And D is calculated as the product of ε_0, ε_r, and E. Therefore, Equation (3) can be formulated as follows:

$$J = \int_0^{P_{max}} EdP = \int_0^{E_{max}} \varepsilon_0 \varepsilon_r EdE \tag{4}$$

Jaffe [13] states that the energy density can be determined by integrating the area that is bounded by the polarisation axis and the polarisation versus electric field (P − E) curve. As shown in Figure 3a–c, the region shaded in blue represents the recovered energy density, denoted as J_{rec}, which is the energy released during the discharging process. Conversely, the orange region indicates the dissipated energy density (J_{loss}), attributable to losses within the dielectric material. The total energy density, J_{tot}, accumulated during the charging process equals the combined area of these two regions, reflecting their integral roles. J_{rec} is calculated using Equation (5):

$$J_{rec} = \int_{P_r}^{P_{max}} EdP \tag{5}$$

Furthermore, the ratio of recoverable energy density (J_{rec}) to the total energy storage density (J_{tot}) is equivalent to the energy efficiency (η), which can be expressed as:

$$\eta = \frac{J_{rec}}{J_{tot}} \times 100\% \tag{6}$$

where $J_{tot} = J_{rec} + J_{loss}$, and J_{loss} is the dissipated energy density, which is depicted as the orange region enclosed by the hysteresis loop.

It is important to note that nonlinear dielectric materials, such as FEs, relaxor ferroelectrics (RFEs), and AFEs, are known to exhibit J_{loss}. AFEs generally exhibit lower J_{loss} and higher E_{max} compared to FEs. According to Jaffe's theory, AFEs have the potential to significantly enhance energy storage capabilities, contingent upon resolving E_{max} challenges. Adequate E_{max} is crucial to enable the AFE-to-FE phase transition [13].

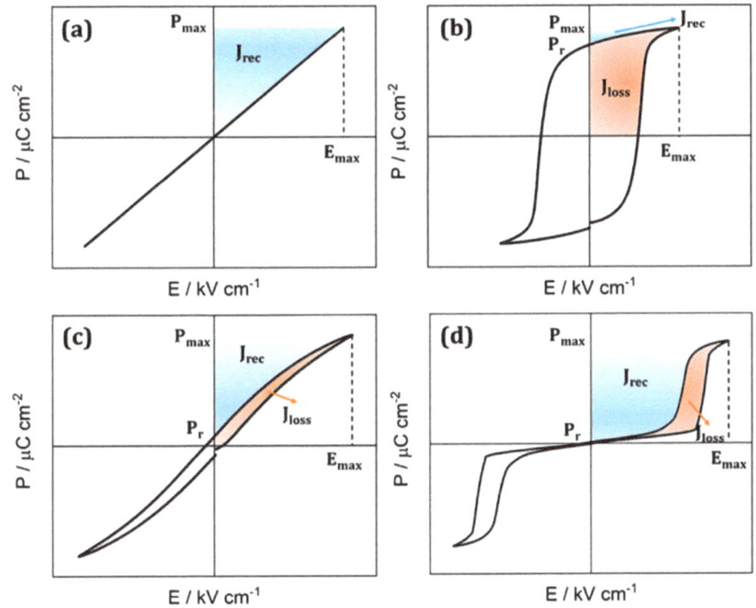

Figure 3. Schematic illustrations showing the polarisation versus electric field (P − E) curve: (**a**) linear dielectrics, (**b**) FEs, (**c**) relaxor ferroelectrics (RFEs), and (**d**) AFEs with the recoverable energy density (J_{rec}) and the dissipated energy density (J_{loss}) indicated in blue and orange areas, respectively.

Figure 3d illustrates that AFEs often display hysteresis loops with a square recoverable energy area instead of a triangular one, a characteristic feature of AFEs [13]. The electric fields at which the AFE-FE transition begins and ends are referred to as the E_F and E_A threshold electric fields, respectively. As a result of their distinctive AFE-FE phase transitions, AFEs have a significant advantage in energy storage over linear dielectrics and FEs, owing to their enormous maximum polarisation (P_{max}) and minimal remnant polarisation (P_r) [70]. The residual polarisation in AFE ceramics is primarily a consequence of system defects and leakage currents. These can be reduced by employing appropriate doping strategies and microstructure optimisation. Modifying the critical phase transition electric field through chemical substitution is also feasible. Furthermore, the degradation resistance of ceramics is affected by various microstructural factors, including particle size, pore presence, secondary phases, and others. These aspects collectively influence the performance and reliability of AFE materials in energy storage applications [71–73].

4. Structural Characteristics of $NaNbO_3$-Based Ceramics

4.1. Crystal Structures of $NaNbO_3$

The structural complexity of $NaNbO_3$ arises from the displacement of Nb within its octahedral coordination and the tilting of the NbO_6 octahedra [74]. As illustrated in Figure 4 [69], at low temperatures (ranging from −100 °C to 360 °C), two phases of $NaNbO_3$ are observed to have very similar energy levels. These phases are designated as the P and Q phases, respectively. The P phase, identified as an AFE phase, possesses an orthorhombic space group Pbcm and is distinguished by antipolar distortions arising from the Nb-O bonds. This phase features a superlattice structure of $\sqrt{2} \times \sqrt{2} \times 4$, relative to the aristo-type perovskite structure. The Q phase, an FE polar phase with the space group $P2_1ma$, features a $\sqrt{2} \times \sqrt{2} \times 2$ supercell and can be induced by applying an electric field. The coexistence of this polar Q phase and the antipolar P phase in $NaNbO_3$ complicates the analysis of double hysteresis loops, as evidenced in references [75–77].

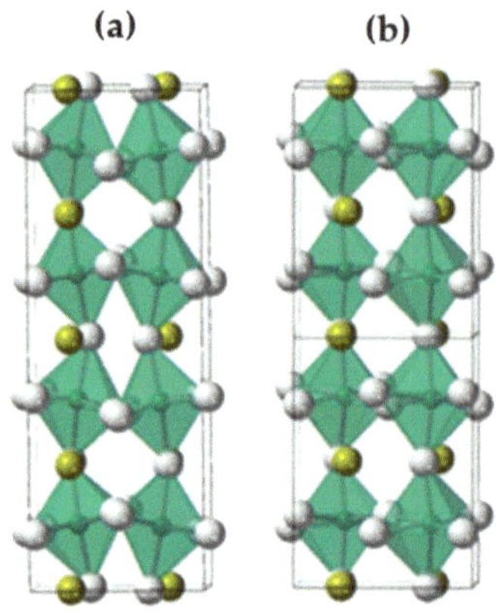

Figure 4. Crystal structures of $NaNbO_3$, with Na^+ in gold, Nb^{5+} in green, and O^{2-} in white: (**a**) the antipolar P phase and (**b**) the polar Q phase [69].

4.2. Phase Transitions in $NaNbO_3$

Seven distinct phases of $NaNbO_3$ can be observed as a function of temperature under ambient pressure, as shown in Figure 5. The high-temperature phase, manifesting at 639 °C, adopts the aristo-type cubic perovskite structure with a cubic Pm3m space group. Upon cooling, this structure transitions through a sequence of phases: tetragonal (P4/mbm), followed by four orthorhombic phases (Ccmm, Pnmm, Pmnm, and Pbcm), and ultimately to a rhombohedral (R3c) phase [23]. The identification of these phases is based on two primary factors: the displacement of Nb^{5+} ions from their octahedral centres and the tilting of oxygen octahedra. With heating, these tilts and displacements diminish, culminating in the restoration of the aristo-type cubic structure [74]. Notably, the rhombohedral N phase exhibits FE, whereas the orthorhombic P and R phases are AFE. All phases existing at temperatures above the R phase are paraelectric, displaying no spontaneous polarisation.

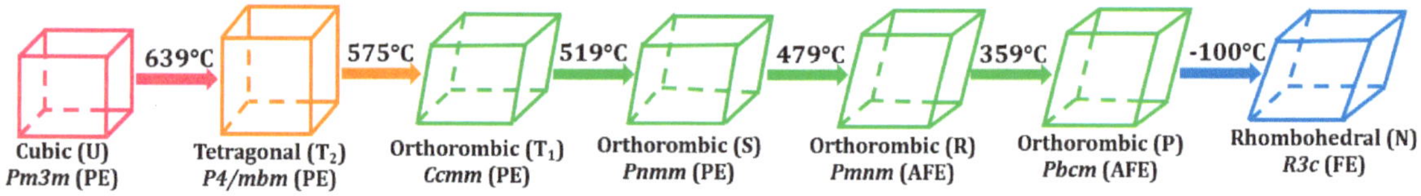

Figure 5. Temperature-dependent phase transitions in $NaNbO_3$.

As opposed to $PbZrO_3$ or $AgNbO_3$ [78,79], which exhibit distinctive double polarisation hysteresis loops at elevated temperatures and ambient temperatures, respectively, pure $NaNbO_3$ undergoes an irreversible phase transition from the AFE state to the FE state [80,81]. Consequently, there is no known existence of steady double polarisation hysteresis loops in this $NaNbO_3$. Research by Cross and Nicholson [82], Ulinzheev et al. [83], and Miller et al. [84] into $NaNbO_3$ crystals has shown that applying electric fields, both parallel and perpendicular to the orthorhombic c axis, also known as the $[110]_{PC}$ direction, induces phase transitions. However, while a double hysteresis loop can initially be observed when a sufficiently strong electric field is applied perpendicular to the orthorhombic c axis [85], subsequent electrical cycles lead to instability, causing the loop to become indistinguishable from a typical FE loop. To date, there are no documented instances of double polarisation hysteresis loops in pure polycrystalline $NaNbO_3$ ceramics, highlighting a significant difference in the behaviour of $NaNbO_3$ compared to $PbZrO_3$ or $AgNbO_3$.

Recent studies reveal that polycrystalline $NaNbO_3$ ceramics may display FE-type polarisation hysteresis loops [86], or display loops that remain unopened [87], as depicted in Figure 6. These phenomena are influenced by factors such as chemical composition [88], starting chemicals [89], and average grain size [90], which affect the stability between the AFE P and the FE Q phase. Koruza et al. [90] found that a phase transition from AFE to FE occurs in $NaNbO_3$ ceramics when the average grain size is below 0.27 mm, due to intragranular stresses.

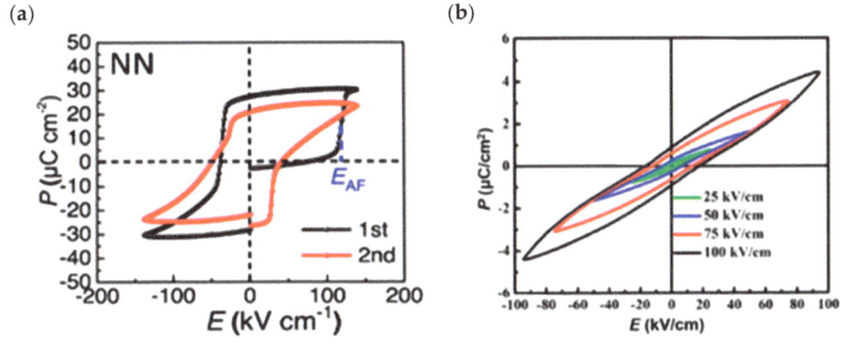

Figure 6. $P-E$ loops of $NaNbO_3$ bulk ceramics, as reported by (**a**) Luo et al. [86] and (**b**) Chao et al. [87].

Li et al. [89] observed that the fraction of P and Q phases varies with different Nb_2O_5 precursors, with a higher fraction from orthorhombic precursors, indicating stress-induced variations in phase amounts. Fan et al. [91] studied the impact of A site nonstoichiometry on phase stability, finding that all samples retained the AFE phase, unaffected by variations in Na content. M.-H. Zhang et al. [92] discovered that sintered $NaNbO_3$ ceramics, using orthorhombic Nb_2O_5 and with an average grain size of 8.9 mm, displayed a 100% AFE phase, highlighting the critical role of grain size in phase composition.

4.3. Phase Manipulation in $NaNbO_3$-Based Ceramics

Reproducing the initial experimental results on AFE in $NaNbO_3$ proved challenging, primarily due to variations in contaminant levels in the samples. These variations subtly affected the relative stability between the AFE and FE phases [82]. Structure analysis of $NaNbO_3$ was challenged by the nearly identical energy profiles of the P (AFE) and Q (FE) phases [77]. The isoenergetic relationship between these phases was confirmed through density functional theory (DFT) experiments [93]. The tolerance factor of $NaNbO_3$, approximately 0.967, prompted the investigation of solid solutions aimed at reducing this factor, thereby favoring the P phase stabilisation [94]. Experimental studies on $xCaZrO_3-(1-x)NaNbO_3$ [95] and $xBiScO_3-(1-x)NaNbO_3$ [94] solid solutions confirmed their AFE nature. Transmission electron microscopy (TEM) analysis of domain configurations revealed that reducing the tolerance factor systematically diminishes the Q phase, potentially stabilising a singular P phase. Although the precise mechanism remains partially understood, it is recognised that incommensurate phases facilitate the electric-field-induced AFE to FE transition. These intermediate phases play a critical role, acting as a transitional bridge during polarisation reversal and the AFE to FE conversion in $NaNbO_3$ solid solutions [96]. Doping has been identified as a stabiliser for the incommensurate phase within the P phase region, a finding supported by analyses of polarisation current density and differential dielectric permittivity anomalies. This incommensurate phase is instrumental in the electric-field-induced polarisation reversal process, bridging the transition from AFE to FE states in $NaNbO_3$-based solid solutions [97].

Considerable effort has been devoted to chemically modified $NaNbO_3$ to achieve dual polarisation loops. Shimizu et al. proposed that the stability of the AFE order could be enhanced by simultaneously reducing the Goldschmidt tolerance factor and polarisability, taking into account the electronegativity differences in the new solid solutions [95]. A study investigated the impact of measurement frequency on the $P-E$ curves for $(0.94-x)NaNbO_3-0.06BaZrO_3-xCaZrO_3$, with $x = 0.04$, as shown in Figure 7a [98].

Increasing frequencies led to a lag in AFE-FE phase transition and domain wall motion, decreasing their polarisation contribution and thus lowering the P_{max}. This was accompanied by increased polarisation hysteresis under an electric field, increasing E_F but decreasing E_A. In another system, $0.7NaNbO_3 - 0.3AgTaO_3$ [99], the $P - E$ loops, displayed in Figure 7b, exhibited distinct behaviour under varying electric fields. Notably, the application of high electric fields over 300 kV/cm resulted in reversible AFE-FE phase shifts, producing double $P - E$ loops with enhanced P_{max}.

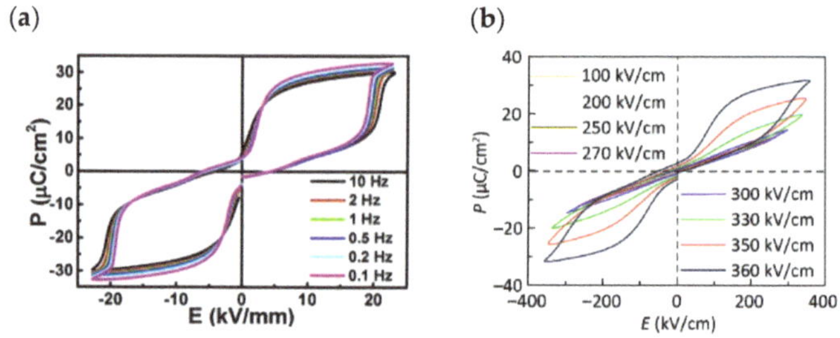

Figure 7. $P - E$ loops of (**a**) $0.9NaNbO_3 - 0.06BaZrO_3 - 0.04CaZrO_3$ [98], (**b**) $Na_{0.7}Ag_{0.3}Nb_{0.7}Ta_{0.3}O_3$ [99].

5. Tailoring Energy Storage Performance in $NaNbO_3$-Based Ceramics

5.1. Incorporation of MnO_2

In 2023, Zhang et al. reported a strategy to enhance the energy storage properties of $NaNbO_3$-based ceramics, specifically $0.95NaNbO_3 - 0.05SrSnO_3(NN5SS)$, through the incorporation of MnO_2. This approach, inspired by techniques used in lead-based materials, involves adding varying amounts of MnO_2 to serve as an electron trap (Equations (7) to (9)) [100]. Such a method has been previously documented to enhance the resistance of other perovskites [101,102]. Notably, it is recognised that Mn exhibits a preference for fewer oxidation states within perovskite structures [103]. The related reactions [100] are as follows:

$$Mn_{Nb}^{x} + e' = Mn_{Nb}' \quad (7)$$

$$Mn_{Nb}' + e' = Mn_{Nb}'' \quad (8)$$

$$Mn_{Nb}'' + e' = Mn_{Nb}''' \quad (9)$$

The modification of $NaNbO_3$ by incorporating $SrSnO_3$ induces an augmentation in the local chemical disorder, thereby stabilising the AFE state [104]. Despite the tolerance factor for $NaNbO_3(NN, 0.965)$ and $0.95NaNbO_3 - 0.05SrSnO_3(NN5SS, 0.964)$ samples being nearly identical, it is insufficient to solely rely on this metric for the stabilisation of AFE order without additional analysis. As shown in Figure 8a,b, NN exhibits FE behaviour, whereas NN5SS displays AFE-type double polarisation loops with a notable high remnant polarisation of 13.9 $\mu C/cm^2$, influenced by leakage current. This phenomenon occurs as the $NaNbO_3 - SrSnO_3$ solid solution demonstrates $n-$type conductivity due to a predominance of excess Sr over Na vacancies, in contrast to the $p-$type conductivity observed in pure $NaNbO_3$ [105].

The inclusion of MnO_2 markedly reduces the dielectric loss, especially in the low-frequency region, as illustrated in Figure 8g, and concurrently elevates the resistance, as shown in Figure 8h. Consequently, the beneficial effect of the diminished quantity of charge carriers on the double hysteresis loops is evidenced in Figure 8c. Unlike the unaltered NN system, the modified samples exhibit a substantially lower remnant polarisation of 3.2 $\mu C/cm^2$, representing a ten-fold reduction compared to the original system. No variation in phase transition behaviour is observed with an increase in Mn modification from 0.1 wt% to 1 wt%. The sole alteration is an increase in the critical field required to initiate the transition [100]. Energy storage efficiency improved from 21% in NN5SS to 33%

in NN5SS_1.0Mn, yet remains low. Enhancing efficiency requires modifying polarisation loop shapes to be much slimmer, reducing hysteresis, and dispersing critical transition fields [106]. Figure 8d,e show that adjusting $SrSnO_3$ substitution levels while maintaining MnO_2 content boosts efficiency significantly from 33% to about 90% in NN9SS_1.0Mn, as emphasised in Figure 8f. In summary, MnO_2 incorporation successfully suppresses mobile charge carriers, drastically lowers P_r, and increases material resistance.

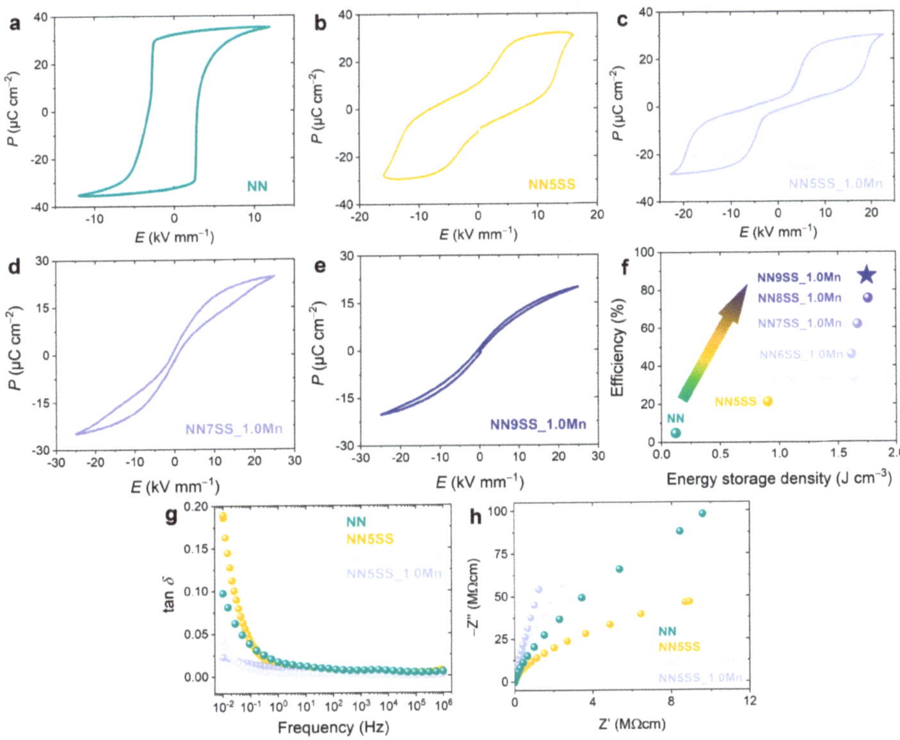

Figure 8. The second electric field cycle at 1 Hz produced P − E loops for (**a**) NN, (**b**) NN5SS, (**c**) NN5SS_0.1Mn and NN5SS_1.0Mn, (**d**) NN7SS_1.0Mn, and (**e**) NN9SS_1.0Mn; (**f**) recoverable energy storage density and efficiency; (**g**) tanδ as a function of frequency; and (**h**) Nyquist plot of the NN, NN5SS, $NN5SS_{0.1Mn}$, and NN5SS_1.0Mn samples [100].

5.2. Chemical Doping

Enhancing the energy storage properties of $NaNbO_3$-based ceramics through single-site doping at either the A-site or B-site is seldom reported. In 2018, Zhou et al. reported an approach to improve the energy storage capabilities of $NaNbO_3$-based ceramics by adding Bi_2O_3 [107]. This strategy potentially increases the E_{max} and reduces the P_r. Following Bi_2O_3 addition, the dielectric constant shows relaxor-like frequency dispersion near room temperature. The high E_{max} of ceramics is attributed to their dense microstructure and uniform chemical composition. When the E_{max} reaches 351 kV/cm, the P_r nearly still zeroes. Moreover, the hybridisation interaction between the 6p orbitals of Bi^{3+} and the 2p orbitals of O^{2-} results in significant polarisation, with peak values exceeding 35 $\mu C/cm^2$. Ceramics with a composition of $Na_{0.7}Bi_{0.1}NbO_3$ achieve an impressive energy storage density of 4.03 J/cm^3 at 250 kV/cm and an energy storage efficiency of 85.4%, as demonstrated in Figure 9a,b [107].

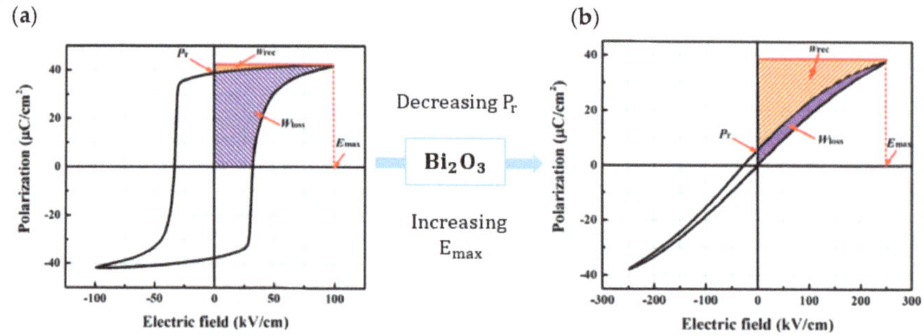

Figure 9. Schematic diagrams illustrating the enhancement of J_{rec} by adding Bi_2O_3 to $NaNbO_3$: (**a**) pure $NaNbO_3$ ceramics exhibit low J_{rec} due to high P_r and small E_{max}; (**b**) $Na_{1-3x}Bi_xNbO_3$ ceramics exhibit high J_{rec} due to low P_r and large E_{max} [107].

In 2022, L. Yang et al. significantly enhanced the energy storage properties of $NaNbO_3$-based ceramics through tantalum (Ta) doping. This modification reduced the dielectric loss and increased volume resistivity by substituting Nb^{5+} cations with Ta^{5+} cations at a concentration in the compound $(Na_{0.8}Bi_{0.1})(Nb_{0.9-x}Ta_xTi_{0.1})O_3$ (where x = 0.15) [108]. Such improvements have also been observed in various niobate ceramics, underscoring the broader applicability of this approach [109,110]. The improved volume resistivity of these ceramics contributes significantly to energy economy and E_{max} in energy storage applications [10]. The leakage current decreased after doping with 15% Ta^{5+} ions due to an increased bandgap, especially under high electric fields. Impedance spectroscopy revealed enhanced resistance both in the grains and at the grain boundaries, leading to an increase in E_{max} from 367 kV/cm to 469 kV/cm. Remarkably, at 450 kV/cm, the ceramics achieved a high recoverable energy density of 6.5 J/cm^3 and an η of 94%, surpassing most $NaNbO_3$-based relaxor ceramics.

Additionally, these ceramic capacitors demonstrated excellent energy storage stability and performance across a wide temperature range (−90 °C to 150 °C) and over 10^5 cycles, as shown in Figure 10a,b [108]. The variation in P − E loops between −10 °C and 0 °C is attributed to adsorbed moisture on the sample interfaces, with a noticeable drop in P_{max} as the measurement temperature increases. These findings underscore the suitability of Ta-doped $NaNbO_3$-based ceramics for practical energy storage applications, thanks to their robust temperature stability and cycle-stable energy storage properties.

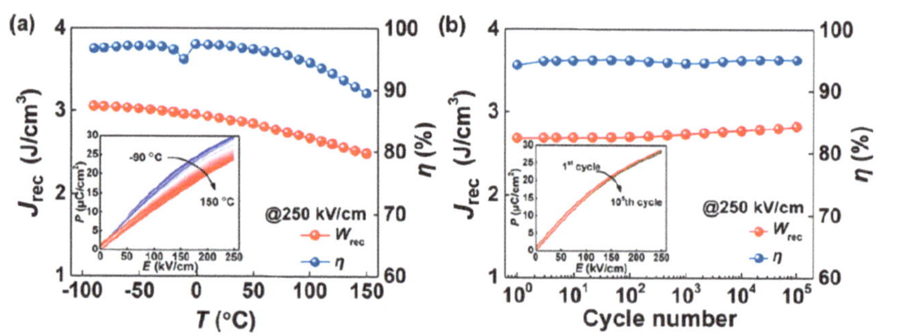

Figure 10. (**a**) The relationship between recoverable energy density and electric field at 250 kV/cm as a function of temperature is illustrated by the P − E loops obtained within an operating temperature range of −90 °C to 150 °C. (**b**) The dependence of recoverable energy density and η on an electric field strength of 250 kV/cm is demonstrated through P − E loops from measurements conducted over several cycles, up to 10^5, at this field strength [108].

5.3. Improving Energy Storage Properties of $NaNbO_3$-Based Ceramics: Comprehensive Strategy Summaries

Recent literature has highlighted a variety of lead-free $NaNbO_3$-based materials notable for their high J_{rec}, as detailed in Table 1 with key sample preparation parameters such as thickness and electrode size. Reducing both thickness and electrode size effectively minimises the likelihood of detrimental defects, thereby increasing the E_{max}. This method facilitates achieving substantial energy storage properties under high electric fields (> 350 kV/cm), making it a preferred method among researchers striving for ultrahigh J_{rec} in materials of identical composition. Furthermore, J_{rec}, being directly correlated with the magnitude of the applied electric field, benefits from higher electric fields. However, to enable a more detailed comparison of energy storage performance across different systems, an analysis that normalises J_{rec} by the E_{max} is indispensable.

Table 1. Summary of energy storage properties of lead-free $NaNbO_3$-based materials with key sample preparation parameters.

$NaNbO_3$-Based Materials	$E_{max}(\frac{kV}{cm})$	$J_{rec}(\frac{J}{cm^3})$	$\frac{J_{rec}}{E_{max}}(\frac{J}{kVcm^2})$	η (%)	Thickness (mm)	Bulk/MLCC	Electrode and Size	Year	Ref.
$Na_{0.7}Ag_{0.3}Nb_{0.7}Ta_{0.3}O_3$	330	3.3	0.01	42.4	0.10	Bulk	Ag, 3.14 mm^2	2023	[99]
$0.8NaNbO_3 - 0.04CaZrO_3 - 0.16Bi_{0.5}Na_{0.5}TiO_3$	400	3.7	0.009	82.1	0.029	MLCC	Pt	2021	[111]
$0.88NaNbO_3 - 0.12Bi(Ni_{0.5}Zr_{0.5})O_3$	500	4.90	0.010	~72.0	0.15 ± 0.01	Bulk	Ag, 0.785 mm^2	2020	[112]
$0.85NaNbO_3 - 0.15Mg_{0.5}Hf_{0.5}O_3$	405	5.0	0.012	83.1	~0.11	Bulk	Au, ~3.14 mm^2	2023	[113]
$0.85NaNbO_3 - 0.15CaZrO_3$	680	5.4	0.008	82	~0.1	Bulk	Ag, ~3.14 mm^2	2023	[114]
$0.7NaNbO_3 - 0.3CaTiO_3$	560	5.52	0.010	83.3	0.1	Bulk	Au	2022	[115]
$0.95NaNbO_3 - 0.05Bi(Mg_{0.5}Sn_{0.5})O_3$	646	6.35	0.010	80.0	0.15 ± 0.01	Bulk	Ag, 0.785 mm^2	2022	[116]
$0.8NaNbO_3 - 0.2Bi(Ni_{0.5}Hf_{0.5})O_3$	380	6.45	0.017	82.72	-	Bulk	-	2024	[117]
$(Na_{0.8}Bi_{0.1})(Nb_{0.75}Ta_{0.15}Ti_{0.1})O_3$	450	6.5	0.014	94.0	0.08	Bulk	Au, 1.77 mm^2	2022	[108]
$Na_{0.7}Bi_{0.1}Nb_{0.9}Ta_{0.1}O_3$	560	6.68	0.012	90.5	0.15	Bulk	Au, 3.14 mm^2	2022	[118]
$0.88NaNbO_3 - 0.12CaZrO_3$	650	8.56	0.013	82.0	~0.1	Bulk	Au, ~3.14 mm^2	2022	[119]
$0.8(0.92NaNbO_3 - 0.08Bi(Ni_{0.5}Zr_{0.5})O_3) - 0.2(Bi_{0.5}Na_{0.5})0.7Sr_{0.3}TiO_3(0.2BNST)$	655	8.6	0.013	83.5	0.094	Bulk	3.14 mm^2	2023	[120]
$0.68NaNbO_3 - 0.32(Bi_{0.5}Li_{0.5})TiO_3$	480	~8.73	~0.018	~80.1	0.12	Bulk	Ag, 3.14 mm^2	2021	[121]
$0.85NaNbO_3 - 0.15Ca_{0.7}Sm_{0.2}TiO_3$	800	9.1	0.011	80.1	0.035	Bulk	Au, 3.14 mm^2	2023	[122]
$0.76NaNbO_3 - 0.24(Bi_{0.5}Na_{0.5})TiO_3$	680	~12.2	~0.018	~69.0	0.15	Bulk	Ag, 7.07 mm^2	2019	[123]
$0.67NaNbO_3 - 0.18(Bi_{0.5}Na_{0.5})TiO_3 - 0.15Bi(Mg_{0.5}Hf_{0.5})O_3$	700	12.65	0.018	88.5	0.0055	MLCC	70 Ag/30 Pd	2024	[124]
$NaNbO_3 - (Bi_{0.8}Sr_{0.2})(Fe_{0.9}Nb_{0.1})O_3$	983	16.5	0.017	83.3	0.06–0.08	Bulk	Ag, 0.785 mm^2	2022	[125]
$0.9NaNbO_3 - 0.1BiFeO_3$	995	18.5	0.019	78.7	0.06–0.08	Bulk	Ag, 0.785 mm^2	2021	[126]

It is worth noting that, compared to current lead-based materials with a remarkable J_{rec} of 11.18 J/cm^3 in $PbZrO_3$-based bulk ceramics and an even larger J_{rec} of 12.6 J/cm^3 in $PbZrO_3$-based multilayer ceramic capacitors (MLCCs) [127,128], lead-free $NaNbO_3$-based materials already exhibit a superior energy density exceeding 18 J/cm^3, as shown in Table 1. To further enhance the J_{rec} of $NaNbO_3$-based ceramics, many strategies have been proposed, such as microstructure modification [129,130], electrical property tuning [82], and phase engineering [131,132].

5.3.1. Microstructure Modification

Examining the grain size of samples facilitates the microstructural tuning of sintered ceramics. Field-emission scanning electron microscopy (FE-SEM) images are indispensable for investigating these microstructures [42,133]. Figure 11a–e present the SEM images of polished samples of $0.955NaNbO_3 - 0.045La(Nb_{0.33}Mg_{0.67})O_3$ with varying additions of MnO_2 (0 wt%, 0.5 wt%, 1 wt%, 1.5 wt%, and 2 wt%) [134]. These images illustrate that all samples exhibit a dense packing with minimal porosity. The addition of MnO_2 resulted in a slight increase in grain size, attributed to the fluid phase facilitating atomic mobility. Notably, the particle size in all samples remained below 1 μm, contributing to an improvement in E_{max}.

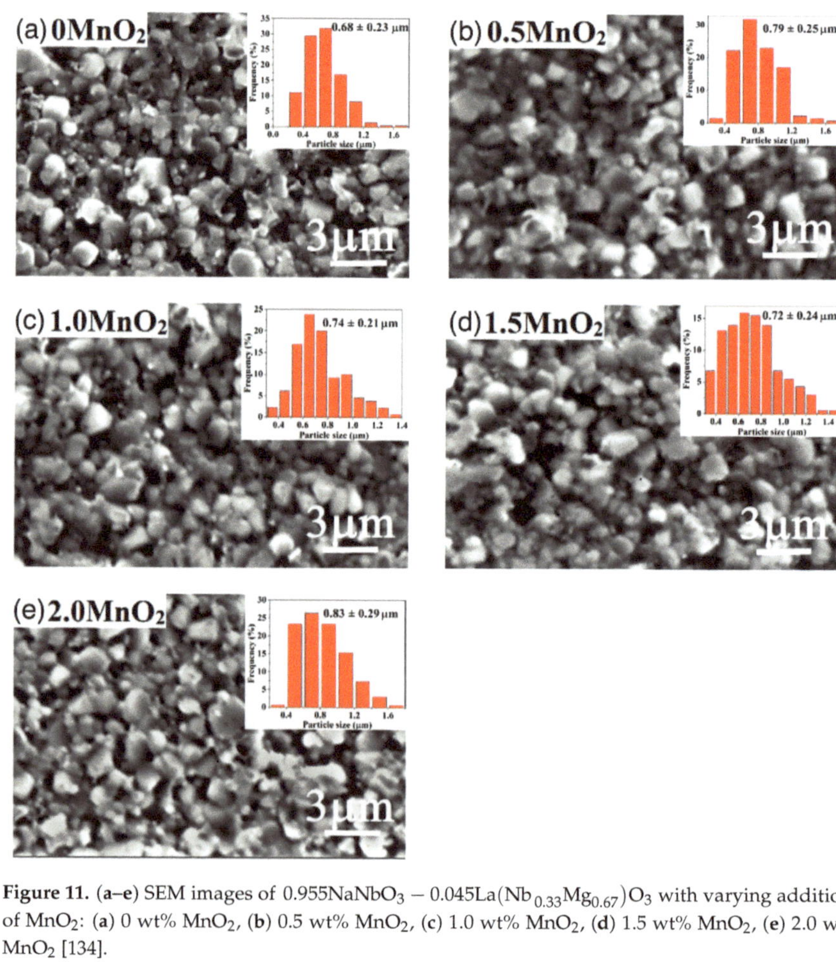

Figure 11. (**a–e**) SEM images of $0.955NaNbO_3 - 0.045La(Nb_{0.33}Mg_{0.67})O_3$ with varying additions of MnO_2: (**a**) 0 wt% MnO_2, (**b**) 0.5 wt% MnO_2, (**c**) 1.0 wt% MnO_2, (**d**) 1.5 wt% MnO_2, (**e**) 2.0 wt% MnO_2 [134].

Recent research has demonstrated that MLCC, which is constructed from a ceramic film with a layered structure [132], exhibits exceptional energy storage performance. This device is characterised by an ultrahigh J_{rec} and superior efficiency, in addition to exhibiting outstanding temperature stability and fatigue resistance [124].

Figure 12a–c present a comparison of $E_{test-max}$, J_{rec}, and η values for $0.82NaNbO_3 - 0.18(Bi_{0.5}Na_{0.5})TiO_3$ (NN–BNT), $0.67NaNbO_3 - 0.18(Bi_{0.5}Na_{0.5})TiO_3 - 0.15Bi(Mg_{0.5}Hf_{0.5})O_3$ (NN–BNT–15BMH) ceramic, and NN–BNT–15BMH MLCCs. The addition of BMH significantly enhances the $E_{test-max}$ in NN-BNT-15BMH ceramic samples. For MLCCs, reducing the dielectric layer thickness to 5.5 μm with NN–BNT–15BMH composition achieves an $E_{test-max}$ of 1100 kV/cm, nearly doubling the value observed in bulk ceramics. Furthermore, NN–BNT–15BMH MLCCs demonstrate an approximate 79.9% increase in J_{rec} compared to their ceramic counterparts [124].

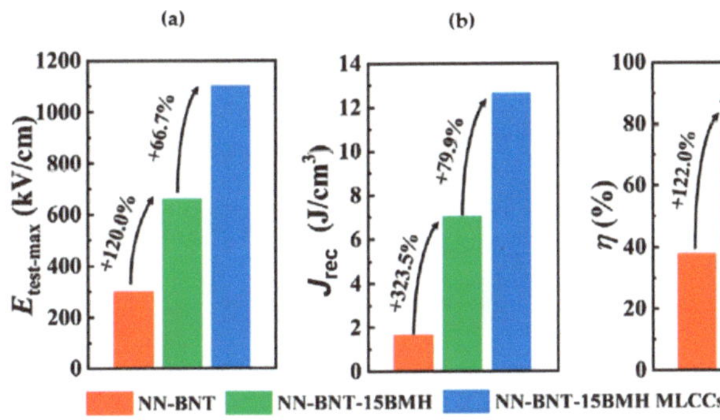

Figure 12. (**a**) Comparison between $E_{test-max}$, J_{rec} and η of NN–BNT, (**b**) NN–BNT–15BMH ceramic, (**c**) NN–BNT–15BMH MLCCs [124].

Additionally, $NaNbO_3$ has been explored for its utility in energy storage materials when integrated into polymer matrices to form ceramic–polymer composites. Pan et al. discovered that high J_{rec} can be achieved in $NaNbO_3$/PVDF composites by optimising the composition of two-dimensional $NaNbO_3$ produced via the molten-salt method [135]. The composite exhibited a remarkable J_{rec} of 13.5 J/cm^3 and a power density of 2.68 MW/cm^3 at an electric field of 4000 kV/cm, underscoring the potential of $NaNbO_3$ as a primary or supplementary component in energy storage applications.

5.3.2. Electrical Property Tuning

The tuning of electrical properties is directly linked to factors such as the band gap, conductivity, and electrical homogeneity. Lanfredi et al. conducted an investigation into the conductive properties of $NaNbO_3$ samples, which were synthesised using a wet chemical approach and subsequently analysed via impedance spectroscopy [136]. This analytical method facilitates the differentiation of grain boundary effects from the intrinsic properties of the grains themselves, thereby providing detailed insights into the resistivity, electrical homogeneity, and relaxation frequency [137,138]. Achieving a higher breakdown strength in $NaNbO_3$ necessitates the establishment of a substantial energy difference between the conduction and valence bands, commonly referred to as the band gap. A larger band gap is indicative of a lower likelihood of conductivity, underscoring the need for the material to possess strong insulating characteristics while maintaining minimal conductivity [82,139,140]. Enhancing electrical homogeneity has been demonstrated to significantly improve the E_{max} and, consequently, the J_{rec} in $BiFeO_3$-based ceramics [141,142]. Although $NaNbO_3$-based ceramics generally exhibit electrical homogeneity, as evidenced by complex impedance spectroscopy data presenting a singular semicircle in NN-BNT-15BMH (as shown in ref. [124]), it is noteworthy that grain refinement within these ceramics markedly increases the number of grain boundaries. This increase in grain boundaries leads to a pronounced rise in electrical resistance. The augmentation in resistance, in turn, enhances the insulating properties of the ceramics, contributing to an elevated E_{max}.

5.3.3. Phase Engineering

It has been reported that forming $NaNbO_3$-based solid solutions is a very promising method for phase engineering, which aims for high energy storage performance generally through inducing a relaxor state. This strategy significantly reduces P_r to achieve slimmer $P - E$ loops and enhances the E_{max}, leading to a higher J_{rec}.

As shown in Figure 13a,b, the compositions of $NaNbO_3$-based materials that employed this strategy exhibit relaxor-like characteristics, including broad dielectric anomalies with discernible frequency dispersion [125,130] and slim polarisation hysteresis loops [115,143,144]. These $NaNbO_3$-based solid solutions exhibit slim loops with remanence significantly reduced—up to ten-fold less—compared to pure $NaNbO_3$, heralding superior energy storage capabilities relative to conventional AFE materials due to the diminished electric-field-induced strain, especially in the absence of pronounced phase transitions that would compel volumetric changes under electric fields. Realising these advancements necessitates precise modifications to the local structure and defect chemistry.

(a) (b)

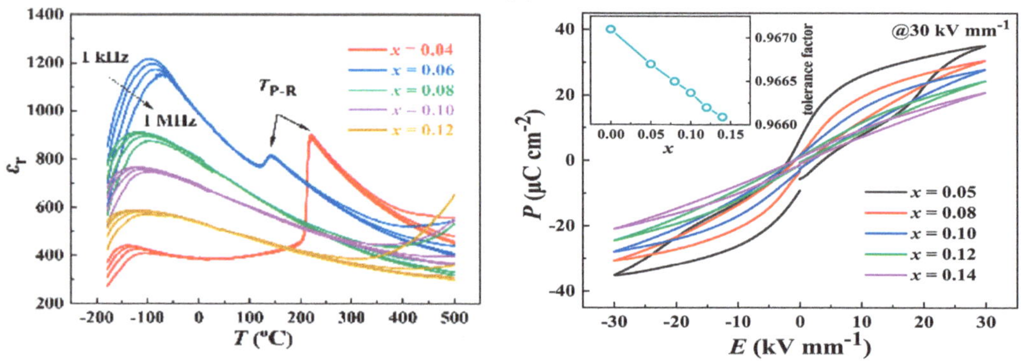

Figure 13. (**a**) Temperature-dependent measurements of dielectric permittivity at various frequencies for AFE samples of $(1 - x)NaNbO_3 - x(Bi_{0.8}Sr_{0.2})(Fe_{0.9}Nb_{0.1})O_3$ [125]; (**b**) $P - E$ loops for $(1 - x)NaNbO_3 - x(Bi_{0.8}Sr_{0.2})(Fe_{0.9}Nb_{0.1})O_3$ ceramics under 30 kV/mm at room temperature [125].

The phenomenon of relaxor behaviour has been identified as a contributor to facilitated polarisation rotation within ceramic materials [145], enhancing energy density through domain evaluation. External electric fields can reorient FE domains due to long-range dipole ordering [146–148]. High-performance piezoelectric materials benefit from phase barriers, which influence FE domain properties, reducing polarisation anisotropy and energy barriers through the creation of small domain sizes and high domain wall densities [149,150]. This reduction is primarily due to spontaneous polarisation (P_s) induced by lattice distortion, which is sensitive to external factors such as temperature, electric fields, ion substitution, and mechanical stresses. Moreover, nanoscale P_s fluctuations result in the formation of polar nanoregions (PNRs), exhibiting relaxor behaviour at the macroscopic level [151]. To enhance energy storage properties, the formation of PNRs has been identified as an effective strategy for increasing the energy efficiency of $NaNbO_3$-based ceramics [100,121]. The relaxor behaviour of the NN9SS_1.0Mn sample is illustrated in Figure 14a, with the presence of randomly distributed PNRs, each a few nm in size, further substantiated by high-resolution transmission electron microscopy (HRTEM) images in Figure 14b. Additionally, small-box modeling of the X-ray pair distribution functions (PDFs), shown in Figure 14c, reveals the local short-range structural order. These findings collectively underscore the emergence of a relaxor state, marked by the development of PNRs due to compositional and microstructural disorder. This relaxor state is instrumental in producing slimmer hysteresis loops and thereby enhancing energy storage efficiency [100].

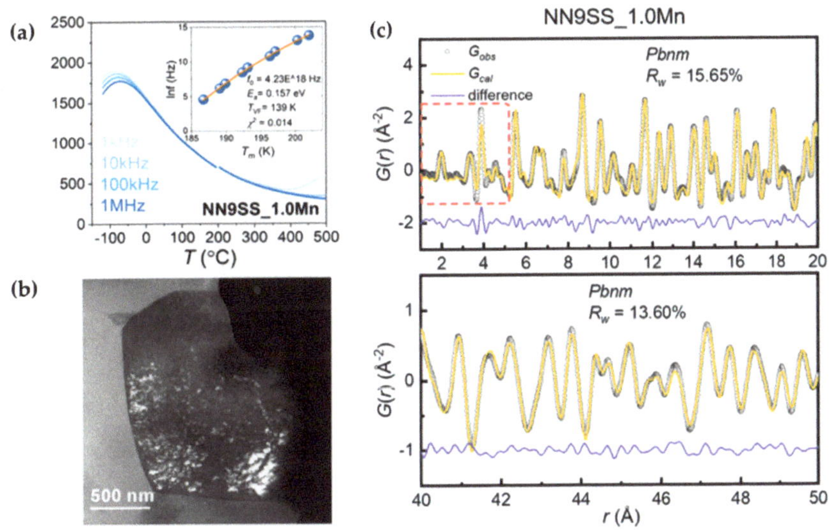

Figure 14. (**a**) Temperature-dependent dielectric permittivity; (**b**) domain morphology of the HRTEM images; (**c**) PDF analysis of relaxor states of the NN9SS_1.0Mn samples [100].

Importantly, $NaNbO_3$-based materials that display relaxorlike behaviours, occasionally termed "relaxor AFEs", present an ambiguous relationship with AFE behaviour, marked by a notable scarcity of superlattice reflection reports indicative of AFEs and the lack of current peaks in the $P-E$ loops that would signify field-induced transitions. A comprehensive analysis of the crystal structure through in situ high-energy synchrotron X-ray diffraction is anticipated to uncover the structural modifications correlating with functional property changes induced by electric field application. It is established that a reversible AFE–FE phase transition underpins the observed double polarisation hysteresis loops. Should phase transitions be unverifiable, leading to an absence of double $P-E$ loops, it may be posited that transitioning the initial state—whether FE or AFE—to a relaxor state could enhance energy storage density, offering a more advantageous approach than categorising them strictly as "relaxor AFE" or "relaxor FE".

6. Summary and Outlook

Over the past decade, there has been significant progress in developing environmentally friendly, lead-free FEs in response to global environmental contamination and health risks. AFE perovskite oxides, distinguished by their intricate structures and a plethora of phase transitions, offer a compelling contrast to traditional FEs. However, despite their intriguing properties and potential applications, the exploration of AFEs, especially those exhibiting reversible AFE-FE transitions and lead-free variants such as $NaNbO_3$, remains nascent. A thorough grasp of crystal structure, phase structure, and microstructure is crucial for developing new AFE compositions. Given the increased complexity of AFE crystal structures compared to FEs, a multi-scale approach is essential to elucidate the fundamental mechanisms of antiferroelectricity.

The development of novel $NaNbO_3$-based AFE materials offers promising insights into the nuanced features of AFEs and their interaction with FE orders, highlighting the need for further theoretical research into the principles of antiferroelectricity in lead-free AFEs. Despite the focus on lead-based materials like $PbZrO_3$, the study of lead-free AFEs, such as $NaNbO_3$, is crucial for a deeper understanding of antiferroelectricity's origins. To finely tune the balance between AFE and FE orders in polycrystalline $NaNbO_3$, a better grasp of phase transition behaviours in single crystals is essential. This understanding could illuminate why the detection of double hysteresis loops is direction-dependent and whether textured ceramics could stabilise the AFE order [82]. However, research on field-induced phase transitions in high-quality alkaline niobate single crystals remains scarce [152].

The crystal structure and the dielectric properties of AFEs are comparable to those of FEs. Furthermore, the presence of variable degrees of competition between AFE and FE phases is almost ubiquitous in many compounds that are like them [153]. $NaNbO_3$ undergoes a convoluted series of phase changes as a function of temperature, which poses difficulties for the precise structural characterisation of the material. Even though various investigations have been conducted, there are still debates regarding this material. In recent years, significant advancements have been made in the development of chemically modified $NaNbO_3$ compositions that exhibit reversible phase transitions. Chemical modification is an effective method that can be utilised to adjust the phase stability of $NaNbO_3$ [154]. Despite these strides, the detailed mechanisms underpinning the structural changes in these materials are still only partially understood. Investigating the nucleation and growth of the field-induced FE phase, along with phase boundary migration during phase transitions in these new ceramics, is of keen interest. These studies are complicated by the processing sensitivities of the materials, including the hygroscopic nature of precursors, limited sintering windows, and the high vapour pressure of alkalis, necessitating further research on processing impacts.

The energy storage capabilities of $NaNbO_3$-based ceramics can be significantly augmented through the induction of relaxor-like characteristics, achieving a $P-E$ loop devoid of hysteresis and thereby maximising energy density. $NaNbO_3$-based materials already exhibit an J_{rec} exceeding 18 J/cm^3, surpassing that of lead-based materials. This significant achievement indicates that lead-free $NaNbO_3$-based ceramics represent a viable alternative to lead-based materials in electrostatic energy storage capacitors. With strategic compositional adjustments and advancements in synthesis techniques, there is strong potential for further enhancing the J_{rec} of $NaNbO_3$-based ceramics. Despite these advancements, considerable challenges persist in resolving fundamental scientific inquiries. These include establishing criteria for the stability assessment between FE and AFE states, tailoring energy states for optimal performance, and precisely delineating the characteristics inherent to AFE materials. Nonetheless, the continuous evolution of experimental and theoretical research methodologies harbours a promising outlook for the development of lead-free AFE materials endowed with superior energy storage capacities.

In the realm of energy storage capacitors, MLCCs boasting higher capacity and compactness are pivotal for practical device applications. To address the demands of high-voltage power electronics, development is required not only in dielectric materials but also in the substitution of expensive electrodes like Ag and Pt with more cost-effective alternatives such as Cu and Ni, ensuring compatibility. Thus, a significant challenge lies in reducing the sintering temperature of $NaNbO_3$-based ceramics without compromising their properties. Additionally, the focus on temperature stability and long-term reliability will be crucial for advancing $NaNbO_3$-based electrostatic energy storage capacitors.

Author Contributions: Writing—original draft preparation, S.N.S.; writing—review and editing, Z.L., D.S. and I.S.; supervision, Z.L.; funding acquisition, Z.L. All authors have read and agreed to the published version of the manuscript.

Funding: This research was supported by the Royal Society of Chemistry Research Fund grant (R23-0577995877).

Institutional Review Board Statement: Not applicable.

Informed Consent Statement: Not applicable.

Data Availability Statement: No new data were created or analyzed in this study.

Conflicts of Interest: The authors declare no conflict of interest.

References

1. Wang, G.; Lu, Z.; Li, Y.; Li, L.; Ji, H.; Feteira, A.; Zhou, D.; Wang, D.; Zhang, S.; Reaney, I.M. Electroceramics for high-energy density capacitors: Current status and future perspectives. *Chem. Rev.* **2021**, *121*, 6124–6172. [CrossRef] [PubMed]
2. Lin, J.; Cao, Y.; Zhu, K.; Yan, F.; Shi, C.; Bai, H.; Ge, G.; Yang, J.; Yang, W.; Li, G. Ultrahigh energy harvesting properties in temperature-insensitive eco-friendly high-performance KNN-based textured ceramics. *J. Mater. Chem. A* **2022**, *10*, 7978–7988. [CrossRef]
3. Yan, F.; Bai, H.; Ge, G.; Lin, J.; Zhu, K.; Li, G.; Qian, J.; Shen, B.; Zhai, J.; Liu, Z. Boosting Energy Storage Performance of Lead-Free Ceramics via Layered Structure Optimization Strategy. *Small* **2022**, *18*, 2202575. [CrossRef] [PubMed]
4. Kusko, A.; Dedad, J. Short-term, long-term, energy storage methods for standby electric power systems. In Proceedings of the Fourtieth IAS Annual Meeting. Conference Record of the 2005 Industry Applications Conference, Hong Kong, China, 2–6 October 2005; pp. 2672–2678.
5. Yao, K.; Chen, S.; Rahimabady, M.; Mirshekarloo, M.S.; Yu, S.; Tay, F.E.H.; Sritharan, T.; Lu, L. Nonlinear dielectric thin films for high-power electric storage with energy density comparable with electrochemical supercapacitors. *IEEE Trans. Ultrason. Ferroelectr. Freq. Control* **2011**, *58*, 1968–1974. [PubMed]
6. Gomez Vidales, A.; Sridhar, D.; Meunier, J.-L.; Omanovic, S. Nickel oxide on directly grown carbon nanofibers for energy storage applications. *J. Appl. Electrochem.* **2020**, *50*, 1217–1229. [CrossRef]
7. Gomez Vidales, A.; Kim, J.; Omanovic, S. $Ni_{0.6-x}Mo_{0.4-x}Ir_x$-oxide as an electrode material for supercapacitors: Investigation of the influence of iridium content on the charge storage/delivery. *J. Solid State Electrochem.* **2019**, *23*, 2129–2139. [CrossRef]
8. Bhalerao, S.; Ambhore, N.; Kadam, M. Polymer matrix composite in high voltage applications: A review. *Biointerface Res. Appl. Chem* **2022**, *12*, 8343–8352.
9. Christen, T.; Carlen, M.W. Theory of Ragone plots. *J. Power Sources* **2000**, *91*, 210–216. [CrossRef]
10. Yang, L.; Kong, X.; Li, F.; Hao, H.; Cheng, Z.; Liu, H.; Li, J.-F.; Zhang, S. Perovskite lead-free dielectrics for energy storage applications. *Prog. Mater. Sci.* **2019**, *102*, 72–108. [CrossRef]
11. Sherrill, S.A.; Banerjee, P.; Rubloff, G.W.; Lee, S.B. High to ultra-high power electrical energy storage. *Phys. Chem. Chem. Phys.* **2011**, *13*, 20714–20723. [CrossRef]
12. Whittingham, M.S. Materials challenges facing electrical energy storage. *Mrs Bull.* **2008**, *33*, 411–419. [CrossRef]
13. Jaffe, B. Antiferroelectric ceramics with field-enforced transitions: A new nonlinear circuit element. *Proc. IRE* **1961**, *49*, 1264–1267. [CrossRef]
14. Ulrich, R.; Schaper, L.; Nelms, D.; Leftwich, M. Comparison of paraeletric and ferroelectric materials for applications as dielectrics in thin film integrated capacitors. *Int. J. Microcircuits Electron. Packag.* **2000**, *23*, 172–181.
15. Park, S.-E.; Shrout, T.R. Ultrahigh strain and piezoelectric behavior in relaxor based ferroelectric single crystals. *J. Appl. Phys.* **1997**, *82*, 1804–1811. [CrossRef]
16. Love, G.R. Energy storage in ceramic dielectrics. *J. Am. Ceram. Soc.* **1990**, *73*, 323–328. [CrossRef]
17. Krohns, S.; Lunkenheimer, P. Ferroelectric polarization in multiferroics. *Phys. Sci. Rev.* **2019**, *4*, 20190015. [CrossRef]
18. Gao, J.; Li, Q.; Zhang, S.; Li, J.-F. Lead-free antiferroelectric $AgNbO_3$: Phase transitions and structure engineering for dielectric energy storage applications. *J. Appl. Phys.* **2020**, *128*, 070903. [CrossRef]
19. Tolédano, P.; Guennou, M. Theory of antiferroelectric phase transitions. *Phys. Rev. B* **2016**, *94*, 014107. [CrossRef]
20. Li, P.-F.; Liao, W.-Q.; Tang, Y.-Y.; Ye, H.-Y.; Zhang, Y.; Xiong, R.-G. Unprecedented ferroelectric–antiferroelectric–paraelectric phase transitions discovered in an organic–inorganic hybrid perovskite. *J. Am. Chem. Soc.* **2017**, *139*, 8752–8757. [CrossRef]
21. Xu, C.; Liu, Z.; Chen, X.; Yan, S.; Cao, F.; Dong, X.; Wang, G. Pulse discharge properties of PLZST antiferroelectric ceramics compared with ferroelectric and linear dielectrics. *AIP Adv.* **2017**, *7*, 115108. [CrossRef]
22. Sawaguchi, E.; Maniwa, H.; Hoshino, S. Antiferroelectric structure of lead zirconate. *Phys. Rev.* **1951**, *83*, 1078. [CrossRef]
23. Yang, D.; Gao, J.; Shu, L.; Liu, Y.-X.; Yu, J.; Zhang, Y.; Wang, X.; Zhang, B.-P.; Li, J.-F. Lead-free antiferroelectric niobates $AgNbO_3$ and $NaNbO_3$ for energy storage applications. *J. Mater. Chem. A* **2020**, *8*, 23724–23737. [CrossRef]
24. Shirane, G.; Pepinsky, R. Phase Transitions in Antiferroelectric $PbHfO_3$. *Phys. Rev.* **1953**, *91*, 812. [CrossRef]
25. Haertling, G. Improved hot-pressed electrooptic ceramics in the (Pb, La)(Zr, Ti) O_3 system. *J. Am. Ceram. Soc.* **1971**, *54*, 303–309. [CrossRef]
26. Yang, Y.; Liu, P.; Zhang, Y.; Kandula, K.R.; Xu, J.; Zhang, G.; Jiang, S. Low electric-field-induced strain and high energy storage efficiency in (Pb, Ba, La)(Zr, Sn, Ti) O_3 antiferroelectric ceramics through regulating the content of La. *Ceram. Int.* **2020**, *46*, 18106–18113. [CrossRef]
27. Dan, Y.; Zou, K.; Chen, G.; Yu, Y.; Zhang, Y.; Zhang, Q.; Lu, Y.; Zhang, Q.; He, Y. Superior energy-storage properties in (Pb, La)(Zr, Sn, Ti) O_3 antiferroelectric ceramics with appropriate La content. *Ceram. Int.* **2019**, *45*, 11375–11381. [CrossRef]
28. Xu, H.; Dan, Y.; Zou, K.; Chen, G.; Zhang, Q.; Lu, Y.; He, Y. Superior energy storage performance in $Pb_{0.97}La_{0.02}(Zr_{0.50}Sn_{0.43}Ti_{0.07})O_3$ antiferroelectric ceramics. *J. Mater. Res. Technol.* **2019**, *8*, 3291–3296. [CrossRef]
29. Isupov, V. Ferroelectric and antiferroelectric perovskites $PbB'\ 0.5\ B''\ 0.5\ O_3$. *Ferroelectrics* **2003**, *289*, 131–195. [CrossRef]
30. Hao, X.; Zhao, Y.; Zhang, Q. Phase structure tuned electrocaloric effect and pyroelectric energy harvesting performance of $(Pb_{0.97}La_{0.02})$(Zr, Sn, Ti) O_3 antiferroelectric thick films. *J. Phys. Chem. C* **2015**, *119*, 18877–18885. [CrossRef]

31. O'Connor, D.; Hou, D.; Ye, J.; Zhang, Y.; Ok, Y.S.; Song, Y.; Coulon, F.; Peng, T.; Tian, L. Lead-based paint remains a major public health concern: A critical review of global production, trade, use, exposure, health risk, and implications. *Environ. Int.* **2018**, *121*, 85–101. [CrossRef]
32. Kumar, A.; Kumar, A.; MMS, C.-P.; Chaturvedi, A.K.; Shabnam, A.A.; Subrahmanyam, G.; Mondal, R.; Gupta, D.K.; Malyan, S.K.; Kumar, S.S. Lead toxicity: Health hazards, influence on food chain, and sustainable remediation approaches. *Int. J. Environ. Res. Public Health* **2020**, *17*, 2179. [CrossRef] [PubMed]
33. Wang, R.; Xu, Z. Recycling of non-metallic fractions from waste electrical and electronic equipment (WEEE): A review. *Waste Manag.* **2014**, *34*, 1455–1469. [CrossRef] [PubMed]
34. Goosey, E.; Goosey, M. The materials of waste electrical and electronic equipment. In *Waste Electrical and Electronic Equipment (WEEE) Handbook*; Elsevier: Amsterdam, The Netherlands, 2019; pp. 231–262.
35. Sun, Z.; Wang, Z.; Tian, Y.; Wang, G.; Wang, W.; Yang, M.; Wang, X.; Zhang, F.; Pu, Y. Progress, outlook, and challenges in lead-free energy-storage ferroelectrics. *Adv. Electron. Mater.* **2020**, *6*, 1900698. [CrossRef]
36. Kumari, P.; Rai, R.; Sharma, S.; Shandilya, M.; Tiwari, A. State-of-the-art of lead free ferroelectrics: A critical review. *Adv. Mater. Lett* **2015**, *6*, 453–484. [CrossRef]
37. Zhang, S.; Malič, B.; Li, J.-F.; Rödel, J. Lead-free ferroelectric materials: Prospective applications. *J. Mater. Res.* **2021**, *36*, 985–995. [CrossRef]
38. Liu, Z.; Lu, T.; Ye, J.; Wang, G.; Dong, X.; Withers, R.; Liu, Y. Antiferroelectrics for energy storage applications: A review. *Adv. Mater. Technol.* **2018**, *3*, 1800111. [CrossRef]
39. Gao, J.; Zhang, Y.; Zhao, L.; Lee, K.-Y.; Liu, Q.; Studer, A.; Hinterstein, M.; Zhang, S.; Li, J.-F. Enhanced antiferroelectric phase stability in La-doped $AgNbO_3$: Perspectives from the microstructure to energy storage properties. *J. Mater. Chem. A* **2019**, *7*, 2225–2232. [CrossRef]
40. Gao, J.; Liu, Q.; Dong, J.; Wang, X.; Zhang, S.; Li, J.-F. Local structure heterogeneity in Sm-doped $AgNbO_3$ for improved energy-storage performance. *ACS Appl. Mater. Interfaces* **2020**, *12*, 6097–6104. [CrossRef] [PubMed]
41. Gao, J.; Zhao, L.; Liu, Q.; Wang, X.; Zhang, S.; Li, J.F. Antiferroelectric-ferroelectric phase transition in lead-free $AgNbO_3$ ceramics for energy storage applications. *J. Am. Ceram. Soc.* **2018**, *101*, 5443–5450. [CrossRef]
42. Lu, Y.; Karaki, T.; Fujii, T. Hydrothermal synthesis of plate-like sodium niobate particles. *Ceram. Int.* **2015**, *41*, S174–S179. [CrossRef]
43. Wang, J.; Wan, X.; Rao, Y.; Zhao, L.; Zhu, K. Hydrothermal synthesized $AgNbO_3$ powders: Leading to greatly improved electric breakdown strength in ceramics. *J. Eur. Ceram. Soc.* **2020**, *40*, 5589–5596. [CrossRef]
44. Vousden, P. The structure of ferroelectric sodium niobate at room temperature. *Acta Crystallogr.* **1951**, *4*, 545–551. [CrossRef]
45. Kittel, C. Theory of antiferroelectric crystals. *Phys. Rev.* **1951**, *82*, 729. [CrossRef]
46. Torres-Pardo, A.; Jiménez, R.; García-González, E.; González-Calbet, J.M. Phase coexistence in $NaNb_{(1-x)}Ta_xO_3$ materials with enhanced dielectric properties. *J. Mater. Chem.* **2012**, *22*, 14938–14943. [CrossRef]
47. Koruza, J.; Malič, B.; Kosec, M. Microstructure evolution during sintering of sodium niobate. *J. Am. Ceram. Soc.* **2011**, *94*, 4174–4178. [CrossRef]
48. Manan, A.; Rehman, M.U.; Ullah, A.; Ahmad, A.S.; Iqbal, Y.; Qazi, I.; Khan, M.A.; Shah, H.U.; Wazir, A.H. High energy storage density with ultra-high efficiency and fast charging–discharging capability of sodium bismuth niobate lead-free ceramics. *J. Adv. Dielectr.* **2021**, *11*, 2150018. [CrossRef]
49. Gao, P.; Liu, Z.; Zhang, N.; Wu, H.; Bokov, A.A.; Ren, W.; Ye, Z.-G. New antiferroelectric perovskite system with ultrahigh energy-storage performance at low electric field. *Chem. Mater.* **2019**, *31*, 979–990. [CrossRef]
50. Chauhan, A.; Patel, S.; Vaish, R.; Bowen, C.R. Anti-ferroelectric ceramics for high energy density capacitors. *Materials* **2015**, *8*, 8009–8031. [CrossRef] [PubMed]
51. Setter, N. What is a ferroelectric—A materials designer perspective. *Ferroelectrics* **2016**, *500*, 164–182. [CrossRef]
52. Berlincourt, D. Transducers using forced transitions between ferroelectric and antiferroelectric states. *IEEE Trans. Sonics Ultrason* **1966**, *13*, 116–124. [CrossRef]
53. Berlincourt, D.; Krueger, H.; Jaffe, B. Stability of phases in modified lead zirconate with variation in pressure, electric field, temperature and composition. *J. Phys. Chem. Solids* **1964**, *25*, 659–674. [CrossRef]
54. Wei, X.-K.; Dunin-Borkowski, R.E.; Mayer, J. Structural phase transition and in-situ energy storage pathway in nonpolar materials: A review. *Materials* **2021**, *14*, 7854. [CrossRef] [PubMed]
55. Zhang, H.; Wei, T.; Zhang, Q.; Ma, W.; Fan, P.; Salamon, D.; Zhang, S.-T.; Nan, B.; Tan, H.; Ye, Z.-G. A review on the development of lead-free ferroelectric energy-storage ceramics and multilayer capacitors. *J. Mater. Chem. C* **2020**, *8*, 16648–16667. [CrossRef]
56. Yang, Z.; Du, H.; Jin, L.; Poelman, D. High-performance lead-free bulk ceramics for electrical energy storage applications: Design strategies and challenges. *J. Mater. Chem. A* **2021**, *9*, 18026–18085. [CrossRef]
57. Engel, G.F. Design and materials of antiferroelectric capacitors for high density power electronic applications. In Proceedings of the CIPS 2016; 9th International Conference on Integrated Power Electronics Systems, Nuremberg, Germany, 8–10 March 2016; pp. 1–7.
58. Hao, X.; Zhai, J.; Kong, L.B.; Xu, Z. A comprehensive review on the progress of lead zirconate-based antiferroelectric materials. *Prog. Mater. Sci.* **2014**, *63*, 1–57. [CrossRef]

59. Tan, X.; Ma, C.; Frederick, J.; Beckman, S.; Webber, K.G. The antiferroelectric↔ ferroelectric phase transition in lead-containing and lead-free perovskite ceramics. *J. Am. Ceram. Soc.* **2011**, *94*, 4091–4107. [CrossRef]
60. Zhou, Z.; Yang, Q.; Liu, M.; Zhang, Z.; Zhang, X.; Sun, D.; Nan, T.; Sun, N.; Chen, X. Antiferroelectric materials, applications and recent progress on multiferroic heterostructures. *Spin* **2015**, *5*, 1530001. [CrossRef]
61. Chen, B.; Hasegawa, T.; Ohta, H.; Katayama, T. Antiferroelectric-to-ferroelectric phase transition in hexagonal rare-earth iron oxides. *J. Mater. Chem. C* **2022**, *10*, 5621–5626. [CrossRef]
62. Prosandeev, S.; Prokhorenko, S.; Nahas, Y.; Yang, Y.; Xu, C.; Grollier, J.; Talbayev, D.; Dkhil, B.; Bellaiche, L. Hidden phases with neuromorphic responses and highly enhanced piezoelectricity in an antiferroelectric prototype. *Phys. Rev. B* **2022**, *105*, L100101. [CrossRef]
63. Goh, Y.; Hwang, J.; Jeon, S. Excellent reliability and high-speed antiferroelectric $HfZrO_2$ tunnel junction by a high-pressure annealing process and built-in bias engineering. *ACS Appl. Mater. Interfaces* **2020**, *12*, 57539–57546. [CrossRef]
64. Ali, T.; Mertens, K.; Olivo, R.; Rudolph, M.; Oehler, S.; Kühnel, K.; Lehninger, D.; Müller, F.; Hoffmann, R.; Schramm, P. Impact of the Nonlinear Dielectric Hysteresis Properties of a Charge Trap Layer in a Novel Hybrid High-Speed and Low-Power Ferroelectric or Antiferroelectric HSO/HZO Boosted Charge Trap Memory. *IEEE Trans. Electron Devices* **2021**, *68*, 2098–2106. [CrossRef]
65. Müller, J.; Polakowski, P.; Mueller, S.; Mikolajick, T. Ferroelectric hafnium oxide based materials and devices: Assessment of current status and future prospects. *ECS J. Solid State Sci. Technol.* **2015**, *4*, N30. [CrossRef]
66. Ali, F.; Abbas, A.; Wu, G.; Daaim, M.; Akhtar, A.; Kim, K.H.; Yang, B. Novel Fluorite-Structured Materials for Solid-State Refrigeration. *Small* **2022**, *18*, 2200133. [CrossRef] [PubMed]
67. Muller, J.; Boscke, T.S.; Schroder, U.; Mueller, S.; Brauhaus, D.; Bottger, U.; Frey, L.; Mikolajick, T. Ferroelectricity in simple binary ZrO_2 and HfO_2. *Nano Lett.* **2012**, *12*, 4318–4323. [CrossRef] [PubMed]
68. Ali, F.; Zhou, D.; Ali, M.; Ali, H.W.; Daaim, M.; Khan, S.; Hussain, M.M.; Sun, N. Recent progress on energy-related applications of HfO_2-based ferroelectric and antiferroelectric materials. *ACS Appl. Electron. Mater.* **2020**, *2*, 2301–2317. [CrossRef]
69. Randall, C.A.; Fan, Z.; Reaney, I.; Chen, L.Q.; Trolier-McKinstry, S. Antiferroelectrics: History, fundamentals, crystal chemistry, crystal structures, size effects, and applications. *J. Am. Ceram. Soc.* **2021**, *104*, 3775–3810. [CrossRef]
70. Xu, H.; Guo, W.; Wang, J.; Ma, Y.; Han, S.; Liu, Y.; Lu, L.; Pan, X.; Luo, J.; Sun, Z. A metal-free molecular antiferroelectric material showing high phase transition temperatures and large electrocaloric effects. *J. Am. Chem. Soc.* **2021**, *143*, 14379–14385. [CrossRef]
71. Kambale, K.; Mahajan, A.; Butee, S. Effect of grain size on the properties of ceramics. *Met. Powder Rep.* **2019**, *74*, 130–136. [CrossRef]
72. Zhang, T.; Lei, Y.; Yin, J.; Du, J.; Yu, P. Effects of pores on dielectric breakdown of alumina ceramics under AC electric field. *Ceram. Int.* **2019**, *45*, 13951–13957. [CrossRef]
73. Haddour, L.; Mesrati, N.; Goeuriot, D.; Tréheux, D. Relationships between microstructure, mechanical and dielectric properties of different alumina materials. *J. Eur. Ceram. Soc.* **2009**, *29*, 2747–2756. [CrossRef]
74. Megaw, H.D. The seven phases of sodium niobate. *Ferroelectrics* **1974**, *7*, 87–89. [CrossRef]
75. Lefkowitz, I.; Łukaszewicz, K.; Megaw, H. The high-temperature phases of sodium niobate and the nature of transitions in pseudosymmetric structures. *Acta Crystallogr.* **1966**, *20*, 670–683. [CrossRef]
76. Xu, Y.; Hong, W.; Feng, Y.; Tan, X. Antiferroelectricity induced by electric field in $NaNbO_3$-based lead-free ceramics. *Appl. Phys. Lett.* **2014**, *104*, 052903. [CrossRef]
77. Mishra, S.; Choudhury, N.; Chaplot, S.; Krishna, P.; Mittal, R. Competing antiferroelectric and ferroelectric interactions in Na Nb O 3: Neutron diffraction and theoretical studies. *Phys. Rev. B* **2007**, *76*, 024110. [CrossRef]
78. Lu, Z.; Sun, D.; Wang, G.; Zhao, J.; Zhang, B.; Wang, D.; Shyha, I. Energy storage properties in Nd-doped $AgNbTaO_3$ lead-free antiferroelectric ceramics with Nb-site vacancies. *J. Adv. Dielectr.* **2023**, *13*, 2242006. [CrossRef]
79. Lu, Z.; Zhu, Y.; Sun, D.; Wang, B.; Wang, D.; Day, S.; Wang, G. Phases study for AgNb (Ta, W)O_3 lead-free antiferroelectric ceramics. *Results Eng.* **2023**, *20*, 101447. [CrossRef]
80. Zhelnova, O.; Fesenko, O. Phase transitions and twinning in $NaNbO_3$ crystals. *Ferroelectrics* **1987**, *75*, 469–475. [CrossRef]
81. Wood, E.; Miller, R.; Remeika, J. The field-induced ferroelectric phase of sodium niobate. *Acta Crystallogr.* **1962**, *15*, 1273–1279. [CrossRef]
82. Cross, L.E.; Nicholson, B.J. The optical and electrical properties of single crystals of sodium niobate. *Lond. Edinb. Dublin Philos. Mag. J. Sci.* **1955**, *46*, 453–466. [CrossRef]
83. Ulinzheev, A.; Leiderman, A.; Smotrakov, V.; Topolov, V.Y.; Fesenko, O. Phase transitions induced in $NaNbO_3$ crystals by varying the direction of an external electric field. *Phys. Solid State* **1997**, *39*, 972–974. [CrossRef]
84. Miller, R.C.; Wood, E.A.; Remeika, J.P.; Savage, A. Na $(Nb_{1-x}V_x)$ O_3 System and "Ferrielectricity". *J. Appl. Phys.* **1962**, *33*, 1623–1630. [CrossRef]
85. Cross, L. Electric double hysteresis in $(K_xNa_{1-x})NbO_3$ single crystals. *Nature* **1958**, *181*, 178–179. [CrossRef]
86. Luo, N.; Ma, L.; Luo, G.; Xu, C.; Rao, L.; Chen, Z.; Cen, Z.; Feng, Q.; Chen, X.; Toyohisa, F. Well-defined double hysteresis loop in $NaNbO_3$ antiferroelectrics. *Nat. Commun.* **2023**, *14*, 1776. [CrossRef] [PubMed]
87. Chao, L.; Hou, Y.; Zheng, M.; Zhu, M. High dense structure boosts stability of antiferroelectric phase of $NaNbO_3$ polycrystalline ceramics. *Appl. Phys. Lett.* **2016**, *108*, 212902. [CrossRef]
88. Dungan, R.; Golding, R. Metastable ferroelectric sodium niobate. *J. Am. Ceram. Soc.* **1964**, *47*, 73–76. [CrossRef]

89. Li, W.; Xia, X.; Zeng, J.; Zheng, L.; Li, G. Significant differences in $NaNbO_3$ ceramics fabricated using Nb_2O_5 precursors with various crystal structures. *Ceram. Int.* **2020**, *46*, 3759–3766. [CrossRef]
90. Koruza, J.; Groszewicz, P.; Breitzke, H.; Buntkowsky, G.; Rojac, T.; Malič, B. Grain-size-induced ferroelectricity in $NaNbO_3$. *Acta Mater.* **2017**, *126*, 77–85. [CrossRef]
91. Fan, Y.; Zhou, Z.; Liang, R.; Zhou, M.; Dong, X. The effect of A-site nonstoichiometry on the microstructure, electric properties, and phase stability of $NaNbO_3$ polycrystalline ceramics. *J. Eur. Ceram. Soc.* **2019**, *39*, 4712–4718. [CrossRef]
92. Zhang, M.-H.; Fulanović, L.; Egert, S.; Ding, H.; Groszewicz, P.B.; Kleebe, H.-J.; Molina-Luna, L.; Koruza, J. Electric-field-induced antiferroelectric to ferroelectric phase transition in polycrystalline $NaNbO_3$. *Acta Mater.* **2020**, *200*, 127–135. [CrossRef]
93. Johnston, K.E.; Griffin, J.M.; Walton, R.I.; Dawson, D.M.; Lightfoot, P.; Ashbrook, S.E. 93Nb NMR and DFT investigation of the polymorphs of $NaNbO_3$. *Phys. Chem. Chem. Phys.* **2011**, *13*, 7565–7576. [CrossRef]
94. Guo, H.; Shimizu, H.; Mizuno, Y.; Randall, C.A. Strategy for stabilization of the antiferroelectric phase (Pbma) over the metastable ferroelectric phase (P21ma) to establish double loop hysteresis in lead-free $(1 - x)$ $NaNbO_3$-$xSrZrO_3$ solid solution. *J. Appl. Phys.* **2015**, *117*, 214103. [CrossRef]
95. Shimizu, H.; Guo, H.; Reyes-Lillo, S.E.; Mizuno, Y.; Rabe, K.M.; Randall, C.A. Lead-free antiferroelectric: X $CaZrO_3$-$(1 - x)$ $NaNbO_3$ system $(0 \leq x \leq 0.10)$. *Dalton Trans.* **2015**, *44*, 10763–10772. [CrossRef] [PubMed]
96. Guo, H.; Shimizu, H.; Randall, C.A. Direct evidence of an incommensurate phase in $NaNbO_3$ and its implication in $NaNbO_3$-based lead-free antiferroelectrics. *Appl. Phys. Lett.* **2015**, *107*, 112904. [CrossRef]
97. Shakhovoy, R.; Raevskaya, S.; Shakhovaya, L.; Suzdalev, A.D.; Raevski, I.; Yuzyuk, Y.I.; Semenchev, A.; El Marssi, M. Ferroelectric Q and antiferroelectric P phases' coexistence and local phase transitions in oxygen-deficient $NaNbO_3$ single crystal: Micro-Raman, dielectric and dilatometric studies. *J. Raman Spectrosc.* **2012**, *43*, 1141–1145. [CrossRef]
98. Qi, H.; Zuo, R.; Xie, A.; Fu, J.; Zhang, D. Excellent energy-storage properties of $NaNbO_3$-based lead-free antiferroelectric orthorhombic P-phase (Pbma) ceramics with repeatable double polarization-field loops. *J. Eur. Ceram. Soc.* **2019**, *39*, 3703–3709. [CrossRef]
99. Ma, L.; Chen, Z.; Luo, G.; Che, Z.; Xu, C.; Shan, D.; Cen, Z.; Feng, Q.; Chen, X.; Fujita, T. High energy storage density in $NaNbO_3$ antiferroelectrics with double hysteresis loop. *J. Mater.* 2023, *in press*. [CrossRef]
100. Zhang, M.-H.; Ding, H.; Egert, S.; Zhao, C.; Villa, L.; Fulanović, L.; Groszewicz, P.B.; Buntkowsky, G.; Kleebe, H.-J.; Albe, K. Tailoring high-energy storage $NaNbO_3$-based materials from antiferroelectric to relaxor states. *Nat. Commun.* **2023**, *14*, 1525. [CrossRef]
101. Kondo, N.; Sakamoto, W.; Lee, B.-Y.; Iijima, T.; Kumagai, J.; Moriya, M.; Yogo, T. Improvement in ferroelectric properties of chemically synthesized lead-free piezoelectric (K, Na)(Nb, Ta)O_3 thin films by Mn doping. *Jpn. J. Appl. Phys.* **2010**, *49*, 09MA04. [CrossRef]
102. Guo, Y.; Xiao, P.; Wen, R.; Wan, Y.; Zheng, Q.; Shi, D.; Lam, K.H.; Liu, M.; Lin, D. Critical roles of Mn-ions in enhancing the insulation, piezoelectricity and multiferroicity of $BiFeO_3$-based lead-free high temperature ceramics. *J. Mater. Chem. C* **2015**, *3*, 5811–5824. [CrossRef]
103. Wechsler, B.; Klein, M.B. Thermodynamic point defect model of barium titanate and application to the photorefractive effect. *JOSA B* **1988**, *5*, 1711–1723. [CrossRef]
104. Ding, H.; Zhang, M.H.; Koruza, J.; Molina-Luna, L.; Kleebe, H.J. Domain morphology of newly designed lead-free antiferroelectric $NaNbO_3$-$SrSnO_3$ ceramics. *J. Am. Ceram. Soc.* **2021**, *104*, 3715–3725. [CrossRef]
105. Villa, L.; Albe, K. Role of doping and defect quenching in antiferroelectric $NaNbO_3$ from first principles. *Phys. Rev. B* **2022**, *106*, 134101. [CrossRef]
106. Berlincourt, D. Transducer using the electric field-forced antiferroelectric-ferroelectric transition. *Ultrasonics* **1968**, *6*, 48–51. [CrossRef]
107. Zhou, M.; Liang, R.; Zhou, Z.; Dong, X. Superior energy storage properties and excellent stability of novel $NaNbO_3$-based lead-free ceramics with A-site vacancy obtained via a Bi_2O_3 substitution strategy. *J. Mater. Chem. A* **2018**, *6*, 17896–17904. [CrossRef]
108. Yang, L.; Kong, X.; Li, Q.; Lin, Y.-h.; Zhang, S.; Nan, C.-w. Excellent energy storage properties achieved in sodium niobate-based relaxor ceramics through doping tantalum. *ACS Appl. Mater. Interfaces* **2022**, *14*, 32218–32226. [CrossRef]
109. Kobayashi, K.; Ryu, M.; Doshida, Y.; Mizuno, Y.; Randall, C.A. Novel High-Temperature Antiferroelectric-Based Dielectric $NaNbO_3$–$NaTaO_3$ Solid Solutions Processed in Low Oxygen Partial Pressures. *J. Am. Ceram. Soc.* **2013**, *96*, 531–537. [CrossRef]
110. Yadav, A.; Fahad, M.; Satapathy, S.; Sarun, P. Effect of tantalum on the temperature dependent electrical characteristics of $NaNb_{1-x}Ta_xO_3$ $(0.0 \leq x \leq 0.3)$ ceramics between 400 and 560 °C. *J. Alloys Compd.* **2019**, *797*, 902–911. [CrossRef]
111. Zhu, L.-F.; Yan, Y.; Leng, H.; Li, X.; Cheng, L.-Q.; Priya, S. Energy-storage performance of $NaNbO_3$ based multilayered capacitors. *J. Mater. Chem. C* **2021**, *9*, 7950–7957. [CrossRef]
112. Chen, H.; Chen, X.; Shi, J.; Sun, C.; Dong, X.; Pang, F.; Zhou, H. Achieving ultrahigh energy storage density in $NaNbO_3$–Bi$(Ni_{0.5}Zr_{0.5})O_3$ solid solution by enhancing the breakdown electric field. *Ceram. Int.* **2020**, *46*, 28407–28413. [CrossRef]
113. Liang, C.; Wang, C.; Zhao, H.; Cao, W.; Huang, X.; Wang, C. Enhanced energy storage performance of $NaNbO_3$-based ceramics via band and domain engineering. *Ceram. Int.* **2023**, *49*, 40326–40335. [CrossRef]
114. Liu, G.; Chen, L.; Qi, H. Energy storage properties of $NaNbO_3$-based lead-free superparaelectrics with large antiferrodistortion. *Microstructures* **2023**, *3*, 2023009.

115. Liu, J.; Li, P.; Li, C.; Bai, W.; Wu, S.; Zheng, P.; Zhang, J.; Zhai, J. Synergy of a stabilized antiferroelectric phase and domain engineering boosting the energy storage performance of $NaNbO_3$-based relaxor antiferroelectric ceramics. *ACS Appl. Mater. Interfaces* **2022**, *14*, 17662–17673. [CrossRef]
116. Chen, H.; Wang, X.; Dong, X.; Pan, Y.; Wang, J.; Deng, L.; Dong, Q.; Zhang, H.; Zhou, H.; Chen, X. Adjusting the energy-storage characteristics of $0.95NaNbO_3$–$0.05Bi(Mg_{0.5}Sn_{0.5})O_3$ ceramics by doping linear perovskite materials. *ACS Appl. Mater. Interfaces* **2022**, *14*, 25609–25619. [CrossRef]
117. Zhang, S.; Li, W.; Zhang, Y.; Tang, X.; Jiang, Y.; Guo, X. Excellent energy density and power density achieved in $NaNbO_3$-based relaxor ferroelectric ceramics. *Mater. Sci. Eng. B* **2024**, *299*, 117025. [CrossRef]
118. Yang, W.; Zeng, H.; Yan, F.; Qian, J.; Zhu, K.; Zhao, K.; Li, G.; Zhai, J. Microstructure-driven excellent energy storage $NaNbO_3$-based lead-free ceramics. *Ceram. Int.* **2022**, *48*, 37476–37482. [CrossRef]
119. Qi, H.; Li, W.; Wang, L.; Chen, L.; Liu, H.; Deng, S.; Chen, J. Large (anti) ferrodistortive $NaNbO_3$-based lead-free relaxors: Polar nanoregions embedded in ordered oxygen octahedral tilt matrix. *Mater. Today* **2022**, *60*, 91–97. [CrossRef]
120. Wang, X.; Chen, H.; Pan, Y.; Dong, Q.; Wang, J.; Chen, X.; Zhou, H. Dielectric ceramics with excellent energy storage properties were obtained by doping $0.92NaNbO_3$-0.08Bi $(Ni_{0.5}Zr_{0.5})O_3$ ceramics. *J. Power Sources* **2023**, *566*, 232934. [CrossRef]
121. Xie, A.; Zuo, R.; Qiao, Z.; Fu, Z.; Hu, T.; Fei, L. $NaNbO_3$-$(Bi_{0.5}Li_{0.5})TiO_3$ lead-free relaxor ferroelectric capacitors with superior energy-storage performances via multiple synergistic design. *Adv. Energy Mater.* **2021**, *11*, 2101378. [CrossRef]
122. Li, H.; Pan, Z.; Chen, X.; Zhao, J.; Tang, L.; Liu, J.; Li, P.; Zhai, J. Stable relaxor ferroelectric phase of $NaNbO_3$-based ceramic with superb energy storage performances. *Mater. Today Phys.* **2023**, *38*, 101208. [CrossRef]
123. Qi, H.; Zuo, R.; Xie, A.; Tian, A.; Fu, J.; Zhang, Y.; Zhang, S. Ultrahigh Energy-Storage Density in $NaNbO_3$-Based Lead-Free Relaxor Antiferroelectric Ceramics with Nanoscale Domains. *Adv. Funct. Mater.* **2019**, *29*, 1903877. [CrossRef]
124. Lv, Z.; Lu, T.; Liu, Z.; Hu, T.; Hong, Z.; Guo, S.; Xu, Z.; Song, Y.; Chen, Y.; Zhao, X. $NaNbO_3$-Based Multilayer Ceramic Capacitors with Ultrahigh Energy Storage Performance. *Adv. Energy Mater.* **2024**, *14*, 2304291. [CrossRef]
125. Jiang, J.; Li, X.; Li, L.; Guo, S.; Zhang, J.; Wang, J.; Zhu, H.; Wang, Y.; Zhang, S.-T. Novel lead-free $NaNbO_3$-based relaxor antiferroelectric ceramics with ultrahigh energy storage density and high efficiency. *J. Mater.* **2022**, *8*, 295–301. [CrossRef]
126. Jiang, J.; Meng, X.; Li, L.; Guo, S.; Huang, M.; Zhang, J.; Wang, J.; Hao, X.; Zhu, H.; Zhang, S.-T. Ultrahigh energy storage density in lead-free relaxor antiferroelectric ceramics via domain engineering. *Energy Storage Mater.* **2021**, *43*, 383–390. [CrossRef]
127. Liu, X.; Li, Y.; Sun, N.; Hao, X. High energy-storage performance of PLZS antiferroelectric multilayer ceramic capacitors. *Inorg. Chem. Front.* **2020**, *7*, 756–764. [CrossRef]
128. Liu, X.; Li, Y.; Hao, X. Ultra-high energy-storage density and fast discharge speed of $(Pb_{0.98-x}La_{0.02}Sr_x)(Zr_{0.9}Sn_{0.1})0.995O_3$ antiferroelectric ceramics prepared via the tape-casting method. *J. Mater. Chem. A* **2019**, *7*, 11858–11866. [CrossRef]
129. Feng, Y.; Zhen, Y.; Jiang, X.; Yang, Z.; Qin, Z.; Yang, W.; Qie, Y.; Geng, H. Optimized energy storage performance in $NaNbO_3$-based ceramics via composition modification and micro-structure control. *Ceram. Int.* **2023**, *49*, 14135–14144. [CrossRef]
130. Jiang, J.; Meng, X.; Li, L.; Zhang, J.; Guo, S.; Wang, J.; Hao, X.; Zhu, H.; Zhang, S.-T. Enhanced energy storage properties of lead-free $NaNbO_3$-based ceramics via A/B-site substitution. *Chem. Eng. J.* **2021**, *422*, 130130. [CrossRef]
131. Qiao, Z.; Li, T.; Qi, H.; Zuo, R. Excellent energy storage properties in $NaNbO_3$-based lead-free ceramics by modulating antiferrodistortive of P phase. *J. Alloys Compd.* **2022**, *898*, 162934. [CrossRef]
132. Yang, J.; Zhao, Y.; Zhu, L.; Hao, X. Enhanced electrocaloric effect of relaxor potassium sodium niobate lead-free ceramic via multilayer structure. *Scr. Mater.* **2021**, *193*, 97–102. [CrossRef]
133. Jenko, D.; Benčan, A.; Malič, B.; Holc, J.; Kosec, M. Electron microscopy studies of potassium sodium niobate ceramics. *Microsc. Microanal.* **2005**, *11*, 572–580. [CrossRef]
134. Zhang, T.; Karaki, T.; Fujii, T. The preparation of MnO_2-doped $NaNbO_3$-based lead-free ceramics with enhanced energy storage performance and attractive electrocaloric effect. *Jpn. J. Appl. Phys.* **2022**, *61*, SB1028. [CrossRef]
135. Pan, Z.; Liu, B.; Zhai, J.; Yao, L.; Yang, K.; Shen, B. $NaNbO_3$ two-dimensional platelets induced highly energy storage density in trilayered architecture composites. *Nano Energy* **2017**, *40*, 587–595. [CrossRef]
136. Lanfredi, S.; Dessemond, L.; Rodrigues, A.C. Effect of porosity on the electrical properties of polycrystalline sodium niobate: I, electrical conductivity. *J. Am. Ceram. Soc.* **2003**, *86*, 291–298. [CrossRef]
137. Lanfredi, S.; Rodrigues, A. Impedance spectroscopy study of the electrical conductivity and dielectric constant of polycrystalline $LiNbO_3$. *J. Appl. Phys.* **1999**, *86*, 2215–2219. [CrossRef]
138. Lanfredi, S.; Carvalho, J.; Hernandes, A.C. Electric and dielectric properties of $Bi_{12}TiO_{20}$ single crystals. *J. Appl. Phys.* **2000**, *88*, 283–287. [CrossRef]
139. Harttar, M.A.M.; Rashid, M.W.A.; Azlan, U.A.A. Physical and electrical properties enhancement of rare-earth doped-potassium sodium niobate (KNN): A review. *Ceram. –Silikáty* **2015**, *59*, 158–163.
140. Lanfredi, S.; Rodrigues, A.C.; Dessemond, L. Effect of porosity on the electrical properties of polycrystalline sodium niobate: II, dielectric behavior. *J. Am. Ceram. Soc.* **2003**, *86*, 2103–2110. [CrossRef]
141. Lu, Z.; Wang, G.; Bao, W.; Li, J.; Li, L.; Mostaed, A.; Yang, H.; Ji, H.; Li, D.; Feteira, A. Superior energy density through tailored dopant strategies in multilayer ceramic capacitors. *Energy Environ. Sci.* **2020**, *13*, 2938–2948. [CrossRef]
142. Wang, G.; Li, J.; Zhang, X.; Fan, Z.; Yang, F.; Feteira, A.; Zhou, D.; Sinclair, D.C.; Ma, T.; Tan, X. Ultrahigh energy storage density lead-free multilayers by controlled electrical homogeneity. *Energy Environ. Sci.* **2019**, *12*, 582–588. [CrossRef]

143. Qi, H.; Zuo, R. Linear-like lead-free relaxor antiferroelectric $(Bi_{0.5}Na_{0.5})TiO_3$–$NaNbO_3$ with giant energy-storage density/efficiency and super stability against temperature and frequency. *J. Mater. Chem. A* **2019**, *7*, 3971–3978. [CrossRef]
144. Dong, X.; Li, X.; Chen, X.; Chen, H.; Sun, C.; Shi, J.; Pang, F.; Zhou, H. High energy storage density and power density achieved simultaneously in $NaNbO_3$-based lead-free ceramics via antiferroelectricity enhancement. *J. Mater.* **2021**, *7*, 629–639. [CrossRef]
145. Tao, H.; Yin, J.; Zhao, C.; Wu, J. Relaxor behavior of potassium sodium niobate ceramics by domain evolution. *J. Eur. Ceram. Soc.* **2021**, *41*, 335–343. [CrossRef]
146. Li, F.; Zhang, S.; Damjanovic, D.; Chen, L.Q.; Shrout, T.R. Local structural heterogeneity and electromechanical responses of ferroelectrics: Learning from relaxor ferroelectrics. *Adv. Funct. Mater.* **2018**, *28*, 1801504. [CrossRef]
147. Yao, F.Z.; Wang, K.; Jo, W.; Webber, K.G.; Comyn, T.P.; Ding, J.X.; Xu, B.; Cheng, L.Q.; Zheng, M.P.; Hou, Y.D. Diffused phase transition boosts thermal stability of high-performance lead-free piezoelectrics. *Adv. Funct. Mater.* **2016**, *26*, 1217–1224. [CrossRef]
148. Zhao, C.; Wu, B.; Wang, K.; Li, J.-F.; Xiao, D.; Zhu, J.; Wu, J. Practical high strain with superior temperature stability in lead-free piezoceramics through domain engineering. *J. Mater. Chem. A* **2018**, *6*, 23736–23745. [CrossRef]
149. Xu, K.; Li, J.; Lv, X.; Wu, J.; Zhang, X.; Xiao, D.; Zhu, J. Superior piezoelectric properties in potassium–sodium niobate lead-free ceramics. *Adv. Mater.* **2016**, *28*, 8519–8523. [CrossRef]
150. Zheng, T.; Wu, H.; Yuan, Y.; Lv, X.; Li, Q.; Men, T.; Zhao, C.; Xiao, D.; Wu, J.; Wang, K. The structural origin of enhanced piezoelectric performance and stability in lead free ceramics. *Energy Environ. Sci.* **2017**, *10*, 528–537. [CrossRef]
151. Tao, H.; Wu, H.; Liu, Y.; Zhang, Y.; Wu, J.; Li, F.; Lyu, X.; Zhao, C.; Xiao, D.; Zhu, J. Ultrahigh performance in lead-free piezoceramics utilizing a relaxor slush polar state with multiphase coexistence. *J. Am. Chem. Soc.* **2019**, *141*, 13987–13994. [CrossRef]
152. Koruza, J.; Liu, H.; Höfling, M.; Zhang, M.-H.; Veber, P. (K, Na) NbO_3-based piezoelectric single crystals: Growth methods, properties, and applications. *J. Mater. Res.* **2020**, *35*, 990–1016. [CrossRef]
153. Liu, X.; Tan, X. Giant strains in non-textured $(Bi_{1/2}Na_{1/2})$ TiO_3-based lead-free ceramics. *Adv. Mater.* **2016**, *28*, 574–578. [CrossRef]
154. Machado, R.; Sepliarsky, M.; Stachiotti, M. Relative phase stability and lattice dynamics of $NaNbO_3$ from first-principles calculations. *Phys. Rev. B* **2011**, *84*, 134107. [CrossRef]

ceramics

Article

Nanosized Tungsten Powder Synthesized Using the Nitridation–Decomposition Method

Qing-Yin He [1], Ben-Li Zhao [1] and Shi-Kuan Sun [2,*]

[1] Foshan Stomatology Hospital & School of Medicine, Foshan University, Foshan 528000, China
[2] School of Materials Science and Hydrogen Energy, Foshan University, Foshan 528000, China
* Correspondence: shikuansun@fosu.edu.cn

Abstract: A facile, one-step nitridation–decomposition method was developed for the synthesis of nanosized tungsten powder with a high surface area. This approach involved the nitridation of WO_3 in NH_3 to form mesoporous tungsten nitride (W_2N), followed by in situ decomposition of W_2N to directly yield single-phase W particles. The phase and morphology evolution during the synthesis were systematically investigated and compared with the carbothermal reduction of WO_3. It was revealed that powdered tungsten product with single-phase particles was obtained after nitridation at 800 °C combined with in situ decomposition at 1000 °C, displaying an average particle size of 15 nm and a large specific surface area of 6.52 m^2/g. Furthermore, the proposed method avoided the limitations associated with intermediate phase formation and coarsening observed in carbothermal reduction, which resulted in the growth of W particles up to ~4.4 μm in size. This work demonstrates the potential of the nitridation–decomposition approach for the scalable and efficient synthesis of high-quality, fine-grained tungsten powder.

Keywords: nanosized tungsten powder; powder processing; morphology evolution; in situ synthesis

Citation: He, Q.-Y.; Zhao, B.-L.; Sun, S.-K. Nanosized Tungsten Powder Synthesized Using the Nitridation–Decomposition Method. *Ceramics* **2024**, *7*, 680–688. https://doi.org/10.3390/ceramics7020044

Academic Editor: Koji Morita

Received: 20 March 2024
Revised: 25 April 2024
Accepted: 2 May 2024
Published: 11 May 2024

1. Introduction

Tungsten (W) is considered to be an important structural material for high-temperature applications. Tungsten possesses a melting point of 3422 °C [1,2], exceeding that of any other pure metal, allowing it to withstand extreme thermal environments. Its high tensile strength of 1725 MPa allows it to resist deformation and breakage under immense loads [3]. Furthermore, tungsten exhibits outstanding corrosion resistance, making it a valuable material for applications involving harsh chemical environments [4]. Coupled with a low thermal expansion coefficient (4.32×10^{-6} K^{-1} at 300 K), it can maintain its dimensions with minimal change under significant temperature changes. This makes it ideal for applications requiring dimensional stability in extreme environments, such as jet engine components and nuclear reactors [5–7]. Moreover, ultrafine tungsten powder is the essential raw material used for the production of ultrafine cemented carbides, widely used in the printed circuit board industry as a component of micro-drilling tools [8,9].

Obtaining high-quality tungsten powder is a critical step in the fabrication of advanced tungsten-based materials [10,11]. The main challenge is focused on densifying the material, often requiring a sintering temperature exceeding 2000 °C. Researchers have explored various methods to synthesize nanoscale tungsten powder, such as high-energy ball milling [12,13], molten salt synthesis [14], and spray conversion processing [15]. Nanocrystalline tungsten powder has been fully densified at temperatures as low as 1400 °C via spark plasma sintering, resulting in an ultrafine-grained microstructure and excellent mechanical performance [16]. However, the most widely-used method is the traditional reduction method [17,18], where tungsten powder is obtained from the reduction of tungsten oxide or ammonium paratungstate (APT) under the flow of hydrogen (H_2). Unfortunately, water vapor produced by the reduction leads to an increase in particle size due to the formation of volatile $WO_2(OH)_2$, which transports tungsten through the gas phase and promotes particle growth [19]. Therefore, reducing the particle size of the starting powder is a key strategy to lower the densification temperature and achieve a controllable microstructure. This necessitates the development of novel synthesis methods that circumvent the limitations of the traditional hydrogen reduction approach.

As previously reported, a nitridation–carburization method was used to synthesize ultrafine tungsten carbide (WC) powder, in which nitridation of WO_3 using NH_3 and carburization using CH_4-H_2 were conducted sequentially [20]. This two-step process involved the sequential nitridation of WO_3 using NH_3 followed by carburization using CH_4-H_2. Interestingly, the nitridation step resulted in the formation of W_2N, which subsequently transformed into WC powder through carburization at 800 °C. A WC powder was formed with an impressively fine particle size of approximately 20–30 nm. The hypothesis suggested that the initial formation of metallic tungsten occurred via the rapid decomposition of W_2N nanoparticles, a process significantly faster than WO_3 reduction. Rather than the growth and coarsening attributed to the existence of $WO_2(OH)_2$, as mentioned previously, the particle size of WC was be maintained at the nanoscale due to the absence of H_2O in the reaction system. Inspired by this work, an in situ synthesis process was proposed to produce metallic tungsten powder, using a one-step nitridation–decomposition method. This method involved the initial preparation of W_2N intermediate through the nitridation of WO_3 in NH_3, followed by in situ decomposition under an inert atmosphere. The phase and morphological evolution throughout the process were studied. As an alternative method, carbothermal reduction offers the advantages of faster reaction times due to the absence of solid–gas reactions and generally low costs. In the present work, we also attempted to produce tungsten powder via carbothermal reduction of WO_3, and the morphology of the final product and reaction mechanism are compared.

2. Experimental Procedures

A sol–gel route was employed to synthesize WO_3 powder, starting from ammonium metatungstate (Zhuzhou Cemented Carbide Group Corp. Ltd., Zhuzhou, China) and anhydrous citric acid (Sinopharm Chemical Reagent Co., Ltd., Shanghai, China). After gelation, drying, and calcination, nanosized WO_3 precursor with an average particle size of approximately 40 nm was obtained for the following process. More details regarding the procedures can be found in a previous study [20].

The prepared WO_3 powder was placed on a graphite boat and heat-treated in a horizontal tube furnace under a constant flow of NH_3 (600 mL/min) at 800 °C for 3 h to facilitate complete nitridation. Following nitridation, the NH_3 flow was immediately switched to Ar (1000 mL/min) and the temperature was elevated to 900 °C and 1000 °C for in situ decomposition of W_2N to metallic W. The intermediate W_2N after ammonolysis of WO_3 at 800 °C was also collected and characterized.

For carbothermal reduction, the molar ratio of WO_3 to carbon black (purity: 99.9%, Beijing Enoch technology Ltd., Beijing, China) was controlled at 1:2.9. The starting mixtures were mixed for 24 h in a polyethene jar using Si_3N_4 balls and ethyl alcohol as a carrier fluid. The slurry was dried in a rotary evaporator and then moved to a graphite element resistance furnace; the temperature was increased to 1300 °C at 10 °C/min followed by a dwell time of 3 h under vacuum.

The phase assemblage of the as-received powders was investigated using X-ray diffraction with Cu Kα radiation at 40 kV and 40 mA (XRD; Rigaku D/MAX-2 550V, Tokyo, Japan). The mass fraction of W in the mixture of W_2N/W was determined based on the relative intensity of the strongest diffraction peak using the K value method [21]. The reference intensity ratio of W_2N (18.34) and W (18.00) was adopted from JCPDS Card nos. 25-1257 and 89-2767, respectively. Transmission electron microscopy (TEM; JEOL 2100F, JEOL Co. Ltd., Tokyo, Japan) and scanning electron microscopy (SEM; LYRA 3 XMU, Tescan, USA) were used to observe the powder morphology, where the powdered samples from both method were dispersed onto a Cu grid. The specific surface area was examined using the BET method (ASAP 2010, Micromeritics, Norcross, USA).

3. Results and Discussion

XRD analysis (Figure 1a) confirmed the formation of pure tungsten product through carbothermal reduction of WO_3 at 1300 °C under vacuum, as evidenced by the presence of only tungsten reflections matching JCPDS Card no. 89-2767. This indicated the absence of residual oxides or unreacted starting materials. However, compared to the initial WO_3 powder with an average particle size of ~40 nm [20], TEM images (Figure 1b) revealed the coarsening of W particles after the reaction. The estimated particle size based on TEM analysis was around 0.8 μm, which was further supported by the measured specific surface area of 0.7 m^2/g calculated using the following equation:

$$S = \frac{6}{D \times \rho} \quad (1)$$

where S denotes the surface area, D represents the particle diameter, and ρ is the density. This corresponded to a theoretical particle diameter of 4.4 μm. Notably, the average particle size observed in SEM images (seen in Figure 1c,d) was determined to be within the range of 1–2 μm. Importantly, both XRD and SEM results confirmed the absence of any graphite phase, indicating a complete reaction and no residual carbon black contamination. The TEM observation (Figure 1b) reveals a finer particle structure compared to the SEM images, primarily attributed to the dispersion of the powdered sample onto a Cu grid during TEM sample preparation.

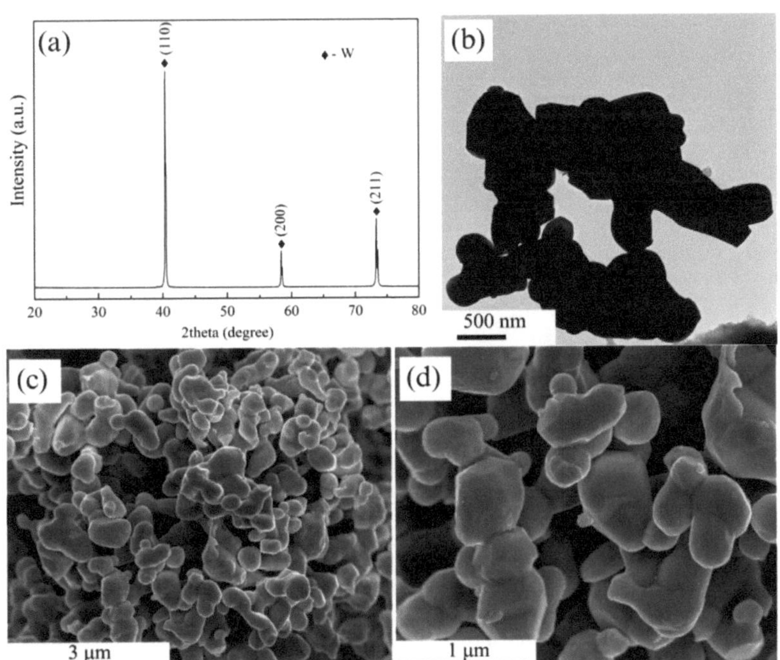

Figure 1. XRD patterns (**a**), TEM (**b**), and SEM observations (**c**,**d**) of the W product after carbothermal reduction of WO_3 at 1300 °C for 3 h under vacuum.

In a previous study, flowing NH_3 was utilized at 600 °C to produce W_2N powder through nitridation of WO_3 [20]. To ensure complete nitridation of WO_3 and prevent oxygen impurities, the nitridation temperature was increased to 800 °C. Figure 2 presented XRD patterns of the nitrided products, which illustrated phase evolution during nitridation–decomposition of WO_3 at various stages. The dominant phase of the product after ammonolysis of WO_3 at 800 °C was W_2N, as evidenced in Figure 2a by characteristic reflections at (111), (200), and (220). However, the presence of W at $2\theta = 40.0°$ as a minor phase in the product suggests simultaneous decomposition of the nitride intermediate due to the relatively high temperature employed. The W mass fraction was calculated to be 0.9 wt. %. In situ decomposition was subsequently performed by switching the flowing gas from NH_3 to Ar and further increasing the heat-treatment temperature to 900 °C and 1000 °C. As shown in Figure 2b,c, the intensity of W peaks gradually increased with increasing temperature, while those of W_2N decreased. Complete decomposition of W_2N to W was achieved at 1000 °C, as confirmed by the disappearance of W_2N peaks in Figure 2e. Importantly, no intermediate phases were observed during the decomposition process (Figure 2a–e).

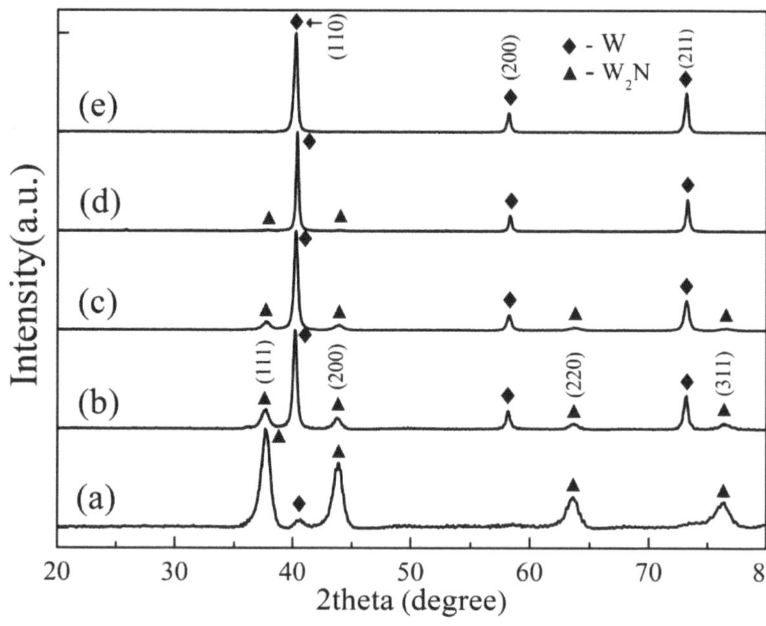

Figure 2. XRD patterns of the intermediate product after nitridation of WO_3 at 800 °C for 3 h (**a**) and in situ nitridation–decomposition at 900 °C for 2 h (**b**) and 6 h (**c**) and at 1000 °C for 2 h (**d**) and 6 h (**e**) in flowing Ar.

The morphology of the W_2N intermediate, obtained from the nitridation of WO_3 at 800 °C for 3 h, is presented in Figure 3a. The particles possessed a near-spherical shape with an average size of approximately 30 nm and exhibited slight agglomeration. Notably, mesopores with a size of approximately 5 nm were observed within the particle structure, as indicated by red arrows in Figure 3b, in contrast to the starting WO_3. The unique formation of mesoporous W_2N was mainly attributed to the nitridation reaction.

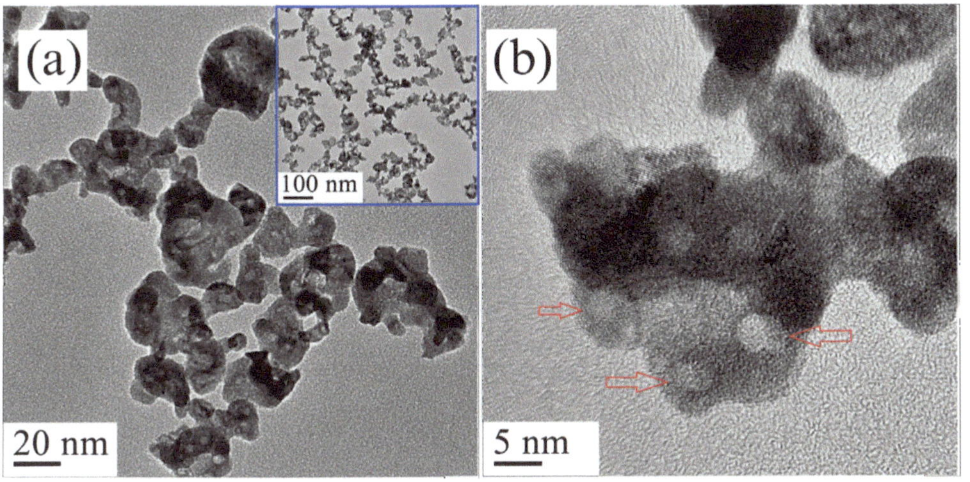

Figure 3. TEM observation of W_2N powders after being heat-treated at 800 °C under low (**a**) and high magnification (**b**).

Tungsten powder derived from the in situ decomposition of W_2N at 1000 °C was analyzed, and the resulting morphology and particle size were investigated. As shown in the TEM observation in Figure 4, the synthesized W powder exhibited a uniform particle size of ~15 nm with narrow distribution. Furthermore, some elongated W particles were observed, possibly originating from the mesoporous W_2N particles. The specific surface area of the W powder synthesized at 1000 °C was measured to be 6.52 m^2/g, indicating a relatively high surface area. In addition, Figure 4b also exhibits that the as-obtained nanosized tungsten powder are bonded together to form some agglomeration. Notably, the particle size of the W product obtained through nitridation–decomposition was smaller than both the raw WO_3 powder and the W product derived from the carbothermal reduction (seen in Figure 1b). This highlighted the potential of this method for synthesizing high-quality, nanosized tungsten powder.

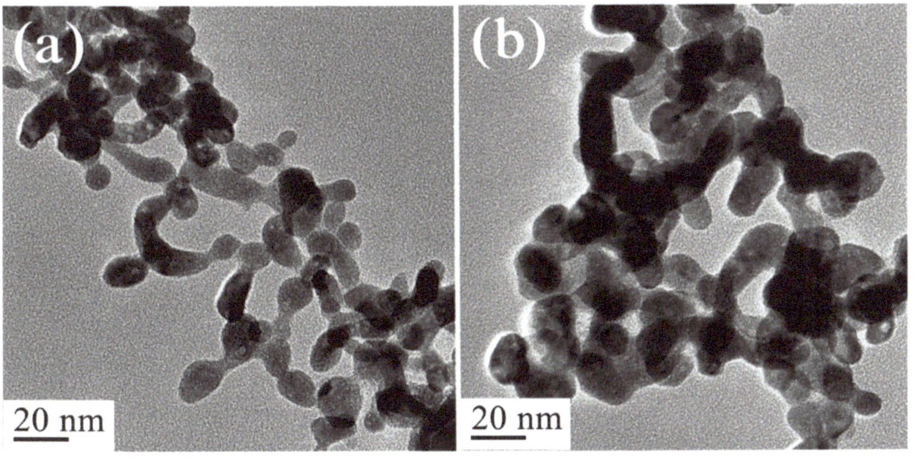

Figure 4. TEM images of W powder after in situ decomposition at 1000 °C in Ar atmosphere showing (**a**) elongated particle, (**b**) agglomeration.

Figure 5 illustrates a proposed mechanism for the morphology changes observed during the nitridation of WO_3. The initial stage involved the formation of a rigid surface layer composed of interconnected W_2N crystallites on the WO_3 particle surface. Due to the great density difference between W_2N (18.1 g/cm^3) and WO_3 (7.28 g/cm^3), a significant volume reduction of up to 67% occurred during the conversion of WO_3 to W_2N. This newly formed nitride layer likely restricted shrinking as the nitridation process continued, consequently, leading to the formation of nanosized pores or cracks within the particles. The W_2N particles formed were comprised of nanocrystalline aggregates, where the mesoporous structure formed within or between these aggregates, which can be observed in Figure 3b. This hypothesis was supported by the variation in the specific surface area. Despite small changes in overall particle size, the W_2N powder exhibits a significantly higher specific surface area (23.63 m^2/g) compared to the starting WO_3 powder (18.10 m^2/g). This increase suggests the development of internal porosity within the W_2N particles, consistent with the proposed mechanism. Due to the crack form within the particle itself, the newly-formed tungsten featured a freshly exposed surface, leading to the formation of soft agglomerates (seen in Figure 4). These agglomerates are expected to be easily broken apart, due to the absence of the water vapor at high temperatures, which would result in the agglomerates of tungsten product derived using the traditional hydrogen reduction method.

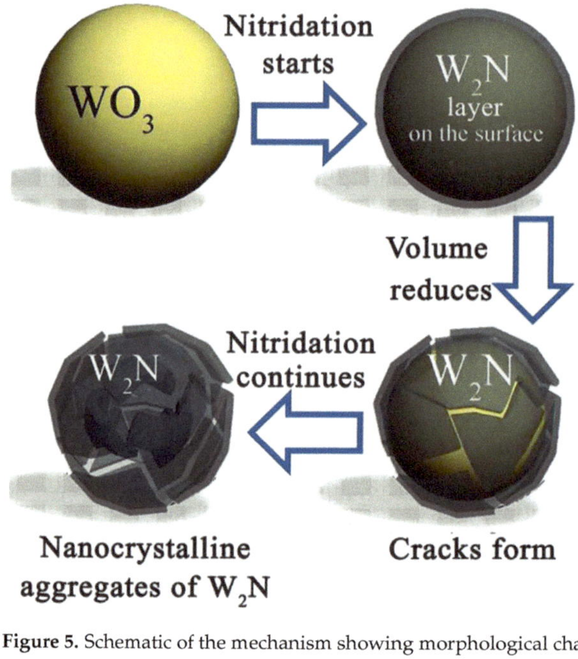

Figure 5. Schematic of the mechanism showing morphological changes of WO_3 during the nitridation process.

Although the overall mechanism of the carbothermal reduction reaction of tungsten oxide can be expressed as $WO_3 + 3C = W + 3CO$, the carbothermal reduction of WO_3 to W proceeds through a multi-step reaction mechanism, as previously reported [22,23], namely $WO_3 \rightarrow WO_{2.72} \rightarrow WO_2 \rightarrow W$. In the initial stage, oxygen removal from WO_3 forms intermediate oxides such as $WO_{2.72}$, creating lattice defects that facilitate further reduction. The subsequent reaction between WO_2 and CO produces tungsten and CO_2:

$$WO_2 + 2CO \rightarrow W + 2CO_2.$$

However, it was confirmed that a competing side reaction involving the spontaneous disintegration of WO_2 occurs simultaneously [24]:

$$2WO_2 \rightarrow W + WO_3 + 0.5\,O_2.$$

WO_2 was heat-treated under vacuum and WO_3 was identified as the gas phase according to the following reaction [25,26]:

$$3/2WO_2(s) \rightarrow 1/x\,(WO_3)_x\,(g) + 1/2W\,(s).$$

During this process, the highly mobile gaseous WO_3 phase underwent sublimation and was deposited onto the surface of WO_2 or carbon black particles as the carbothermal reduction was occurring. The resulting gaseous $(WO_3)_x$ species contributed to gas-phase diffusion, potentially promoting tungsten particle coarsening via surface energy reduction. Notably, achieving minimal carbon impurities often required non-stoichiometric ratios like WO_3:C = 1:2.9 (used in this study), which presented challenges in precise control due to activity variations in the carbon source.

Unlike carbothermal reduction, the decomposition of W_2N offered a direct route to metallic W with a significantly smaller particle size. This faster process avoided the growth of W particles through chemical vapor transport of tungsten-containing species, as confirmed by the absence of intermediate phases during decomposition. Notably, the direct conversion from nanosized W_2N particles led to inherently smaller W particles compared to the coarsening observed in carbothermal reduction. Table 1 lists the average particle size of W powder obtained through both the nitridation–decomposition approach and carbothermal reduction. The comparison clearly exhibits the particle refinement and advantage for the proposed approach in the present work.

Table 1. Comparison of particle size and surface area of W powder from difference approaches.

Approach	Particle Size (TEM)	BET Surface Area	Feature
Nitridation–decomposition	~15 nm	0.7 m^2/g	Coarsening
Carbothermal reduction	0.8 μm	6.52 m^2/g	Nano-aggregates; mesoporous

Synthesizing high-quality, fine-grained tungsten posed a significant challenge due to the limitations of existing methods. This work directly addressed this need by presenting a novel route for W_2N decomposition, which overcame the drawbacks of particle coarsening originating from intermediate species [26]. Our findings not only offer potential improvements in the densification of tungsten but also open doors for applications demanding a high surface area and precise control over tungsten nanostructures, for instance, in the fields of catalysis [27], nanocomposites [28], and microelectronics [29]. Importantly, further exploration of this method and optimizing reaction parameters and precursor properties hold immense promise for developing an even more efficient and scalable synthesis of fine-grained tungsten. This paves the way for broader adoption of this valuable material in advanced technological applications across various fields.

4. Conclusions

This work successfully demonstrated the potential of an advanced nitridation–carburization method for synthesizing high-quality, fine-grained tungsten powder. Compared to traditional carbothermal reduction, the proposed method yielded W particles with a significantly smaller and more uniform size distribution (~15 nm) and a considerably higher surface area (6.52 m^2/g). This approach offers intriguing possibilities for fabricating advanced tungsten materials, particularly for applications requiring precise control over microstructure and high surface area, such as plasma-facing components in fusion reactors. Furthermore, the presented synthesis method aligns well with the growing demand for tailored powder materials with specific properties, driven by advancements in fabrication technologies like 3D printing. The rapid processing time and superior product quality suggest the potential for high material throughput upon scaling up the method. Future research should explore the feasibility of integrating this approach with existing production facilities in the cemented carbide industry, aiming to mature and adapt the technology for large-scale production of fine-grained tungsten powder.

Author Contributions: Conceptualization, Q.-Y.H. and S.-K.S.; methodology, B.-L.Z. and S.-K.S.; data curation, Q.-Y.H. and B.-L.Z.; writing—original draft preparation, Q.-Y.H. and S.-K.S.; writing—review and editing, Q.-Y.H., B.-L.Z., and S.-K.S.; project administration, Q.-Y.H. and S.-K.S.; funding acquisition, Q.-Y.H. and S.-K.S. All authors contributed to the article. All authors have read and agreed to the published version of the manuscript.

Funding: This study was support by the Science and Technology Innovation Project of Foshan (Grant No. 2220001005552) and the "14th Five-Year Plan" Medical Training Specialist Project of Foshan. SKS acknowledges the Guangdong Key Platform and Programs of the Education Department of Guangdong Province for funding under grant No. 2021ZDZX1003 and Guangdong Science and Technology Project under grant No. 2021B1212050004. ZBL acknowledges the Refined Management of Public Hospital Logistics (2023) Project from National Institute of Hospital Administration, China for funding under grant No. GYZ2023HQ28.

Data Availability Statement: The original contributions presented in the study are included in the article, further inquiries can be directed to the corresponding author.

Conflicts of Interest: The authors declare that the research was conducted in the absence of any commercial or financial relationships that could be construed as a potential conflicts of interest.

References

1. Klopp, W.D. A review of chromium, molybdenum, and tungsten alloys. *J. Less Common. Met.* **1975**, *42*, 261–278. [CrossRef]
2. Lassner, E.; Schubert, W.-D. (Eds.) The element tungsten. In *Tungsten: Properties, Chemistry, Technology of the Element, Alloys, and Chemical Compounds*; Springer US: Boston, MA, USA, 1999; pp. 1–59.
3. Gumbsch, P.; Riedle, J.; Hartmaier, A.; Fischmeister, H.F. Controlling factors for the brittle-to-ductile transition in tungsten single crystals. *Science* **1998**, *282*, 1293–1295. [CrossRef] [PubMed]
4. Nelson, A.T.; O'Toole, J.A.; Valicenti, R.A.; Maloy, S.A. Fabrication of a tantalum-clad tungsten target for lansce. *J. Nucl. Mater.* **2012**, *431*, 172–184. [CrossRef]
5. Knaster, J.; Moeslang, A.; Muroga, T. Materials research for fusion. *Nat. Phys.* **2016**, *12*, 424–434. [CrossRef]
6. Abernethy, R.G. Predicting the performance of tungsten in a fusion environment: A literature review. *Mater. Sci. Technol.* **2017**, *33*, 388–399. [CrossRef]
7. Omole, S.; Lunt, A.; Kirk, S.; Shokrani, A. Advanced processing and machining of tungsten and its alloys. *J. Manuf. Mater. Process* **2022**, *6*, 15. [CrossRef]
8. Upadhyaya, G.S. *Cemented Tungsten Carbides Production, Properties and Testing*; Noyes Publications: New York, NY, USA; Westwood, NJ, USA, 1998.
9. Sun, J.; Zhao, J.; Huang, Z.; Yan, K.; Shen, X.; Xing, J.; Gao, Y.; Jian, Y.; Yang, H.; Li, B. A review on binderless tungsten carbide: Development and application. *Nanomicro Lett.* **2019**, *12*, 13. [CrossRef] [PubMed]
10. Naito, M.; Abe, H.; Kondo, A.; Yokoyama, T.; Huang, C.C. Smart powder processing for advanced materials. *KONA* **2009**, *27*, 130–143. [CrossRef]
11. Appleyard, D. Powering up on powder technology. *Met. Powder Rep.* **2015**, *70*, 285–289. [CrossRef]
12. Liang, Y.X.; Wu, Z.M.; Fu, E.G.; Du, J.L.; Wang, P.P.; Zhao, Y.B.; Qiu, Y.H.; Hu, Z.Y. Refinement process and mechanisms of tungsten powder by high energy ball milling. *Int. J. Refract. Met. Hard Mater.* **2017**, *67*, 1–8. [CrossRef]
13. Wu, Z.M.; Liang, Y.X.; Fu, E.G.; Du, J.L.; Wang, P.P. The process and mechanisms for the transformation of coarse grain to nanoscale grain in tungsten by ball milling. *Powder Technol.* **2018**, *326*, 222–227. [CrossRef]
14. Zhang, S.; Wen, Y.; Zhang, H. Low temperature preparation of tungsten nanoparticles from molten salt. *Powder Technol.* **2014**, *253*, 464–466. [CrossRef]
15. Kim, B.K.; Lee, G.G.; Ha, G.H.; Lee, D.W. Spray Drying Metal Salts, Desalting, Dehydrating, Milling Metal Oxide Powder, Forming into Green Body, Sintering to Produce Oxide for Use as Alloy Precursor. U.S. Patent No. 5842108A, 14 September 1998.
16. Senthilnathan, N.; Annamalai, A.R.; Venkatachalam, G. Microstructure and mechanical properties of spark plasma sintered tungsten heavy alloys. *Mater. Sci. Eng. A* **2018**, *710*, 66–73. [CrossRef]
17. Luidold, S.; Antrekowitsch, H. Hydrogen as a reducing agent: State-of-the-art science and technology. *JOM* **2007**, *59*, 20–26. [CrossRef]

18. Wu, C. Preparation of ultrafine tungsten powders by in-situ hydrogen reduction of nano-needle violet tungsten oxide. *Int. J. Refract. Met. Hard Mater.* **2011**, *29*, 686–691. [CrossRef]
19. Venables, D.S.; Brown, M.E. Reduction of tungsten oxides with hydrogen and with hydrogen and carbon. *Thermochim. Acta* **1996**, *285*, 361–382. [CrossRef]
20. Sun, S.-K.; Kan, Y.-M.; Zhang, G.-J.; Wang, P.-L. Ultra-fine tungsten carbide powder prepared by a nitridation–carburization method. *J. Am. Ceram. Soc.* **2010**, *93*, 3565–3568. [CrossRef]
21. Madsen, I.C.; Scarlett, N.V.Y.; Kleeberg, R.; Knorr, K. Quantitative phase analysis. In *International Tables for Crystallography*; Gilmore, C.J., Kaduk, J.A., Schenk, H., Eds.; International Union of Crystallography: Chester, UK, 2019; pp. 344–373.
22. Shveikin, G.P.; Kedin, N.A. Products of carbothermal reduction of tungsten oxides in argon flow. *Russ. J. Inorg. Chem.* **2014**, *59*, 153–158. [CrossRef]
23. Gruner, W.; Stolle, S.; Wetzig, K. Formation of CO_x species during the carbothermal reduction of oxides of Zr, Si, Ti, Cr, W, and Mo. *Int. J. Refract. Met. Hard Mater.* **2000**, *18*, 137–145. [CrossRef]
24. Blackburn, P.E.; Hoch, M.; Johnston, H.L. The vaporization of molybdenum and tungsten oxides. *J. Phys. Chem.* **1958**, *62*, 769–773. [CrossRef]
25. Ostermann, M.; Dalbauer, V.; Schubert, W.-D.; Haubner, R. Preparation of nano-crystalline tungsten powders from gaseous $WO_2(OH)_2$. *Tungsten* **2022**, *4*, 60–66. [CrossRef]
26. Schubert, W.D.; Lassner, E. Production and characterization of hydrogen-reduced submicron tungsten powders—Part 1: State of the art in research, production and characterization of raw materials and tungsten powders. *Int. J. Refract. Met. Hard Mater.* **1991**, *10*, 133–141. [CrossRef]
27. Can, F.; Courtois, X.; Duprez, D. Tungsten-Based Catalysts for Environmental Applications. *Catalysts* **2021**, *11*, 703. [CrossRef]
28. Wahlberg, S. Tungsten-Based Nanocomposites by Chemical Methods. Ph.D. Dissertation, KTH Royal Institute of Technology, Stockholm, Sweden, 2014.
29. Inberg, A.; Ginsburg, E.; Shacham-Diamand, Y.; Croitoru, N.; Seidman, A. Electroless and sputtered silver–tungsten thin films for microelectronics applications. *Microelectron. Eng.* **2003**, *65*, 197–207. [CrossRef]

ceramics

MDPI

Article

The Origin of the Low-Temperature Minimum of Electrical Resistivity in Strontium Ferromolybdate Ceramics

Gunnar Suchaneck [1,*], Evgenii Artiukh [2] and Gerald Gerlach [1]

[1] Solid-State Electronics Laboratory, TU Dresden, 01062 Dresden, Germany; gerald.gerlach@tu-dresden.de
[2] Cryogenic Research Division, SSPA "Scientific-Practical Materials Research Centre of NAS of Belarus", 220072 Minsk, Belarus; sirfranzferdinand@yandex.ru
* Correspondence: gunnar.suchaneck@tu-dresden.de

Abstract: In this work, we analyze the electrical behavior of strontium ferromolybdate below room temperature. We demonstrate that in SFMO ceramics, SFMO thin films deposited by pulsed laser deposition including (100) and (111) textured thin films, as well as in nonstoichiometric SFMO ceramics, an intergrain tunneling mechanism of charge carrier conduction leads to a decrease in resistivity with increasing temperature in the low-temperature region. This intergrain tunneling can be attributed to fluctuation-induced tunneling. On the other hand, bulk metallic resistivity of the grains, which increases with temperature, becomes dominant at higher temperatures and magnetic fluxes. The interplay of these conduction mechanisms leads to a resistivity minimum, i.e., a resistivity upturn below the temperature of minimum resistivity. Several mechanisms have been discussed in the literature to describe the low-temperature upturn in resistivity. Based on available literature data, we propose a revised model describing the appearance of a low-temperature resistivity minimum in SFMO ceramics by an interplay of fluctuation-induced tunneling and metallic conductivity. Additionally, we obtained that in the region of metallic conductivity at higher temperatures and magnetic fluxes, the pre-factor R_m of the temperature-dependent term of metallic conductivity written as a power law decreases exponentially with the temperature exponent m of this power law. Here, the value of m is determined by the charge scattering mechanism.

Keywords: strontium ferromolybdate; electrical resistivity; low-temperature electrical behavior

Citation: Suchaneck, G.; Artiukh, E.; Gerlach, G. The Origin of the Low-Temperature Minimum of Electrical Resistivity in Strontium Ferromolybdate Ceramics. *Ceramics* **2024**, 7, 491–503. https://doi.org/10.3390/ceramics7020032

Academic Editors: Dawei Wang and Fayaz Hussain

Received: 12 February 2024
Revised: 19 March 2024
Accepted: 28 March 2024
Published: 1 April 2024

1. Introduction

Strontium ferromolybdate ($Sr_2FeMoO_{6-\delta}$, SFMO) double perovskites are promising candidates for magnetic electrode materials for room-temperature spintronic applications, because they have a half-metallic character (with theoretically 100% electron spin polarization), present a high Curie temperature of about 420 K (magnets should be operated in their ordered magnetic state below Curie temperature), and show a low-field magnetoresistance [1].

Modern artificial intelligence (AI) text generators such as ChatGPT are based on neural networks. The conventional complementary metal–oxide–semiconductor (CMOS) neuron and synapse designs require numerous transistors and feedback mechanisms and would be unsuitable for developing modern AI systems. Since the early developments of neural network theory, magnetic materials have been used for modeling brain-like systems. The main advantage of spintronics compared to other resistive memories for neuromorphic computing is the possibility to induce complex and tunable resistance dynamics through spin torque. Like other memory cells, they can switch between two fixed states, allowing them to emulate synapses. Thus, neuromorphic spintronics aims to develop spintronic hardware devices and circuits with brain-inspired principles [2], i.e., spintronics is a promising approach to neuromorphic computing as it potentially enables energy-efficient and area-efficient embedded applications by mimicking key features of biological synapses and neurons with a single device instead of using multiple electronic components [3,4].

The main reasons for the still missing wide application of SFMO is the low reproducibility of its electrical and magnetic properties originating in the formation of point defects and grain boundaries with a composition different from the bulk of the grains, as well as its aging in contact with air and moisture.

The charge transport mechanisms of perovskites are strongly dependent on synthesis conditions and, thus, on the microstructure of the material (single-crystalline, polycrystalline, grain size, nature of grain boundaries, and others) formed by its synthesis. SFMO ceramics obtained by solid-state reactions using one and the same procedure were insulating, metallic, or in an intermediate state in dependence on the precursors and annealing temperature and time [5]. In SFMO, the transport properties are dominated by spin-polarized intergrain tunneling through insulating grain boundaries [6]. Therefore, the transport properties will be different in ceramics, thin films, and single crystals. Single crystals possess a 4 mm symmetry while ceramics have a ∞m one. Columnar thin films will be in an intermediate state between 4 mm and ∞m symmetry. Consequently, the interpretation of grain size parameters will be very different for each of these structures. Under specific synthesis conditions, SFMO ceramics consist of $SrMoO_4$ (SMO) intergrain energy barriers between conductive bulks of SFMO grains induced by a small oxygen excess during material fabrication [7]. Previously, we have shown that an appropriate thermal treatment leads to the formation of dielectric $SrMoO_4$ shells at the surface of SFMO nanograins. The formation of thin insulating $SrMoO_4$ surface layers suppresses the metallic conductivity owing to the appearance of intergrain fluctuation-induced tunneling [8]. The ease of forming SMO shells around SFMO grains during synthesis simply by excess oxygen during synthesis makes SFMO an easy-to-fabricate, natural core–shell material.

Generally, several mechanisms have been discussed to interpret the low-temperature resistivity upturn that was visible in experiments: (i) inelastic scattering of electrons by impurity ions in impure metals [9], (ii) Kondo-like effects associated with dilute magnetic impurities in a nonmagnetic host [10–14], (iii) quantum interference effects arising from electron–electron interactions and weak localization [15–22], (iv) electron–electron interaction driven by Coulomb forces in a strongly correlated system [23], (v) spin-polarized tunneling via grain boundaries in ceramic manganites [24], (vi) the competition of two mechanisms—one contribution decreasing with temperature combined with another contribution increasing with temperature [10,25], and (vii) two spin channels in SFMO connected in parallel, where the spin-down channel is metallic (with a non-zero density of states at the Fermi level), and the spin-up channel (with a gap in the band structure) behaves like a semiconductor [26].

Several reports have considered an interplay of two mechanisms or a crossover between them [13,16,27]. However, a satisfactory description has not yet emerged and is still under discussion.

The Kondo effect mentioned above was discovered by Jun Kondo in 1964 [10]. It describes the scattering of conduction electrons in a metal due to the resonant interaction between conduction charge carriers and the spin of localized magnetic impurity ions. Kondo applied third-order perturbation theory to the problem of scattering of s-orbital conduction electrons by d-orbital electrons localized at impurities. Kondo's approach predicted a logarithmical increase in the scattering rate and the resulting part of the resistivity as the temperature approaches absolute zero. This results in an anomalous upturn in resistivity at low temperatures. Thereby, the steepness of the upturn is proportional to the impurity concentration. Considering an Fe^{3+}-O-Fe^{4+} Kondo lattice in SFMO [28], the resistivity upturn, which occurs in polycrystalline SFMO ceramics in the absence of a magnetic field [29], was reproduced. However, the authors in [28] took into account only four data points of [29]. Considering all data points of [29], the agreement with the experimental data above 240 K would become worse. In $La_{2/3}Sr_{1/3}MnO_3$, the Kondo resistivity contributions are expected to collapse at a magnetic flux density of about 13 T [27]. We excluded the Kondo effect as a possible cause since a frustrated spin-glass-like structure is present at the surfaces of weakly connected ferromagnetic grains in cold-pressed polycrystalline SFMO [30]. In this case, the spin degrees of freedom are completely frozen and, thus, internal degrees of freedom are absent. The frustrated spin-glass structure leads to a strong exchange bias effect, which is seen in the field-cooled magnetoresistance measurements [31]. Here, the presence of a high field irreversibility starting below 235 K between the field-cooled and zero-field-cooled curves at a magnetic flux of 1 T is attributed to a possible spin-glass component at the distorted surfaces of nanoparticles obtained earlier in $NiFe_2O_4$ [32] and $La_{2/3}Sr_{1/3}MnO_3$ [33]. Other signatures for a frustrated spin-glass structure are a wider hysteresis in magnetoresistance compared to that in magnetization [34,35] and the shift of the low-temperature peak of the imaginary part of AC susceptibility to higher temperatures with the increase in frequency [35,36].

Weak localization is a physical effect, which occurs in disordered electronic systems at very low temperatures. The origin of weak localization is quantum interference of back-scattered electrons. The effect manifests itself as a correction, $\Delta\sigma$, to the conductivity (or correspondingly the resistivity) of a metal or semiconductor arising in the case that the mean free path l is in the order of the Fermi wavelength $\lambda_F = 2\pi/k_F$ of the carrier wave functions with k_F denoting the Fermi wave vector, i.e., $k_F l \sim 1$. Calculations of k_F were carried out assuming an electron density of $n_e = 1.1 \times 10^{22}$ cm^{-3} [37], yielding $k_F = (3\pi ne)^{1/3} = 4.7 \times 10^9$ m^{-1} and $\lambda_F = 0.913$ nm. The experimentally obtained low-temperature mean free path amounts to l(4 K) = 1.11 nm [38] (p. 72), in good agreement with a value of $l = \hbar k_F / n_e e^2 \rho_0 = 0.975$ nm where $\hbar$ is the Planck constant expressed in J s radian^{-1} and ρ_0 the residual resistivity amounting to $\rho_0 = 1.8 \times 10^{-6}$ Ωm [39]. A strong magnetic field suppresses the contribution of weak localization [16]. In $La_{2/3}Sr_{1/3}MnO_3$ thin films consisting of ZrO_2 nanoparticles, weak localization vanishes at about 9 T [27]. In ferromagnetic systems, including also ferrimagnetic SFMO, the nuclei already experience a magnetic field without applying any external magnetic field. This is the hyperfine field created by the electrons at the nuclei [40]. It describes the hyperfine interaction between the magnetic moment of the nucleus and the magnetic moment of the electrons in the solid. In SFMO, the hyperfine magnetic field amounts to about 30 T [41]. Thus, weak localization may be excluded from consideration. Recently, we claimed the absence of weak localization in SFMO since fluctuation-induced tunneling as well as adiabatic small polaron hopping do not favor quantum interference. We concluded that the resistivity upturn behavior of SFMO cannot be explained by weak localization [17].

The contribution of electron–electron interactions to the conductivity is similar to that due to weak localization. In this case, the inelastic diffusion length should be replaced by the thermal diffusion length $L_T = (\hbar D/kT)^{1/2}$ and the coefficients are slightly modified [42]. Therefore, the large hyperfine magnetic field will also erase the electron–electron interaction correction to resistivity. In $La_{2/3}(Sr,Ca)_{1/3}MnO_3$ epitaxial thin films deposited on a (001) $LaAlO_3$ substrate where the appearance of a resistivity minimum was attributed to the electron–electron interaction of the strongly correlated system, the resistivity upturn below the temperature of minimum resistivity T_{min} was largely unaffected by externally applied fields up to 8 T. This is contrary to data of polycrystalline SFMO ceramics [29].

Spin-polarized intergrain tunneling in manganites was discovered by Hwang et al. in 1996 [43]. The low-temperature resistivity minimum obtained in ceramic manganites was found to explain charge carriers' tunneling between antiferromagnetically coupled grains [21]. The tunneling resistance between two FM grains was described by the theory of tunneling conduction through the FM metal/nonmagnetic barrier/FM metal (fbf) junction using the phenomenological expression [44]

$$\rho(T,B) = \frac{\rho_{\text{fbf}}}{1 + P^2 \langle \cos\theta_{ij} \rangle}, \quad (1)$$

where P is the degree of the spin polarization of the current carriers and θ_{ij} is the angle between the magnetization directions of the grains i and j:

$$\cos\theta_{ij} \underset{B=0}{=} -L(|J|/kT). \quad (2)$$

Here, $L(x)$ = coth(x) − $1/x$ is the Langevin function and J the antiferromagnetic interaction constant. Since the temperature dependence of the in-grain and grain boundary magnetization is dominated by the spin-wave $T^{3/2}$ terms at low temperatures, ρ_{fbf} was taken as

$$\rho_{\text{fbf}} = \rho_0 + \rho_{1.5}T^{1.5}, \quad (3)$$

where ρ_0 and ρ_1 are parameters independent of B. The model parameters are then defined from the requirement that in the case B = 0, Equation (1) fits the experimental data for $\rho(T,B = 0)$ in the range 4.3–50 K. For $La_{0.5}Pb_{0.5}MnO_3$, the fitting parameters that reflect the behavior of the experimentally obtained resistivity minimum were P = 0.697, $|J|/k$ = 155 K, $\rho_0 = 1.45 \times 10^{-3}$ Ωm, and $\rho_{1.5} = 1.026 \times 10^{-6}$ ΩmK$^{-3/2}$. In the case of SFMO, the parameter $|J|/k$ is significantly lower than for manganites amounting to 47.6 K [37]. This considerably lowers the depth of the resistivity minimum. Also, the resistivity parameters $\rho_0 = 2 \times 10^{-6}$ Ωm [39] and $\rho_{1.5} \approx 3 \times 10^{-6}$ ΩmK$^{-3/2}$ (estimated below) are different, leading to a curve that increases with temperature. A resistivity minimum is obtained only in the parameter range $\rho_1 \approx$ 1.5–8 × 10^{-9} ΩmK$^{-3/2}$, which does not correspond to experimental data. Therefore, we disregard spin-polarized tunneling as the origin of the resistivity minimum.

Spin-polarized tunneling between antiferromagnetically coupled grains in ceramic manganites causes a resistivity upturn, which shifts to lower temperature with an increasing magnetic field and disappears above a critical field value amounting to 1.5 T for $La_{0.5}Pb_{0.5}MnO_3$ [24]. The application of a magnetic field increases the probability of electron tunneling through dielectric grain boundaries, thus decreasing the resistivity and recovering the metallic conductivity of the nanograins. This leads to a decrease in the upturn temperature as obtained for SFMO in [29]. It decreases from 180 K at 0.2 T to about 40 K at 7 T. Since the low-field magnetoresistance in polycrystalline perovskites is governed by spin-polarized tunneling across grain boundaries [43], this provides a sufficient low-field magnetoresistance promising for device application [1].

The decrease in the temperature T_{min} of minimum resistivity with an increasing magnetic field also corresponds to the predictions of a model consisting of a series connection of elastic and inelastic scattering mechanisms [21]. The elastic contribution was assumed to arise from the scattering of holes by magnons without a spin flip whereas the inelastic contribution was calculated as quantum correction to resistivity, taking the inverse of the total scattering length as the geometric mean of the inverse phase coherence length and the inverse magnetic length. We have analyzed the disappearance of the resistivity upturn in SFMO with increasing magnetic flux density from resistivity data of polycrystalline SFMO ceramics taken from [29]. Here, an exponential law, $T_{min} \propto B^{-0.4}$, appeared, which is unsuitable for defining a critical magnetic field.

An example of the competition of two resistivity contributions with opposite temperature dependencies is on one hand thermally activated hopping and on the other hand excitation to the mobility edge [45] proposed for $La_{1-x}Sr_xMnO_3$ [46]. For SFMO, another model was considered consisting of a semiconductor-like thermally activated resistivity in one spin channel in parallel to metallic conductivity in the other spin channel [26].

In this work, we propose a modified model describing the appearance of a low-temperature resistivity upturn in SFMO ceramics by an interplay of two conductivity mechanisms: (i) fluctuation-induced tunneling and (ii) metallic conductivity. It is based on the thorough analysis and evaluation of literature data of conductivity behavior of SFMO.

2. Materials and Methods

The resistivity of SFMO thin films produced by pulsed laser deposition below 30 K increases strictly logarithmically with decreasing temperature [38] (p. 64). Such a behavior may be attributed either to the Kondo effect, weak localization, or electron–electron interaction [14]. However, we have ruled out all these mechanisms for strontium ferromolybdate above.

Near room temperature, two possible conduction mechanisms of nano-sized granular SFMO were obtained that cannot be distinguished by experimental data [8]. The first one is the Mott variable-range hopping describing a carrier transport in a disordered semiconductor or in an amorphous solid by charge carrier hopping between two spatially separated states in an extended temperature range [45]. For three-dimensional conductance when the Coulomb interaction between the carrier is negligible, the resistivity has a characteristic temperature dependence of

$$\rho(T) = \rho_0 \exp\left(\frac{T_0}{T}\right)^{1/4}, \tag{4}$$

with a characteristic temperature T_0 in the order of a few 10^6 K. The value of T_0 defines the hopping activation energy and the mean hopping distance. Alternatively to Mott variable-range hopping, the conductivity in this temperature region may be modeled by a $\ln\rho \propto T^{-1/2}$ dependence derived for (i) two-dimensional conductance in the Mott variable-range hopping model [45]; (ii) Efros–Shklovskii variable-range hopping, which accounts for the Coulomb gap interaction of electrons in localized states that results in the creation of the Coulomb gap near the Fermi energy [47]; or (iii) the low-field conductivity of fine metallic particles dispersed in a dielectric matrix [48]. Note that at higher temperatures, the Coulomb gap can be disregarded and Mott's variable-range hopping comes into effect.

Above room temperature, the resistivity of SFMO ceramics annealed in a vacuum was separated into three regions [49]: (i) from 300 K up to the Curie temperature of about 405 K where the electrical resistivity increases with temperature characteristic for metallic behavior; (ii) above 405 K up to approximately 590 K where resistivity decreases with temperature, which is attributed to a B-site disorder-induced weak Anderson localization of the electrical carriers; and (iii) from 590 K up to 900 K where the resistivity of the material becomes metallic again. On the other hand, the decrease in resistivity in SFMO above the resistivity maximum around Curie temperature was attributed to adiabatic small polaron hopping [17] and not to localization effects as originally suggested [49].

Another report of the electrical resistivity of SFMO [50] indicates metallic conduction behavior below 420 K, the localization of the carriers in the temperature range 420–820 K, and reversion to metallic conduction behavior between 820 K and 1120 K.

In the absence of a magnetic field, the temperature dependence of the conductivity of SFMO ceramics [29] was well described by the fluctuation-induced tunneling (FIT) model [51], e.g., by the presence of conducting grains separated by nano-sized energy barriers where large thermal voltage fluctuations occur when the capacitance of an intergrain junction is in the order of 0.1 fF. Nano-sized energy barriers with a barrier width of a few nanometers and a barrier area of several hundred square micrometers usually occur at grain surfaces in cold-pressed ceramics or ceramics thermally treated under oxidation conditions.

Below 500 K, monocrystalline, half-metallic SFMO exhibits a resistivity, $\rho(T)$, increasing with temperature as

$$\rho(T) = \rho_0 + R_l T^l, \tag{5}$$

where $\rho_0 = 1.8 \times 10^{-6}$ Ωm is the residual metallic resistivity, $l = 2$ near room temperature and $R_2 = 2.16 \times 10^{-11}$ ΩmK^{-2} [37,39]. A low-temperature upturn of the resistivity is missing in single-crystal SFMO due to the absence of intergrain boundaries characteristic for ceramics. The T^2 dependence suggests that resistivity is dominated by either electron–electron scattering [39,52] or spin-wave scattering [39]. At higher temperatures up to the Curie temperature T_C, other charge scattering mechanisms come into play that can be considered in Equation (1) by additional terms:

$$\rho(T) = \rho_0 + R_2 T^2 + \sum_l R_l T^l. \tag{6}$$

A term with $l = 2.5$ represents the combination of electron–electron, electron–phonon, and electron–magnon scattering [53], an $l = 3$ term stands for the scattering with anomalous single magnons in half-metallic systems [54], an $l = 3.5$ one stands for the scattering with spin-waves for low temperatures [55,56], and an $l = 4.5$ term describes electron–magnon scattering. The latter was derived in the double-exchange theory at low temperatures [57] but was experimentally observed at elevated temperatures of 200–350 K [52]. Alternatively, the $l = 4.5$ term was attributed to spin-wave scattering [58]. An $l = 5$ term was attributed to acoustic phonon scattering [59]. Additionally, there is a term related to optical phonon scattering:

$$\rho(T) \propto \frac{\omega_s}{\sinh^2(\hbar\omega_s/2kT)}, \tag{7}$$

where ω_s is the average frequency of the softest optical mode, which is consistent with small polaron coherent motion involving relaxation [60], and k is the Boltzmann constant.

Unfortunately, none of these models fit to the experimental data of SFMO thin films in the whole temperature range. However, according to [26], the electron–electron scattering model seems to be appropriate in a broader temperature range. In general, the metallic resistivity of SFMO can be modeled by the equation

$$\rho(T) = \rho_0 + R_m T^m, \tag{8}$$

where m is a fitting parameter that represents an averaged value of the above-mentioned contributions of charge scattering mechanisms characterized by the parameter l.

The FIT model describes conducting grains separated by energy barriers that are subjected to large thermal fluctuations, e.g., carbon-polyvinylchloride composites, consisting of aggregates of carbon spheres [51]. The tunneling occurs between large metallic grains across insulating barriers with width w and area A. It is specified by two parameters, temperature T_1 and a normalized temperature, T_0. T_1 characterizes the electrostatic energy of a parabolic potential barrier:

$$kT_1 = \frac{A \cdot w \cdot \varepsilon_0 E_0^2}{2}. \tag{9}$$

Here, ε_0 is the vacuum permittivity. The characteristic electric field strength E_0 is determined by the barrier height V_0:

$$E_0 = \frac{4V_0}{e \cdot w}. \tag{10}$$

with e the electron charge. The normalized temperature T_0 relates T_1 to the tunneling constant:

$$T_0 = T_1 \cdot \left(\frac{\pi\chi w}{2}\right)^{-1}, \tag{11}$$

with the reciprocal localization length of the wave function

$$\chi = \sqrt{\frac{2m_e V_0}{\hbar^2}}, \tag{12}$$

where m_e is the electron mass. The resulting electrical conductivity in this model is then given by [51]:

$$\sigma(T) = \sigma_{0,FIT} \exp\left(-\frac{T_1}{T_0 + T}\right). \tag{13}$$

Note that the original derivation of Equation (13) considers conductivity, meaning the inverse of resistivity. For SFMO, the characteristic field strength E_0 of the fluctuation-induced tunneling model is nearly independent of temperature and amounts to 4.07×10^4 Vm^{-1} [61], while V_0 is in the order of 10 meV and w in the range of 1–3 nm [8,61]. Equation (13) demonstrates that the FIT model links the temperature-independent tunneling conductivity [62] with a temperature-dependent conductivity obtained for tunneling of spin-polarized electrons in granular metal films [63]. The FIT model was recently applied to intergrain tunneling in polycrystalline SFMO ceramics [64], in half-metallic double-perovskite $Sr_2BB'O_6$ (BB'—FeMo, FeRe, CrMo, CrW, CrRe) ceramics [65], in Ba_2FeMoO_6 thin films [66], and in nano-sized SFMO ceramics fabricated by the citrate-gel technique [8]. Surprisingly, we obtained recently [17] that the FIT model satisfactorily describes conductivity behavior of polycrystalline SFMO ceramics in the absence of a magnetic field [29]. In this work, we extend this analysis to other literature data and include the presence of a magnetic field.

3. Results and Discussion

In [26], SFMO was considered as a system with two spin channels connected in parallel. The spin-down channel is metallic and the spin-up channel with a gap in the band structure behaves like a semiconductor. The band gap of the SFMO spin-up band is small enough so that the temperature can excite electrons to the conduction band. The total resistivity is formed by the resistivity of the semiconductive spin channel including some residual resistivity from impurities as well as defects and the resistivity of the metallic spin channel. The E_g values reported in [26] were less than 10 meV, except for a film with higher saturation magnetization with E_g = 47.3 meV. The latter was deposited at slightly higher pressure (11 Pa compared to 9 Pa) and in a more oxidizing atmosphere with a reduced 5%H_2/Ar flow. These E_g values are significantly smaller than the majority of spin-up band gap in SFMO, amounting to about 0.8 eV [67]. They are in the order of the parameter Δ describing the spin-dependent tunneling for Co-Al-O insulating granular film [68] and lie in the order of the barrier V_0 of fluctuation-induced tunneling [8,61]. Therefore, we postulate that the low-temperature conductivity mechanism with high probability originates from intergranular tunneling. In contrast, the conductivity at mediate temperature can be attributed to metallic conductivity.

In the following, we will consider a brick model consisting of cube-shaped grains with metallic conductivity covered at the surface by grain boundaries creating nano-sized intergrain tunneling barriers. Here, two types of grain boundaries appear, namely one with a normal vector parallel to the applied field (perpendicular boundaries) and another one with a normal vector perpendicular to the applied field (parallel boundaries). The corresponding DC equivalent circuit is an intragrain resistance in parallel with the intergrain resistance of parallel grain boundaries and a series connection of this parallel circuit with the intergrain resistance of perpendicular grain boundaries [69]. In our case, the series intergrain resistance of perpendicular grain boundaries may be neglected since—with regard to a very small barrier height in the order of 10 meV—a bias of already a few mV sufficiently increases the tunneling current and, thus, decreases the barrier resistance. Contrarily, the resistance of parallel boundaries remains high since the normal vector of the grain boundary is along an equipotential line. Following [69], we obtain a temperature-dependent conductivity of

$$\sigma(T) = [\rho_0 + R_m T^m]^{-1} + \frac{2w}{d}\sigma_{0,FIT} \exp\left(-\frac{T_1}{T_0 + T}\right), \tag{14}$$

where w is the barrier width and d the grain size amounting for nano-sized SFMO ceramics fabricated by the citrate-gel method; w = 1.24 nm and d = 75 nm [8]. Unfortunately, only a few publications provide values of both w and d. When matching the FIT model and metallic conductivities, the inverse of the metallic residual resistivity corresponds to the FIT model conductivity for temperature and barrier height tending to zero. Thus, we introduce for the sake of simplicity an effective model resistivity, $\rho_0 = 1/\sigma_{0,FIT}$, which is valid for $2w/d \approx 1$. In the following, we are considering a resistivity minimum arising from the competition of two contributions, one of which is fluctuation-induced tunneling dominating the resistivity at low temperatures and low magnetic fields and the other one is metallic conductivity leading to a growing conductivity with increasing temperature. To confirm this, we will now analyze available literature data with respect to the parameters of electrical conductivity. Figure 1 shows resistivity data of polycrystalline SFMO ceramics fabricated by the solid-state reaction technique [29]. The solid lines are curves fitting these data to the model of this work, Equation (14). Table 1 compiles the model parameters of various SFMO materials. Corresponding fits for SFMO thin films deposited by pulse laser deposition are depicted in Figure 2.

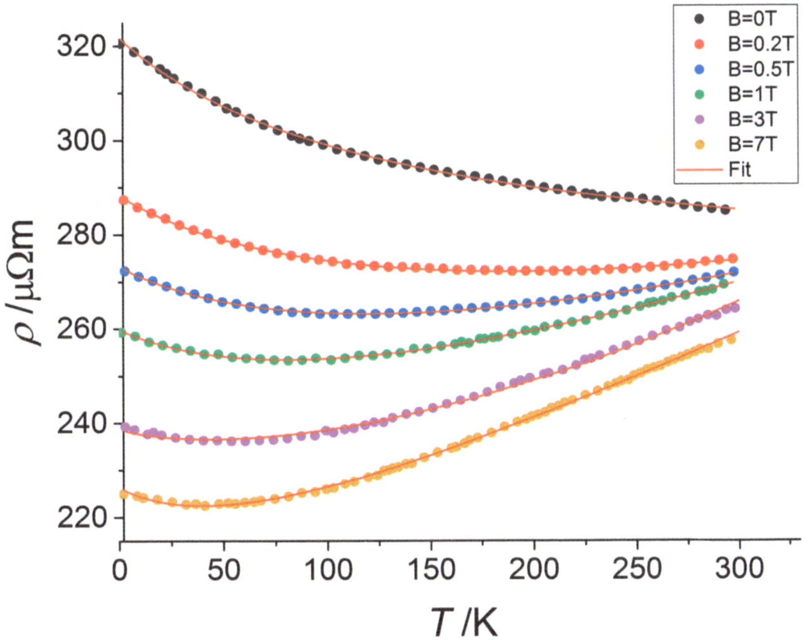

Figure 1. Fit of resistivity data of polycrystalline SFMO ceramics fabricated by the solid-state reaction technique [29] to Equation (14).

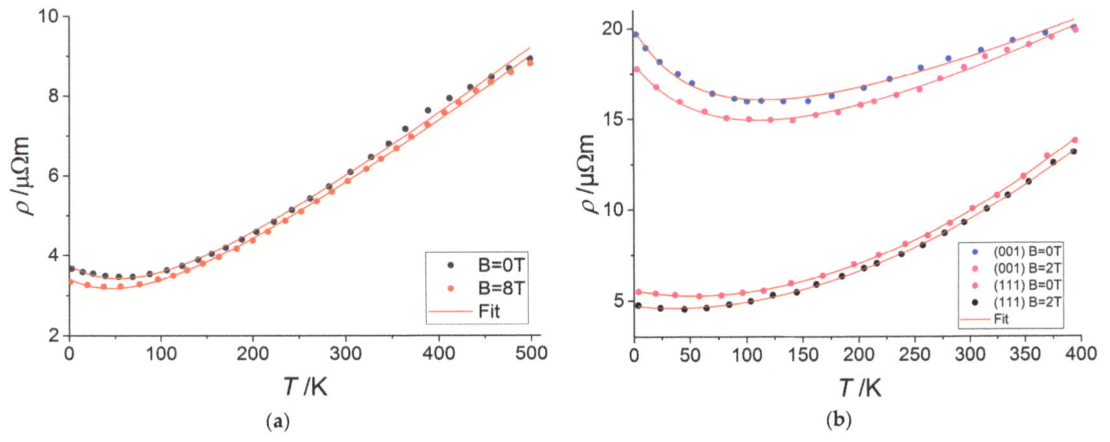

Figure 2. Fit of resistivity data (**a**) from [71] and (**b**) from [73] of ordered SFMO thin films deposited by pulsed laser deposition to Equation (14).

Table 1. Fit parameters for Equation (14) of various SFMO materials.

SFMO Material	B, T	T_1, K	T_0, K	$1/\rho_0$, S/m	m	R_m, ΘmK^{-m}	Ref.
Sintered [1] polycrystalline SFMO ceramics	0	362.2	180.7	0.28	1.24	7.84×10^{-4}	[29]
	0.2	341.9	102.5	0.34	1.41	5.13×10^{-4}	
	0.5	265.4	67.7	0.36	1.44	6.50×10^{-4}	
	1	323.7	77.6	0.38	1.59	2.43×10^{-4}	
	3	362.3	75.5	0.42	1.75	9.82×10^{-5}	
	7	323.1	62.0	0.44	1.80	9.28×10^{-5}	
SFMO ceramics [1]	0	1541.7	492.6	14.6	3	1.19×10^{-9}	[70]
Thin film [2]	0	1272.5	509.3	26.0	2.72	9.76×10^{-9}	[71]
	8	2003.8	831.6	28.2	2.40	5.20×10^{-8}	
Thin film [2]	0	88.0	84.5	42.1	2.9	1.93×10^{-9}	[72]

[1] Sintered 2 h at 1200 °C in 1 % H_2/Ar, [2] deposited by pulse laser deposition.

Figure 3 demonstrates that model Equation (14) is also valid for the data of ordered, nonstoichiometric SFMO ceramics prepared by a solid-state reaction at 900 °C and sintered at 1280 °C for 12 h in a stream of 5%H_2/Ar [74].

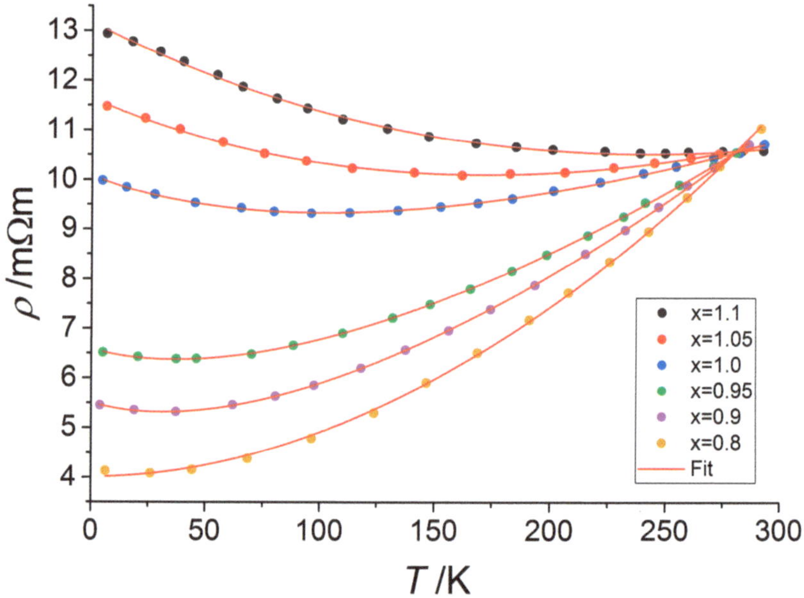

Figure 3. Fit of resistivity data from [74] of nonstoichiometric $Sr_2Fe_xMo_{2\text{-}x}O_6$ ceramics prepared by solid-state reaction and sintered for 12 h at 1280 °C in a stream of 5% H_2/Ar to Equation (14).

The ratio T_1/T_0 defines a value of the product χw of the reciprocal localization length of the wave function χ and barrier width w. Since the value of the tunnel barrier height V_0 depends on the nature and configuration of chemical bonds in the intergrain area, the differences in V_0 for different SFMO materials should be small. For given values of V_0, the value of χ can be calculated by means of Equation (12). The barrier width w then defines T_1/T_0 via Equation (11). In the first case, the magnetic flux dependence on V_0 should be taken into account that follows the relation [75]

$$V_0(B) = V_0(0) - \beta B + \gamma B^2 \quad (15)$$

with $\beta = 1.16$ meV/T and $\gamma = 0.04$ meV/T^2 [76]. This results in a ratio, T_1/T_0, that decreases slightly as the magnetic flux increases. This is in qualitative agreement with the ratios T_1/T_0

derived from experimental data [71,73] (cf. Table 1). On the other hand, this contradicts the data derived from [29], where T_1/T_0 increases with magnetic flux.

For metallic conductivity, the values of R_m and m in Equation (8) are correlated [77]. In our case, this is illustrated in Figure 4, showing an exponential decrease in R_m with m. Note that Figure 4 yields for m = 1.5 a value of $\rho_{1.5} \approx 3 \times 10^{-6}$ ΩmK$^{-3/2}$ which is required in Equation (3).

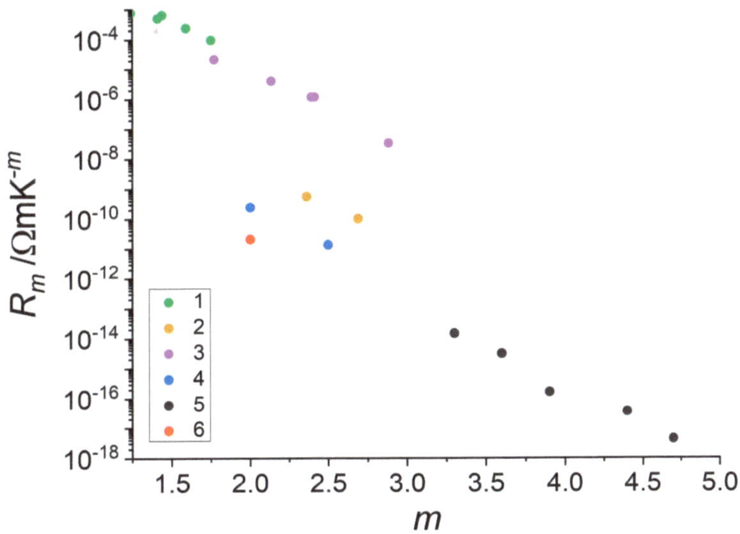

Figure 4. Correlation between pre-factor R_m and exponent m in Equation (8). 1—[29], 2—[71], 3—[74], 4—[77], 5—[78], 6—[79].

4. Conclusions

In this work, we propose a novel explanation of the low-temperature resistivity minimum in ceramic strontium ferromolybdate ceramics possessing nano-sized intergrain barriers that is in agreement with given experimental data from various publications. According to our findings, the resistivity upturn is not caused by a semiconductor–metallic transition, but it can be attributed to intergrain fluctuation-induced tunneling dominating at low temperatures the conductivity of the considered SFMO ceramics. Based on known analytical approaches, a modified model, Equation (14), of the total conductivity of granular ceramics was derived. This model is shown to be valid for cold-pressed SFMO ceramics as well as SFMO ceramics thermally treated under oxidation conditions, SFMO thin films deposited by pulsed laser deposition including (100) and (111) textured thin films, and nonstoichiometric SFMO ceramics. In the region of metallic conductivity at elevated temperatures and higher magnetic fluxes, the pre-factor R_m of the temperature-dependent term of metallic conductivity written as a power law decreases exponentially with the exponent m of this power law. Here, the value of m is determined by the charge scattering mechanism.

Author Contributions: Conceptualization, G.S.; methodology, G.S.; software, E.A.; validation, E.A. and G.G.; formal analysis, E.A.; investigation, G.S. and E.A.; resources, G.G.; data curation, E.A.; writing—original draft preparation, G.S.; writing—review and editing, E.A. and G.G.; visualization, E.A.; supervision, G.G.; project administration, G.S. and G.G.; funding acquisition, G.S. and G.G. All authors have read and agreed to the published version of the manuscript.

Funding: This work was funded by the EU project H2020-MSCA-RISE-2017-778308-SPINMULTIFILM.

Institutional Review Board Statement: Not applicable.

Informed Consent Statement: Not applicable.

Data Availability Statement: The original contributions presented in the study are included in the article; further inquiries can be directed to the corresponding author.

Conflicts of Interest: The authors declare no conflicts of interest.

References

1. Suchaneck, G.; Kalanda, N.; Artsiukh, E.; Gerlach, G. Challenges in $Sr_2FeMoO_{6-\delta}$ thin film deposition. *Phys. Status Solidi B* **2020**, *257*, 1900312. [CrossRef]
2. Grollier, J.; Querlioz, D.; Stiles, M.D. Spintronic nanodevices for bioinspired computing. *Proc. IEEE* **2024**, *104*, 2024–2039. [CrossRef] [PubMed]
3. Mishra, R.; Kumar, D.; Yang, H. Oxygen-migration-based spintronic device emulating a biological synapse. *Phys. Rev. Appl.* **2019**, *11*, 054065. [CrossRef]
4. Zhou, J.; Zhao, T.; Shu, X.; Liu, L.; Lin, W.; Chen, S.; Shi, S.; Yan, X.; Liu, X.; Chen, J. Spin–orbit torque-induced domain nucleation for neuromorphic computing. *Adv. Mater.* **2021**, *33*, 2103672. [CrossRef] [PubMed]
5. Chmaissem, O.; Kruk, R.; Dabrowski, B.; Brown, D.E.; Xiong, X.; Kolesnik, S.; Jorgensen, J.D.; Kimball, C.W. Structural phase transition and the electronic and magnetic properties of Sr_2FeMoO_6. *Phys. Rev. B Condens. Matter Mater. Phys.* **2000**, *62*, 14197–14206. [CrossRef]
6. Kim, T.H.; Uehara, M.; Cheong, S.-W.; Lee, S. Large room-temperature intergrain magnetoresistance in double perovskite $SrFe_{1-x}(Mo\ or\ Re)_xO_3$. *Appl. Phys. Lett.* **1999**, *74*, 1737–1739. [CrossRef]
7. Niebieskikwiat, D.; Caneiro, A.; Sánchez, R.D.; Fontcuberta, J. Oxygen-induced grain boundary effects on magnetotransport properties of $Sr_2FeMoO_{6+\delta}$. *Phys. Rev. B Condens. Matter Mater. Phys.* **2001**, *64*, 180406. [CrossRef]
8. Suchaneck, G.; Kalanda, N.; Artiukh, E.; Yarmolich, M.; Sobolev, N.A. Tunneling conduction mechanisms in strontium ferromolybdate ceramics with strontium molybdate dielectric intergrain barriers. *J. Alloys Comp.* **2020**, *860*, 158526. [CrossRef]
9. Takayama, H. Electron-phonon interaction in impure metals. *Z. Phys.* **1973**, *263*, 329–340. [CrossRef]
10. Kondo, J. Resistance minimum in dilute magnetic alloys. *Prog. Theor. Phys.* **1964**, *32*, 37–49. [CrossRef]
11. Matsushita, Y.; Bluhm, H.; Geballe, T.H.; Fisher, I.R. Evidence for charge Kondo effect in superconducting Tl-doped PbTe. *Phys. Rev. Lett.* **2005**, *94*, 157002. [CrossRef] [PubMed]
12. Hien-Hoang, V.; Chung, N.-K.; Kim, H.J. Electrical transport properties and Kondo effect in $La_{1-x}Pr_xNiO_{3-\delta}$ thin films. *Sci. Rep.* **2021**, *11*, 5391. [CrossRef] [PubMed]
13. Syskakis, E.; Choudalakis, G.; Papastaikoudis, C. Crossover between Kondo and electron–electron interaction effects in $La_{0.75}Sr_{0.20}MnO_3$ manganite doped with Co impurities? *J. Phys. Condens. Matter* **2003**, *15*, 7735–7749. [CrossRef]
14. Wang, Y.; Xie, C.; Li, J.; Du, Z.; Cao, L.; Han, Y.; Zu, L.; Zhang, H.; Zhu, H.; Zhang, X.; et al. Weak Kondo effect in the monocrystalline transition metal dichalcogenide $ZrTe_2$. *Phys. Rev. B Condens. Matter Mater. Phys.* **2021**, *103*, 174418. [CrossRef]
15. Lee, P.A.; Ramakrishnan, T.V. Disordered electronic systems. *Rev. Mod. Phys.* **1985**, *57*, 287–337. [CrossRef]
16. Ziese, M. Searching for quantum interference effects in $La_{0.7}Ca_{0.3}MnO_3$ films on $SrTiO_3$. *Phys. Rev. B Condens. Matter Mater. Phys.* **2003**, *68*, 132411. [CrossRef]
17. Suchaneck, G.; Artiukh, E. Absence of weak localization effects in strontium ferromolybdate. *Appl. Sci.* **2023**, *13*, 7096. [CrossRef]
18. Maritato, L.; Adamo, C.; Barone, C.; De Luca, G.M.; Galdi, A.; Orgiani, P.; Petrov, A.Y. Low-temperature resistivity of La0.7Sr0.3MnO3 ultra thin films: Role of quantum interference effects. *Phys. Rev. B Condens. Matter Mater. Phys.* **2006**, *73*, 094456. [CrossRef]
19. Kumar, D.; Sankar, J.; Narayan, J.; Singh, R.K.; Majumdar, A.K. Low-temperature resistivity minima in colossal magnetore-sistive $La_{0.7}Ca_{0.3}MnO_3$ thin films. *Phys. Rev. B Condens. Matter Mater. Phys.* **2002**, *65*, 094407. [CrossRef]
20. Herranz, G.; Sánchez, F.; Fontcuberta, J.; Laukhin, V.; Galibert, J.; García-Cuenca, M.V.; Ferrater, C.; Varela, M. Magnetic field effect on quantum corrections to the low-temperature conductivity in metallic perovskite oxides. *Phys. Rev. B Condens. Matter Mater. Phys.* **2005**, *72*, 014457. [CrossRef]
21. Auslender, M.; Kar'kin, A.E.; Rozenberg, E. Low-temperature resistivity minima in single-crystalline and ceramic $La_{0.8}Sr_{0.2}MnO_3$: Mesoscopic transport and intergranular tunneling. *J. Appl. Phys.* **2001**, *89*, 6639–6641. [CrossRef]
22. Kalanda, N.; Demyanov, S.; Yarmolich, M.; Petrov, A.; Sobolev, N. Electric transport characteristics of $Sr_2FeMoO_{6-\delta}$ ceramics with structurally inhomogeneous surfaces. In *Workshop Book, Proceedings of the 2nd International Workshop on Advanced Magnetic Oxides (IWAMO), Aveiro, Portugal, 24–25 November 2021*; Department of Physic of the University of Aveiro: Aveiro, Portugal, 2021; p. 24.
23. Zhang, J.; Xu, Y.; Yu, L.; Cao, S.; Zhao, Y. Resistivity minimum and the electronic strongly correlation characteristic for $La_{2/3}Sr_{1/3}MnO_3$ thin film. *Physica B* **2008**, *403*, 1471–1473. [CrossRef]
24. Rozenberg, E.; Auslender, M.; Felner, I.; Gorodetsky, G. Low-temperature resistivity minimum in ceramic manganites. *J. Appl. Phys.* **2000**, *88*, 2578–2582. [CrossRef]
25. Barman, A.; Ghosh, M.; Biswas, S.; De, S.K.; Chatterjee, S. Electrical and magnetic properties of $La_{0.7-x}Y_xSr_{0.3}MnO_3$ ($0 \leq x \leq 0{:}2$) perovskite at low temperature. *J. Phys. Condens. Matter* **1998**, *10*, 9799–9811. [CrossRef]
26. Saloaro, M.; Majumdar, S.; Huhtinen, H.; Paturi, P. Absence of traditional magnetoresistivity mechanisms in Sr_2FeMoO_6 thin films grown on $SrTiO_3$, MgO and $NdGaO_3$ substrates. *J. Phys. Condens. Matter* **2012**, *24*, 366003. [CrossRef] [PubMed]

27. Gao, Y.; Cao, G.; Zhang, J.; Habermeier, H.-U. Intrinsic and precipitate-induced quantum corrections to conductivity in $La_{2/3}Sr_{1/3}MnO_3$ thin films. *Phys. Rev. B Condens. Matter Mater. Phys.* **2012**, *85*, 195128. [CrossRef]
28. Koo, J.H.; Kim, J.-H.; Yu, D.G.; Cho, G. Magnetoresistance in double-perovskite structure for Sr_2FeMoO_6. *Mod. Phys. Lett. B* **2007**, *21*, 1593–1598. [CrossRef]
29. Kobayashi, K.I.; Kimura, T.; Sawada, H.; Terakura, K.; Tokura, Y. Room-temperature magnetoresistance in an oxide material with an ordered double-perovskite structure. *Nature* **1998**, *395*, 677–680. [CrossRef]
30. Nag, A.; Jana, S.; Middley, S.; Ray, S. The many facets of tunneling magnetoresistance in Sr_2FeMoO_6. *Indian J. Phys.* **2017**, *91*, 883–893. [CrossRef]
31. Middey, S.; Jana, S.; Ray, S. Surface spin-glass and exchange bias in nanoparticle. *J. Appl. Phys.* **2010**, *108*, 043918. [CrossRef]
32. Kodama, R.H.; Berkowitz, A.E.; McNiff, E.J., Jr.; Foner, S. Surface spin disorder in $NiFe_2O_4$ nanoparticles. *Phys. Rev. Lett.* **1996**, *77*, 394–397. [CrossRef] [PubMed]
33. Zhu, T.; Shen, B.G.; Sun, J.R.; Zhao, H.W.; Zhan, W.S. Surface spin-glass behavior in $La_{2/3}Sr_{1/3}MnO_3$ nanoparticles. *Appl. Phys. Lett.* **2001**, *78*, 3863–3865. [CrossRef]
34. Sarma, D.D.; Ray, S.; Tanaka, K.; Kobayashi, M.; Fujimori, A.; Sanyal, P.; Krishnamurthy, H.R.; Dasgupta, C. Intergranular magnetoresistance in Sr_2FeMoO_6 from a magnetic tunnel barrier mechanism across Grain Boundaries. *Phys. Rev Lett.* **2007**, *98*, 20157205. [CrossRef] [PubMed]
35. Ray, S.; Middey, S.; Jana, S.; Banerjee, A.; Sanyal, P.; Rawat, R.; Gregoratti, L.; Sarma, D.D. Origin of the unconventional magnetoresistance in Sr_2FeMoO_6. *Europhys. Lett.* **2011**, *94*, 47007. [CrossRef]
36. Poddar, A.; Bhowmik, R.N.; Muthuselvam, I.P.; Das, N. Evidence of disorder induced magnetic spin glass phase in Sr_2FeMoO_6 double perovskite. *J. Appl. Phys.* **2009**, *106*, 073908. [CrossRef]
37. Tomioka, Y.; Okuda, T.; Okimoto, Y.; Kumai, R.; Kobayashi, K.; Tokura, Y. Magnetic and electronic properties of a single crystal of ordered double perovskite. *Phys. Rev. B Condens. Matter Mater. Phys.* **2000**, *61*, 422–427. [CrossRef]
38. Westerburg, W. Spinpolarisierter Transport in Epitaktischen Manganoxid- und Doppelperowskitschichten. PhD Thesis, Department of Physics in the Johannes Gutenberg-University Mainz, Mainz, Germany, 2000. [CrossRef]
39. Yanagihara, H.; Salamon, M.B.; Lyanda-Geller, Y.; Xu, S.; Moritomo, Y. Magnetotransport in double perovskite (formula presented) role of magnetic and nonmagnetic disorder. *Phys. Rev. B Condens. Matter Mater. Phys.* **2001**, *64*, 214407. [CrossRef]
40. Akai, H.; Akai, M.; Blügel, S.; Drittler, B.; Ebert, H.; Terakura, K.; Zeller, R.; Dederichs, P.H. Theory of hyperfine interactions in metals. *Progr. Theor. Phys. Suppl.* **1990**, *101*, 11–77. [CrossRef]
41. Yarmolich, M.; Kalanda, N.; Demyanov, S.; Fedotova, J.; Bayev, V.; Sobolev, N.A. Charge ordering and magnetic properties in nanosized Sr_2FeMoO_{6-} powders. *Phys. Status Solidi B* **2016**, *253*, 2160–2166. [CrossRef]
42. Thompson, R.S.; Li, D.; Witte, C.M.; Jia, G.; Lu, J.G. Weak localization and electron-electron interactions in indium-doped ZnO nanowires. *Nano Lett.* **2009**, *9*, 3991–3995. [CrossRef]
43. Hwang, H.Y.; Cheong, S.-W.; Ong, N.P.; Batlogg, B. Spin-polarized intergrain tunneling in $La_{2/3}Sr_{1/3}MnO_3$. *Phys. Rev. Lett.* **1996**, *77*, 2041–2044. [CrossRef] [PubMed]
44. Slonczewski, J.C. Conductance and exchange coupling of two ferromagnets separated by a tunneling barrier. *Phys. Rev B* **1989**, *39*, 6995–7002. [CrossRef] [PubMed]
45. Mott, N.F.; Davis, E.A. *Electronic Processes in Non-Crystalline Materials*, 2nd ed.; Clarendon Press: Oxford, UK, 1979.
46. Allub, R.; Alascio, B. Magnetization and conductivity $La_{1-x}Sr_xMnO_3$-type crystals. *Phys. Rev. B Condens. Matter Mater. Phys.* **1997**, *55*, 14113–14116. [CrossRef]
47. Efros, A.L.; Shklovskii, B.I. Coulomb gap and low temperature conductivity of disordered systems. *J. Phys. C Solid State Phys.* **1975**, *8*, L49–L51. [CrossRef]
48. Sheng, P.; Abeles, B.; Arie, Y. Hopping conductivity in granular metals. *Phys. Rev. Lett.* **1973**, *31*, 44–47. [CrossRef]
49. Niebieskikwiat, D.; Sánchez, R.; Caneiro, A.; Morales, L.; Vásquez-Mansilla, M.; Rivadulla, F.; Hueso, L. High-temperature properties of the double perovskite: Electrical resistivity, magnetic susceptibility, and ESR. *Phys. Rev. B Condens. Matter Mater. Phys.* **2000**, *62*, 3340–3345. [CrossRef]
50. Zhang, L.; Zhou, Q.; He, Q.; He, T. Double-perovskites $A_2FeMoO_{6-\delta}$ (A = Ca, Sr, Ba) as anodes for solid oxide fuel cells. *J. Power Sources* **2010**, *195*, 6356–6366. [CrossRef]
51. Sheng, P.; Sichel, E.K.; Gittleman, J.I. Fluctuation-induced tunneling conduction in carbon-polyvinylchloride composites. *Phys. Rev. Lett.* **1978**, *40*, 1197–2000. [CrossRef]
52. Snyder, G.J.; Hiskes, R.; DiCarolis, S. Intrinsic electrical transport and magnetic properties of $La_{0.67}Ca_{0.33}MnO_3$ and $La_{0.67}Sr_{0.33}MnO_3$ MOCVD thin films and bulk material. *Phys. Rev. B Condens. Matter Mater. Phys.* **1996**, *53*, 14434–14444. [CrossRef] [PubMed]
53. Schiffer, P.; Ramirez, A.P.; Bao, W.; Cheong, S.-W. Low temperature magnetoresistance and the magnetic phase diagram of $La_{1-x}Ca_xMnO_3$. *Phys. Rev. Lett.* **1995**, *75*, 3336–3339. [CrossRef]
54. Akimoto, T.; Moritomo, Y.; Nakamura, A.; Furukawa, N. Observation of anomalous single-magnon scattering in half-metallic ferromagnets by chemical pressure control. *Phys. Rev. Lett.* **2000**, *85*, 3914–3917. [CrossRef] [PubMed]
55. Dyson, F.J. Thermodynamic behavior of an ideal ferromagnet. *Phys. Rev.* **1956**, *102*, 1230–1244. [CrossRef]
56. Zhang, H.; Chen, X.J.; Zhang, C.L.; Almasan, C.C.; Habermeier, H.U. Spin-wave scattering at low temperatures in manganite films. *Phys. Rev. B Condens. Matter Mater. Phys.* **2003**, *67*, 134405. [CrossRef]

57. Kubo, K.; Ohatata, N.A. Quantum theory of double exchange. I. *J. Phys. Soc. Japan* **1972**, *33*, 21–32. [CrossRef]
58. Urushibara, A.; Moritomo, Y.; Arima, T.; Asamitsu, A.; Kido, G.; Tokura, Y. Insulator-metal transition and giant magnetoresistance in $La_{1-x}Sr_xMnO_3$. *Phys. Rev. B Condens. Matter Mater. Phys.* **1995**, *51*, 14103. [CrossRef] [PubMed]
59. Li, G.; Zhou, H.D.; Feng, S.J.; Fan, X.J.; Li, X.G.; Wang, Z.D. Competition between ferromagnetic metallic and paramagnetic insulating phases in manganites. *J. Appl. Phys.* **2002**, *92*, 1406–1410. [CrossRef]
60. Zhao, G.M.; Smolyaninova, V.; Prellier, W.; Keller, H. Electrical transport in the ferromagnetic state of manganites: Small-polaron metallic conduction at low temperatures. *Phys. Rev. Lett.* **2000**, *84*, 6086–6089. [CrossRef] [PubMed]
61. Fisher, B.; Chashka, K.B.; Patlagan, L.; Reisner, G.M. Intergrain tunneling in granular Sr_2FeMoO_6 studied by pulsed high currents. *Phys. Rev. B Condens. Matter Mater. Phys.* **2003**, *68*, 13442. [CrossRef]
62. Simmons, J.G. Generalized formula for the electric tunnel effect between similar electrodes separated by a thin insulating film. *J. Appl. Phys.* **1963**, *34*, 1793–1803. [CrossRef]
63. Helman, J.S.; Abeles, B. Tunneling of spin-polarized electrons and magnetoresistance in granular Ni films. *Phys. Rev. Lett.* **1976**, *37*, 1429–1432. [CrossRef]
64. Fisher, B.; Genossar, J.; Chashka, K.B.; Patlagan, L.; Reisner, G.M. Remarkable power-law temperature dependencies of inter-grain conductivity. *Solid State Commun.* **2006**, *137*, 641–644. [CrossRef]
65. Fisher, B.; Genossar, J.; Chashka, K.B.; Patlagan, L.; Reisner, G.M. Intergrain tunnelling in the half-metallic double-perovskites $Sr_2BB'O_6$ (BB′ = FeMo, FeRe, CrMo, CrW, CrRe). *EPJ Web Conf.* **2014**, *75*, 01001. [CrossRef]
66. Granville, S.; Farrell, I.L.; Hyndman, A.R.; McCann, D.M.; Reeves, R.J.; Williams, G.V.M. Indications of spin polarized transport in Ba_2FeMoO_6 thin films. *arXiv* **2017**, arXiv:1707.01208.
67. Sarma, D.D.; Mahadevan, P.; Saha-Dasgupta, T.; Ray, S.; Kumar, A. Electronic structure of Sr_2FeMoO_6. *Phys. Rev. Lett.* **2000**, *85*, 2549–2552. [CrossRef] [PubMed]
68. Mitani, S.; Takahashi, S.; Takanashi, K.; Yakushiji, K.; Maekawa, S.; Fujimori, H. Enhanced magnetoresistance in insulating granular systems: Evidence for higher-order tunneling. *Phys. Rev. Lett.* **1998**, *81*, 2799–2802. [CrossRef]
69. Haile, S.M.; West, D.L.; Campbell, J. The role of microstructure and processing on the proton conducting properties of gadolinium-doped barium cerate. *J. Mater. Res.* **1998**, *13*, 1576–1595. [CrossRef]
70. Niebieskikwiat, D.; Prado, F.; Caneiro, A.; Sánchez, R.D. Antisite defects versus grain boundary competition in the tunneling magnetoresistance of the Sr_2FeMoO_6 double perovskite. *Phys. Rev. B Condens. Matter Mater. Phys. B.* **2004**, *70*, 132412. [CrossRef]
71. Westerburg, W.; Reisinger, D.; Jakob, G. Epitaxy and magnetotransport of Sr_2FeMoO_6 thin films. *Phys. Rev. B Condens. Matter Mater. Phys.* **2000**, *62*, R767–R770. [CrossRef]
72. Venimadhav, A.; Sher, F.; Attfield, J.P.; Blamire, M.G. Oxygen assisted deposition of Sr_2FeMoO_6 thin films on $SrTiO_3$. *J. Magn. Magn. Mater.* **2004**, *269*, 101–105. [CrossRef]
73. Manako, T.; Izumi, M.; Konishi, Y.; Kobayashi, K.-I.; Kawasaki, M.; Tokura, Y. Epitaxial thin films of ordered double perovskite Sr_2FeMoO_6. *Appl. Phys. Lett.* **1999**, *74*, 2215–2217. [CrossRef]
74. Liu, G.Y.; Rao, G.H.; Feng, X.M.; Yang, H.F.; Ouyang, Z.W.; Liu, W.F.; Liang, J.K. Metal–semiconductor transition in non-stoichiometric double perovskite $Sr_2Fe_xMo_{2-x}O_6$. *Phys. B* **2003**, *334*, 229–233. [CrossRef]
75. López-Mir, L.; Frontera, C.; Aramberri, H.; Bouzehouane, K.; Cisneros-Fernández, J.; Bozzo, B.; Balcells, L.; Martínez, B. Anisotropic sensor and memory device with a ferromagnetic tunnel barrier as the only magnetic element. *Sci. Rep.* **2018**, *8*, 861. [CrossRef] [PubMed]
76. Suchaneck, G.; Artiukh, E. Intergranular magnetoresistance of strontium ferromolybdate ceramics caused by spin-polarized tunneling. *Open Ceram.* **2021**, *7*, 100171. [CrossRef]
77. Suchaneck, G.; Artiukh, E.; Gerlach, G. Resistivity and tunnel magnetoresistance in double-perovskite strontium ferromolybdate ceramics. *Phys. Status Solidi B* **2022**, *259*, 2200012. [CrossRef]
78. Deniz, H.; Preziosi, D.; Alexe, M.; Hesse, D. Coherent Fe-rich nano-scale perovskite oxide phase in epitaxial Sr_2FeMoO_6 films grown on cubic and scandate substrates. *J. Appl. Phys.* **2017**, *121*, 023906. [CrossRef]
79. Suchaneck, G.; Artiukh, E. Nonstoichiometric strontium ferromolybdate as an electrode material for solid oxide fuel cells. *Inorganics* **2022**, *10*, 230. [CrossRef]

ceramics

Article

Zero-Temperature Coefficient of Resonant Frequency in $[(Mg_{0.6}Zn_{0.4})_{0.95}Co_{0.05}]_{1.02}TiO_{3.02}$-$Ca_{0.6}(La_{0.9}Y_{0.1})_{0.2667}TiO_3$ Ultra-Low-Loss Composite Dielectrics

Yuan-Bin Chen * and Jie Peng *

School of Electronics and Electrical Engineering, Zhaoqing University, Zhaoqing 526061, China
* Correspondence: n2890103@outlook.com (Y.-B.C.); kit.j.peng@gmail.com (J.P.)

Abstract: Investigating the microwave dielectric properties of ceramics prepared through the conventional solid-state route, such as $x[(Mg_{0.6}Zn_{0.4})_{0.95}Co_{0.05}]_{1.02}TiO_{3.02}$-$(1-x)Ca_{0.6}(La_{0.9}Y_{0.1})_{0.2667}TiO_3$, reveals notable characteristics. $[(Mg_{0.6}Zn_{0.4})_{0.95}Co_{0.05}]_{1.02}TiO_{3.02}$ shows a permittivity (ε_r) of approximately 20, a high quality factor ($Q \times f$) ranging between 250,000 and 560,000 GHz, and a temperature coefficient of resonant frequency (τ_f) of approximately −65 ppm/°C. To enhance the temperature stability, $Ca_{0.6}(La_{0.9}Y_{0.1})_{0.2667}TiO_3$ featuring a τ_f value of +374 ppm/°C was incorporated into the $[(Mg_{0.6}Zn_{0.4})_{0.95}Co_{0.05}]_{1.02}TiO_{3.02}$ composition. τ_f demonstrated an increase with rising $Ca_{0.6}(La_{0.9}Y_{0.1})_{0.2667}TiO_3$ content, reaching zero at x = 0.95. A ceramic composition of $0.95[(Mg_{0.6}Zn_{0.4})_{0.95}Co_{0.05}]_{1.02}TiO_{3.02}$-$0.05Ca_{0.6}(La_{0.9}Y_{0.1})_{0.2667}TiO_3$, incorporating 3wt.% $BaCu(B_2O_5)$ as sintering aids, exhibited outstanding microwave dielectric properties: ε_r~22.5, Q × f~195,000 (at 9 GHz), and τ_f~0.1ppm/°C, with a sintering temperature at 950 °C. This material is proposed as a prospective candidate for 6G band components and GPS antennas.

Keywords: $[(Mg_{0.6}Zn_{0.4})_{0.95}Co_{0.05}]_{1.02}TiO_{3.02}$; $Ca_{0.6}(La_{0.9}Y_{0.1})_{0.2667}TiO_3$; microwave dielectric properties; temperature coefficient of resonant frequency

Citation: Chen, Y.-B.; Peng, J. Zero-Temperature Coefficient of Resonant Frequency in $[(Mg_{0.6}Zn_{0.4})_{0.95}Co_{0.05}]_{1.02}TiO_{3.02}$-$Ca_{0.6}(La_{0.9}Y_{0.1})_{0.2667}TiO_3$ Ultra-Low-Loss Composite Dielectrics. *Ceramics* **2024**, *7*, 466–477. https://doi.org/10.3390/ceramics7020030

Academic Editors: Dawei Wang and Fayaz Hussain

Received: 15 January 2024
Revised: 18 March 2024
Accepted: 22 March 2024
Published: 26 March 2024

1. Introduction

The distinct electrical properties of ceramic dielectric resonators (DRs) have spurred a revolution in the wireless communications industry by minimizing the dimensions and cost of oscillator and filter components within the microwave circuit [1–3]. The integration of DRs enables the downsizing of microwave components, with requirements including a low dielectric loss, high permittivity, and a closee-zero τ_f [4–8]. The product of the quality factor (Q) and the frequency (f), denoted as Q × f, exhibits an inverse relationship with the dielectric loss tangent (tanδ) in dielectric materials. Specifically, a higher value of Q × f signifies a lower dielectric loss, which is particularly crucial for the efficiency of ceramics used in microwave dielectrics [9]. These materials have become the preferred choice for 6G communications owing to their exceptional low-loss properties [10–12].

$MgTiO_3$-based ceramics, known for their low dielectric loss, have been the subject of intensive research. The $MgTiO_3$-$CaTiO_3$ ceramic, composed of a blend of $MgTiO_3$ (ε_r approximately 17, Q × f approximately 160,000 (at 9 GHz), and τ_f approximately −50 ppm/°C) [13] and $CaTiO_3$ (ε_r~170, Q × f~3600 (at 7 GHz), and τ_f~800 ppm/°C) [14], has been utilized in dielectric resonators and patch antennas. A ceramic composition of 0.95 $MgTiO_3$-$0.05CaTiO_3$ exhibits ε_r~21, a Q × f~56,000 (at 7 GHz), and a zero τ_f value, but necessitates high sintering temperatures of 1400~1450 °C [15]. In industrial settings, there is a pressing need to reduce the sintering temperature in order to enhance energy efficiency, lower production costs, and mitigate potential adverse effects on the material properties during the fabrication process [16–19]. Given the close similarity in ionic radii between Zn^{2+} (0.083 nm) and Mg^{2+} (0.078 nm), the strategic substitution of Zn^{2+}

ions with Mg^{2+} ions emerges as a viable approach for crafting (Mg, Zn)TiO_3 compositions. The incorporation of Zn into $MgTiO_3$ to create a solid solution, (Mg, Zn)TiO_3, has proven advantageous not only because it maintains the structural integrity but also because it reduces the required sintering temperature. This not only streamlines the manufacturing process but also holds promise for optimizing material properties in practical industrial applications. For instance, $(Mg_{0.7}Zn_{0.3})TiO_3$, sintered at 1200 °C, exhibits ε_r~19.8, Q × f~142,000 GHz, and τ_f~−66 ppm/°C [20]. Additionally, incorporating Co into $(Mg_{0.7}Zn_{0.3})$, as seen in $(Mg_{0.7}Zn_{0.3})_{0.95}Co_{0.05}TiO_3$, results in excellent dielectric properties, with ε_r~20, Q × f~163,560 GHz, and τ_f~−65 ppm/°C, even when sintered at a relatively low temperature of 1200 °C [21]. Non-stoichiometric ceramic $Mg_{1+\delta}$ $TiO_{3+\delta}$ compositions were initially reported by Huang [22]. The Q × f value exhibited a nonlinear variation, reaching a peak of approximately 357,600 GHz when δ reached 0.02, beyond which it exhibited a decline. This noteworthy peak in the Q × f value is attributed to the distinct $MgTiO_3$ phase, underscoring the influence of non-stoichiometry on the microwave properties of the ceramic material. The exploration of these non-stoichiometric variations holds the potential for tailoring and optimizing microwave characteristics in diverse applications.

In this paper, we explore the substitution of Zn^{2+} and Co^{2+} ions for Mg^{2+} ions, resulting in the formation of $[(Mg_{0.6}Zn_{0.4})_{0.95}Co_{0.05}]_{1.02}TiO_{3.02}$ (MZCT). The ionic radius of Mg (0.78 Å) closely resembles that of Zn^{2+} (0.83 Å) and Co^{2+} (0.82 Å). Consequently, the substitution greatly enhanced the Q × f value, which establishes it as a highly dependable method for producing materials with consistent dielectric properties. The dielectric properties in the microwave applications were assessed through the analysis of X-ray diffraction (XRD) patterns, densification, and microstructures.

We investigated the partial substitution of Y^{3+} for La^{3+} in $Ca_{0.6}La_{0.2667}TiO_3$ ceramics. This substitution is viable due to the smaller ionic radius of Y^{3+} (0.106 Å) compared to that of La^{3+} (0.122 Å), allowing for the formation of $Ca_{0.6}(La_{0.9}Y_{0.1})_{0.2667}TiO_3$ (CLYT). A solid-state synthesis method was employed in this study. Examining the correlations between the surface morphology and quantity of CLYT ceramics sintered at 1350 °C revealed the following properties: ε_r = 111, Q × f = 23,100 (GHz), and τf = 374.6 ppm/°C. The formation of $BaCu(B_2O_5)$ (BCB) was observed at 700 °C, with subsequent melting occurring above 850 °C [23–25]. Due to its low melting temperature and advanced microwave dielectric properties, BCB proved effective as a sintering aid for microwave dielectric materials [22].

In this work, BCB powder was employed to effectively reduce the sintering temperature during the fabrication of the xMZCT-(1−x)CLYT ceramics. The impact of BCB on the microwave dielectric properties of the ceramics was investigated.

2. Experiments

MZCT was synthesized using the solid-state mixed oxide route, employing high-purity oxide powders (>99.9%) such as $MgCO_3{\cdot}0.2708H_2O$ as starting materials. The decomposition reaction of $MgCO_3{\cdot}0.2708H_2O$ at temperatures between 600 and 700 °C for 5 h can be represented as follows [26]:

$$MgCO_3{\cdot}0.2708H_2O \rightarrow MgO + CO_2 + xH_2O$$

The ceramic preparation process commenced with an initial firing at 650 °C to eliminate moisture content and ensure a stable starting point. MZCT ceramics were fabricated through a conventional solid-state reaction utilizing high-purity oxide powders (MgO, ZnO, CoO, and TiO_2). The raw materials were precisely weighed and subjected to a thorough mixing process through ball milling with agate media in distilled water for an extensive period of 24 h. Following meticulous mixing, resulting mixtures underwent a drying phase before being carefully calcined at a temperature of 1000 °C for 4 h.

Samples of CLYT ceramics were fabricated through a conventional solid-state reaction utilizing high-purity oxide powders ($CaCO_3$, La_2O_3, Y_2O_3, and TiO_2). The process involved mixing, milling for 24 h, drying, and grinding, followed by calcination at 1000 °C for 2 h. XRD analyses were employed to identify the crystal phases of the calcined powders. For the synthesis of BCB, $BaCO_3$ (>99%), CuO (>99%), and B_2O_3 (>99%) were mixed for 4 h in a nylon jar with zirconia balls and then dried and calcined at 700 °C for 3 h. Subsequently, xMZCT-(1−x)CLYT powders were dry-mixed with an agate mortar and pestle before being wet-mixed using distilled water for 12 h. To facilitate the formation of well-structured ceramic samples, a finely powdered mixture was prepared, incorporating 3 wt% of a 10 wt% of an aqueous solution of polyvinyl alcohol (PVA 500, Showa, Japan) solution as a binder. This mixture was meticulously compressed into pellets, with each one measuring 1.1 cm in diameter and 0.5 cm in thickness, utilizing a pressure of 200 MPa to ensure a compact and uniform structure.

Following pelletization, the specimens underwent a sintering process in ambient air, for a duration of 4 h at temperatures ranging from 900 to 1000 °C. Throughout this thermal treatment, both the heating and cooling rates were carefully controlled at 10 °C/min to prevent thermal stress and ensure a gradual transformation. To assess the structural evolution, an X-ray diffraction pattern analysis was employed. This analytical technique was applied to identify the crystalline phases present in both the calcined powder and the sintered ceramics, providing valuable insights into the structural changes and the formation of desired phases during the manufacturing process. Scanning electron microscopy (SEM, Philips XL-40FEG, Eindhoven, The Netherlands) was used in the microstructure analysis of the sintered surface. Additionally, dispersive spectroscopy (EDS) was utilized to identify the presence of any second phases. The Archimedes method was used to measure the bulk densities of the sintered pellets. The Hakki–Coleman [27] dielectric resonator method, adapted and refined by Courtney [28], was applied to determine the permittivity and the quality factor value at the microwave frequencies under both TE011 and TE01 modes.

For measurement purposes, the dielectric resonator was meticulously placed between two brass plates, establishing a reliable and controlled environment to assess the microwave properties with precision. A comprehensive measurement system that utilized the synergistic capabilities of an HP8350B sweep oscillator and an HP8757D network analyzer for precision analysis was employed. The test set was positioned atop a thermostat, spanning a temperature spectrum in the range of 25 to 80 °C. The τ_f value can be calculated using Equation (1).

$$\tau_f = \frac{f_{80} - f_{25}}{f_{25}(80 - 25)} \times 10^6 (\text{ppm}/^\circ\text{C}) \quad (1)$$

where f_{80} and f_{25} represent the resonant frequencies at 80 and 25 °C, respectively.

3. Results and Discussion

Figure 1 depicts the X-ray diffraction (XRD) pattern of 0.95MZCT-0.05CLYT, augmented with a 3wt.% BCB dopant, recorded at room temperature. The crystal structures of MZCT and CLYT were revealed to be rhombohedral (ICDD-PDF #01-073-7752) and orthorhombic (ICDD-PDF #00-022-0153), respectively. Remarkably, the XRD patterns reveal the persistent presence of the MZCT phase in these specimens, exhibiting negligible alternations with varying sintering temperatures within the range of 900–1000 °C. The XRD patterns affirm the dominance of MZCT as the major crystalline phase, accompanied by a minor phase of CLYT. The inability to achieve a solid solution in the xMZCT-(1−x)CLYT ceramic system is attributed to inherent structural disparities. Notably, the XRD patterns of the 0.95MZCT-0.05CLYT ceramic exhibit minimal fluctuations across the sintering temperature spectrum of 900–1000 °C. In Figure 2, SEM images of the 0.95MZCT-0.05CLYT ceramic, doped with 3wt.% BCB and subjected to various sintering temperatures, are presented. At 900 °C, porous microstructures are evident in the grains. Grain growth is initiated at 950 °C and significantly escalates at 975 °C. Beyond 975 °C, inhomogeneous grain growth becomes apparent, potentially impacting the microwave dielectric properties of the ceramics.

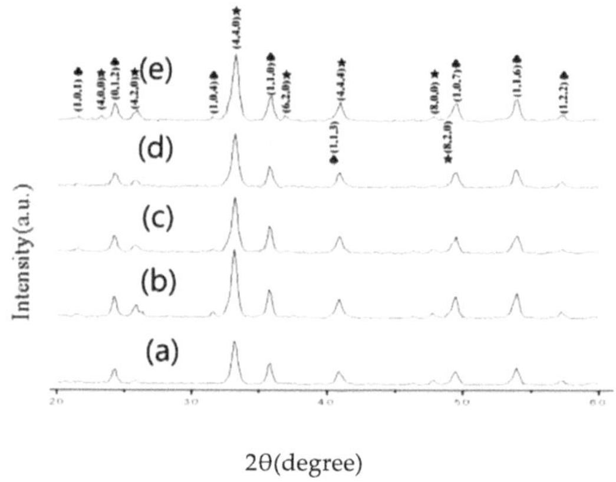

Figure 1. X-ray diffraction patterns of 0.95MZCT-0.05CLYT ceramics doped with 3wt.% BCB and sintered at various temperatures for 4 h: (**a**) 900 °C, (**b**) 925 °C, (**c**) 950 °C, (**d**) 975 °C, and (**e**) 1000 °C (♣: MZCT; ★: CLYT).

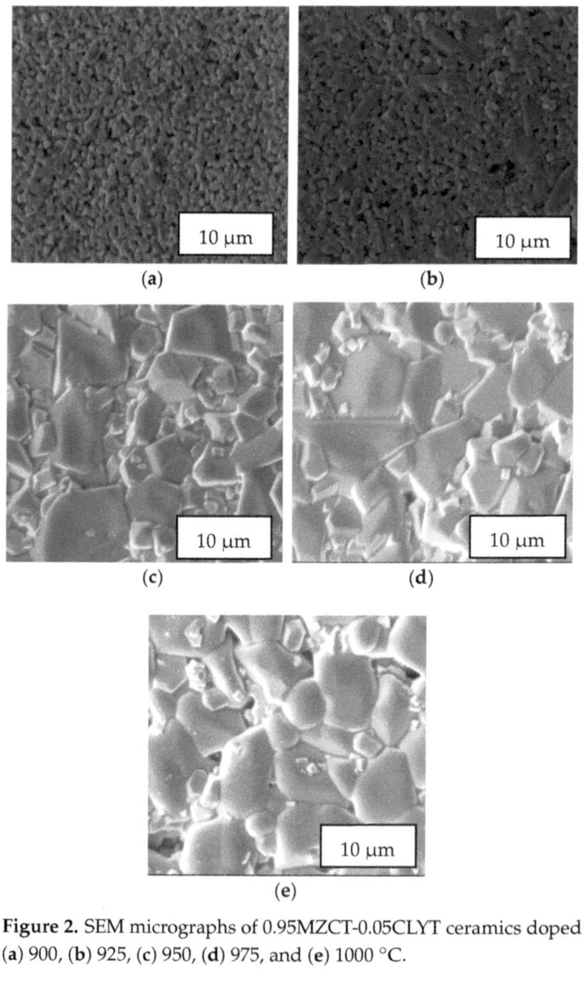

Figure 2. SEM micrographs of 0.95MZCT-0.05CLYT ceramics doped with 3wt.% BCB and sintered at (**a**) 900, (**b**) 925, (**c**) 950, (**d**) 975, and (**e**) 1000 °C.

Energy-dispersive X-ray (EDX) analysis, complemented by SEM, was employed to discern each individual grain in the 0.95MZCT-0.05CLYT ceramics sintered at 950 °C, and the results, along with the corresponding spots A–B, are presented in Figure 3. The grain morphology of the well-developed 0.95MZCT-0.05CLYT ceramics can be classified into two types: the larger grains (spot A), indicated as the Mg–Ti phase, were confirmed to be MZCT, while the smaller cubic-shaped grains (spot B) were identified as CLYT. The EDX findings align with the XRD results obtained from the 0.95MZCT-0.05CLYT ceramics. In contrast to the pure MZCT, CLYT exhibits a lower sintering temperature. This discrepancy arises from the smaller grain size of CLYT compared to that of MZCT, and the incorporation of CLYT into MZCT contributes to the improved densification of the ceramics. Details of the EDX data for spots A-B are shown in Table 1, offering insights into the elemental composition at specific regions. The well-developed grain morphology shows the distinctive characteristics of the dual-phase ceramics.

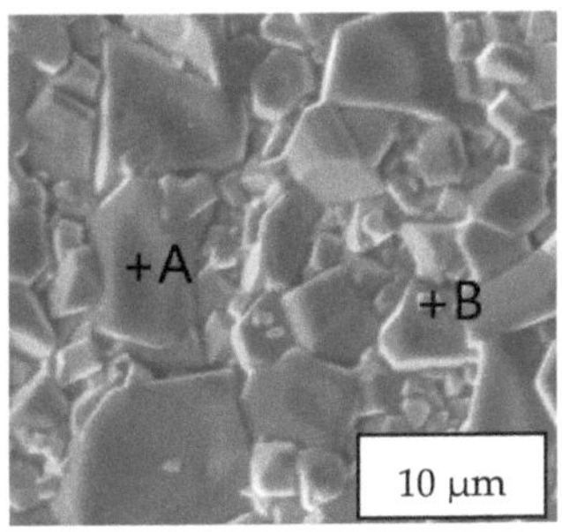

Figure 3. The marks of SEM for the 0.95MZCT-0.05CLYT ceramics doped with 3wt.% BCB and sintered at 950 °C.

Table 1. EDX data of 0.95MZCT-0.05CLYT ceramics doped with 3wt.% BCB for areas A-B.

Elements	Mg	Zn	Co	Ti	Ca	La	Y	O
Area A	20.11 (±0.01)	1.23 (±0.02)	1.9 (±0.03)	23.21 (±0.02)	0	0	0	53.55 (±0.02)
Area B	0	0	0	12.75 (±0.01)	9.89 (±0.01)	11.98 (±0.02)	12.1 (±0.01)	53.28 (±0.01)

Figure 4 illustrates the bulk density, relative density, and porosity of 3wt.% BCB-doped xMZCT-(1−x)CLYT ceramics sintered at various temperatures for 4 h. As the temperature rose, the bulk density exhibited an upward trend, reaching a peak value of 3.86 g/cm^3 at 950 °C, after which it gradually declined. The reduction in density is attributed to the abnormal grain growth, as depicted in Figure 2. The variation in permittivity mirrored the density changes, showing an increase with the sintering temperature. Following its zenith at 1150 °C, the permittivity experienced a subsequent decrease. With an increase in sintering temperature, the relative density also increased, as shown in Figure 4b, and the porosity also decreased, as shown in Figure 4c. When the temperature rose to 950 degrees C, the relative density of the xMZCT-(1−x)CLYT ceramics could reach more than 95%, and the porosity reached the minimum value, which was less than 5%.

Figure 5 depicts the permittivity curves of the xMZCT-(1−x)CLYT ceramics at different sintering temperatures over a 4 h period. The relationship between the ε_r values and sintering temperature (presented in Table 2) mirrors the trend observed between density and sintering temperature, given that higher density corresponds to lower porosity. The permittivity exhibited a slight increase with an increase in the sintering temperature. The permittivity of the xMZCT-(1−x)CLYT ceramics showed a progression from 17.27 to 26.2 as the sintering temperature increased from 900 to 950 °C. A peak value of 26.2 was achieved for the 0.85MZCT-0.15CLYT ceramics sintered at 950 °C for 4 h. It is well known that the permittivity of microwave dielectric ceramics is influenced by both the mixing rule and density.

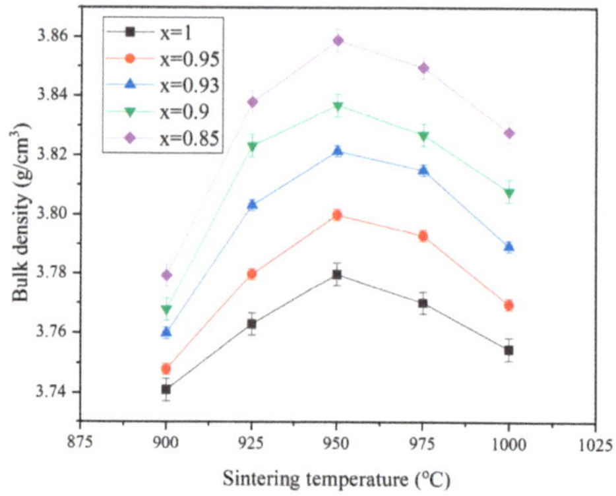

(**a**)

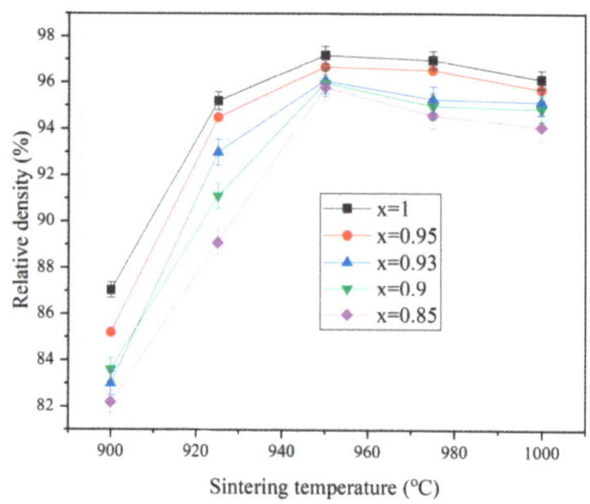

(**b**)

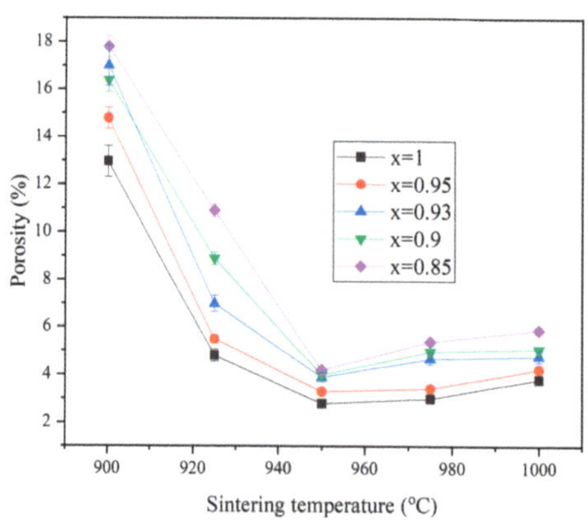

(**c**)

Figure 4. (**a**) Bulk density, (**b**) relative density, and (**c**) porosity of xMZCT-(1-x)CLYT ceramics with 3wt.% BCB as a function of the sintering temperature.

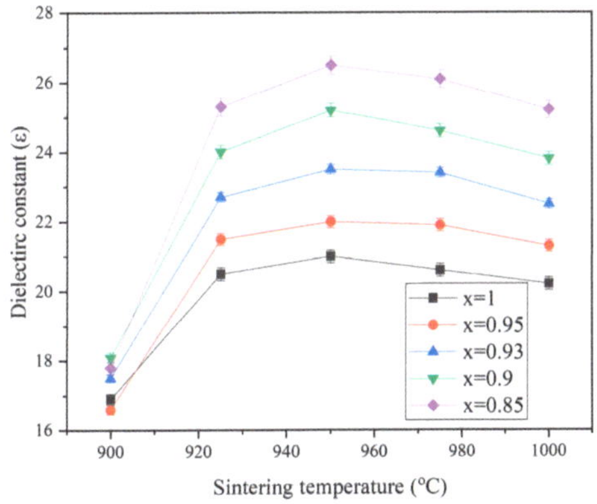

Figure 5. Permittivity curves of xMZCT-(1−x)CLYT ceramics doped with 3wt.% BCB at different sintering temperatures for 4 h.

Table 2. Microwave dielectric properties of the xMZCT-(1-x)CLYT ceramic system doped with 3 wt.% BCB and sintered at 950 °C for 4 h.

x Value	Bulk Density (g/cm^3)	ε_r	Q × f (GHz)	τ_f (ppm/°C)
1	3.78 (±0.04)	20.95 (±0.05)	250,000 (±6700)	−63 (±0.4)
0.95	3.8 (±0.03)	22.5 (±0.01)	190,000 (±3810)	0 (±0.01)
0.93	3.81 (±0.03)	23.5 (±0.02)	180,000 (±1220)	0.1 (±0.05)
0.9	3.83 (±0.02)	25.2 (±0.03)	172,000 (±1800)	10 (±0.3)
0.85	3.88 (±0.03)	26.7 (±0.01)	156,000 (±1610)	21 (±0.5)

It is also known that the permittivity of microwave dielectric ceramics is affected by their ion polarizability. As mentioned above, in xMZCT-(1−x)CLYT, the ε_r value of CLYT increases. To elucidate the effect of CLYT Zn substituting MZCT on the dielectric constant, the ion polarizability of xMZCT-(1−x) CLYT was estimated using the Clausius–Mossotti equation:

$$\varepsilon_r = \frac{3V_m + 8\pi\alpha_m}{3V_m - 4\pi\alpha_m} \quad (2)$$

where ε_r, V_m, and α_m are the relative permittivity, molar volume, and macroscopic polarizability, respectively. Using experimental relative permittivity data and unit cell volume data, the macroscopic polarizability αm was calculated. The theoretical polarizability data in Table 1 show that, with an increase in xMZCT-(1−x)CLYT content, there is almost an S-shaped increase, while the unit cell volume increases with an increase in x. The relative dielectric constant increases as αm increases. When the αm value approaches 3Vm/4π, the relative dielectric constant increases rapidly. It has been reported that the macroscopic polarizability of a complex system with ideal symmetry can be determined by the summation of the polarizabilities of the constituent cations, meaning that:

$$\alpha_m = \Sigma\alpha(\text{ions}) \quad (3)$$

The theoretical polarizability (expressed as αm (theoretical)) value was calculated according to Equation (3). This was compared with the "experimental" polarizability, which was denoted as αm(exp) and determined using the Clausius–Mossotti equation. It should be noted that the αm(exp) of the xMZCT-(1−x)CLYT end member is greater than the αm (theoretical) value, and the αm(exp) value is larger than the αm (theoretical) value. Shannon showed that, as the cation size changes, deviations in the additivity of ion polarizability occur when cation compression or vibration occurs in structural sites. The lower αm(exp) value of xMZCT-(1−x)CLYT may be due to the compression effect caused by large difference in polarizability MZCT and CLYT ions.

The logarithmic mixing rule, proposed by Lichterecker in 1926, is an intermediate form between serial and parallel mixing [29]:

$$\log \varepsilon_m = v_h \log \varepsilon_h + v_l \log \varepsilon_l \tag{4}$$

where ε_h and ε_l represent the relative permittivity of the high- and dielectric phases, respectively. V_h and V_l denote the volume fractions of the high- and low-dielectric phases, respectively, ensuring that their sum is equal to 1 (V_h + V_l = 1). The value ε_m signifies the effective permittivity of the composite.

Microwave dielectric loss primarily arises from factors such as the lattice vibration mode, porosity, secondary-phase impurities and lattice defects. The role of relative density in governing dielectric losses is noteworthy, echoing findings in other microwave dielectric materials. It is widely acknowledged that the factors diminishing the dielectric Q value fall into two categories: natural loss and external loss. Natural loss stems from the interaction between polar phonon vibrations and microwave electric fields in crystals. External loss encompasses phenomena like order–disorder transformations, pore density, grain size, oxygen vacancies, and impurity phases in ceramics. The intrinsic Q value establishes the upper limit for a defect-free single crystal, quantitatively described in the microwave frequency range by the classical damping oscillator model. In this model, under the one-phonon absorption approximation, a reciprocal relationship between Q × f and the permittivity emerges [30]:

$$Q \times f \propto \varepsilon^{-1} \tag{5}$$

where the frequency f is confined within the vicinity of the phonon engine frequency for the validity of the estimation, typically around 10^{12} Hz at room temperature. Nevertheless, a set of experiments has shown that extending the application of Equation (5) from the microwave frequency to the megahertz frequency (1–4 orders of magnitude lower than the optical phonon resonance frequency) at room temperature yields a satisfactory dielectric Q magnitude for well-crafted ceramics. However, the results show that Q × f solely depends on ε_r [30]:

$$Q \times f \propto \varepsilon_r^{-0.6} \tag{6}$$

This implies an inverse relationship between the values of Q × f and ε_r, where, as ε_r increases, Q × f decreases, and vice versa. This phenomenon is most likely attributed to external factors. Numerous authors acknowledge that porosity within a dielectric has an adverse impact on the Q × f values, and the extent of this impact varies depending on the type of dielectric. For low-dielectric Q × f ceramics of the order of 10^3 GHz, the impact of porosity on dielectric Q can be described as $Q = Q_o(1 - 1.5P)$, where Q_o represents the intrinsic dielectric Q measured using microwave reflectance spectroscopy, and P denotes porosity. However, in the case of high Q × f ceramics of the order of 10^5–10^6 GHz, such as polycrystalline Al_2O_3 ceramics, even a small amount of porosity significantly diminishes the dielectric Q, as described by Equation (7) [30,31]:

$$\frac{1}{Q} = (1 - P)\frac{1}{Q_o} + A'P(\frac{P}{1 - P})^{2/3} \tag{7}$$

where Qo represents the full-density dielectric quality factor (1.565 × 10^{-5}), A' is a constant (=6.3 × 10^{-3}), and P is the porosity. Figure 6 illustrates the Q × f value of the xMZCT-(1−x)CLYT ceramics sintered at various temperatures. The Q × f value exhibited an initial increase with rising sintering temperature, reaching a peak value before subsequently declining. Specifically, the peak of the Q × f value was achieved for the xMZCT-(1−x)CLYT ceramics at 950 °C, ranging from 256,000 to 150,000 depending on the value of x. The reduction in the Q × f value is attributed to the abnormal grain growth observed at higher temperatures, as shown in Figure 2.

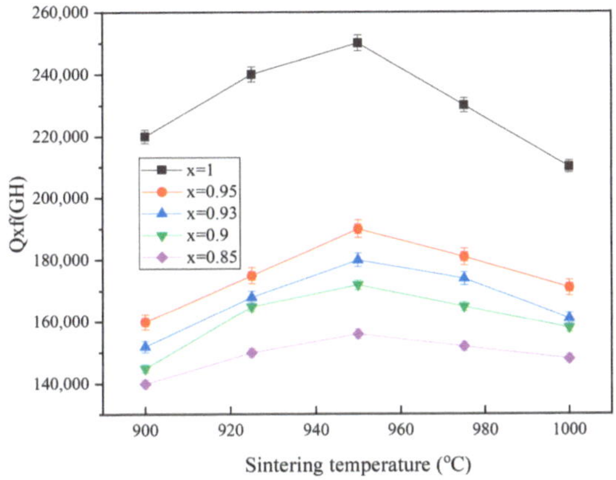

Figure 6. Q × f values of xMZCT-(1−x)CLYT ceramics doped with 3wt.% BCB as a function of the sintering temperature.

The linear thermal expansion coefficient is a fixed parameter in ceramics, and it directly influences the temperature coefficient of capacitance by its correlation with the temperature dependence of the permittivity. The relationship between the temperature coefficient of resonant frequency (τ_f), τ_c, and the thermal expansion coefficient (α_L) is stated in Equation (8) [32]:

$$\tau_f = \left(\frac{\tau_c}{2} + \alpha_L\right) \quad (8)$$

Figure 7 illustrates the temperature coefficient of the resonant frequency τ_f for the xMZCT-(1−x)CLYT ceramic system doped with 3wt.% BCB. τ_f is known to be influenced by the composition, additives, and second phase of the material. Given that the τ_f values for MZCT and CLYT are −65 and +320 ppm/°C, respectively, an increase in CLYT content shifts the τ_f value towards a more positive range. This suggests that, by adjusting the amount of CLYT content, a τ_f value of zero can be achieved. A τ_f value of zero was attained with x = 0.95 when sintering was performed at 950 °C for 4 h.

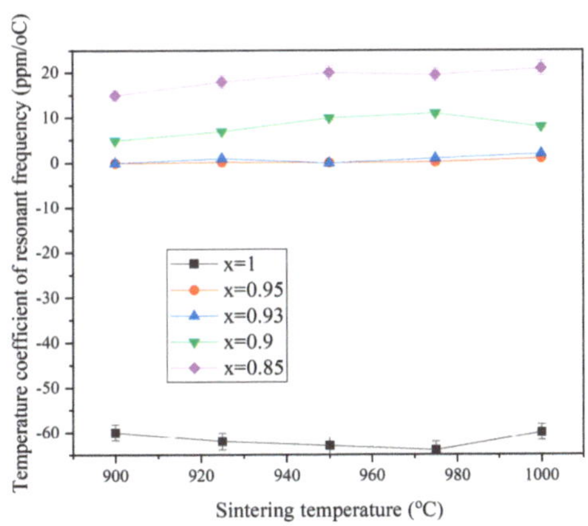

Figure 7. τ_f values of the xMZCT-(1−x)CLYT ceramic system with 3 wt.% BCB sintered at different temperatures for 4 h.

Table 2 presents the microwave dielectric properties of the xMZCT-(1−x)CLYT ceramic system doped with 3 wt.% BCB and sintered at 950 °C for 4 h. As the value of x varied from 0.95 to 0.85, τ_f underwent a transition from 0 to 21 ppm/°C. The curve exhibited a crossing point at zero, indicating that a τ_f value of zero can be achieved by appropriately adjusting the x value in the xMZCT-(1−x)CLYT ceramic system. However, the Q × f value decreased with decreasing MZCT content, as the CLYT ceramic had a low Q × f value of 23,000 GHz. The microwave dielectric properties of MZCT ceramics were pre-examined, and the findings are elucidated in Table 2. Notably, as the sintering temperature was elevated to 950 °C, significant improvements in key parameters were observed. The measured Q × f reached an impressive 190,000 GHz, indicating an enhanced quality factor and resonant frequency characteristics. Simultaneously, the permittivity (ε_r) exhibited a notable increase, reaching 22.5, showcasing the material's improved response to electric fields. Additionally, the temperature coefficient of resonant frequency (τ_f) was determined to be 0 ppm/°C, highlighting the stability of the ceramics' microwave properties over a range of temperatures. In an effort to refine the microwave dielectric properties of the xMZCT-(1−x)CLYT ceramics, subtle modifications were introduced. Specifically, minute substitutions were made, with Co^{2+} or Zn^{2+} replacing Mg^{2+} to create MZCT, and Y^{3+} taking the place of La^{3+} to form CLYT in the experimental setup.

There may be some uncertainty due to measurement errors in the experiments. These errors, shown in Figures 4–7 and Tables 1 and 2, demonstrate the uncertainty caused by measurement differences.

4. Conclusions

The examination of the microwave properties of the xMZCT-(1−x)CLYT ceramic, doped with 3wt.% BCB, revealed the successful achievement of ultra-low dielectric loss. Specifically, the 0.95MZCT-0.05CLYT ceramic exhibited outstanding microwave properties, even when sintered at low temperatures. This was achieved through the strategic replacement of partial Zn^{2+} and Co^{2+} ions in $MgTiO_3$, resulting in the formation of the $[(Mg_{0.6}Zn_{0.4})_{0.95}Co_{0.05}]_{1.02}$ ceramic. Within the xMZCT-(1−x)CLYT system, a distinctive mix of phases emerged, with MZCT constituting the predominant crystalline phase and CLYT having a minor presence.

This composition displayed strong dielectric properties, featuring a permittivity ε_r of around 20, a noteworthy quality factor (Q × f) that reached 256,000 GHz, and a temperature coefficient of resonant frequency (τ_f) of approximately −65 ppm/°C when sintered at 1200 °C. The intricacies of this composition were further explored at the specific sintering temperature of 950 °C. Under this condition, the system demonstrated exceptional microwave dielectric properties. The ε_r reached 22.5, showing its efficient propagation of electromagnetic waves. The Q × f soared to an impressive 190,000 GHz (measured at 9 GHz), indicating minimal signal loss during transmission. Most notably, the system achieved a τ_f value of zero, implying outstanding temperature stability.

The ultra-low temperature coefficient and high quality factor of the newly discovered 0.95MZCT-0.05CLYT ceramic make it an ideal material for application in the field of 6G band components and GPS antennas. Achieving such good properties at a relatively low sintering temperature aligns with the contemporary drive for energy-efficient and sustainable material processing.

Author Contributions: Conceptualization, Y.-B.C. and J.P.; methodology, Y.-B.C. and J.P.; software, Y.-B.C. All authors have read and agreed to the published version of the manuscript.

Funding: This research received no external funding.

Data Availability Statement: The data presented in this study are available on request from the corresponding author.

Conflicts of Interest: The authors declare no conflict of interest.

References

1. Huang, C.-L.; Wang, J.-J.; Li, B.-J.; Lee, W.-C. Effect of B_2O_3 additives on sintering and microwave dielectric behaviors of $0.66Ca(Mg_{1/3}Nb_{2/3})O_3$-$0.34CaTiO_3$ ceramics. *J. Alloys Compd.* **2008**, *461*, 440–446. [CrossRef]
2. Huang, C.-L.; Li, G.-J.; Wang, J.-J. Microwave dielectric properties of $(1-x)(Mg_{0.95}$Zn0.05)TiO3–x(Na0.5La0.5)TiO3 ceramic system. *J. Alloys Compd.* **2009**, *472*, 497–501. [CrossRef]
3. Hsu, C.-H. Improved high-Q microwave dielectric resonator using ZnO-doped $La(Co_{1/2}Ti_{1/2})O_3$ ceramics. *J. Alloys Compd.* **2008**, *464*, 412–417. [CrossRef]
4. Xiao, M.; Sun, H.-R.; Zhou, Z.-Q.; Zhang, P. Bond ionicity lattice energy, bond energy, and microwave dielectric properties of $Ca_{1-x}Sr_xWO_4$ ceramics. *Ceram. Int.* **2018**, *44*, 20686–20691. [CrossRef]
5. Kim, E.-S.; Kim, S.-H.; Lee, B.-I. Low-temperature sintering and microwave dielectric properties of $CaWO_4$ ceramics for LTCC applications. *J. Eur. Ceram. Soc.* **2006**, *26*, 2101–2104. [CrossRef]
6. Krzmanc, M.-M.; Logar, M.; Budic, B.; Suvorov, D. Dielectric and microstructural study of the $SrWO_4$, $BaWO_4$, and $CaWO_4$ scheelite ceramics. *J. Am. Ceram. Soc.* **2011**, *94*, 2464–2472. [CrossRef]
7. Kim, E.S.; Kim, S.H. Effects of structural characteristics on microwave dielectric properties of (1-x)$CaWO_4$-x$LaNbO_4$ ceramics. *J. Electroceram.* **2006**, *17*, 47–77. [CrossRef]
8. Hu, X.-Q.; Jiang, J.; Wang, J.-Z.; Gan, L.; Zhang, T.-J. A new additive-free microwave dielectric ceramic system for LTCC applications: (1-x)$CaWO_4$-x$(Li_{0.5}Sm_{0.5})WO_4$. *J. Mater. Sci. Mater. Electron.* **2020**, *31*, 2544–2550. [CrossRef]
9. Bian, J.-J.; Ding, Y.-M. Structure, sintering behavior, and microwave dielectric properties of (1-x)$CaWO_4$-x$YLiF_4$ $(0.02<x<0.10)$ ceramics. *Mater. Res. Bull.* **2015**, *67*, 245–250.
10. Zhang, S.; Su, H.; Zhang, H.-W.; Jing, Y.-L.; Tang, X.-L. Microwave dielectric properties of $CaWO_4$-Li_2TiO_3 ceramics added with LBSCA glass for LTCC applications. *Ceram. Int.* **2016**, *42*, 15242–15246. [CrossRef]
11. Jeon, C.-J.; Kim, E.-S. Low-temperature sintering of $0.85CaWO_4$-$0.15LaNbO_4$ ceramics. *Ceram. Int.* **2008**, *34*, 921–924. [CrossRef]
12. Liao, Q.-W.; Wang, Y.-L.; Jiang, F.; Guo, D. Ultra-low fire glass-free $Li_3FeMo_3O_{12}$ microwave dielectric ceramics. *J. Am. Ceram. Soc.* **2014**, *97*, 2394–2396. [CrossRef]
13. Shen, C.-H.; Huang, C.-L. Microwave dielectric properties and sintering behaviors of $(Mg_{0.95}$Ni0.05)TiO3–CaTiO3 ceramic system. *J. Alloys Compd.* **2009**, *472*, 451–455. [CrossRef]
14. Kell, R.C.; Greenham, A.C.; Olds, G.C.E. High-Permittivity Temperature-Stable Ceramic Dielectrics with Low Microwave Loss. *J. Am. Ceram. Soc.* **1973**, *56*, 352–354. [CrossRef]
15. Huanga, C.-L.; Houa, J.L.; Panb, C.-L.; Huang, C.-Y.; Pengd, C.-W.; Weid, C.-H.; Huang, Y.-H. Effect of ZnO additive on sintering behavior and microwave dielectric properties of 0.95MgTiO3-0.05CaTiO3 ceramics. *J. Alloys Compd.* **2008**, *450*, 359–363. [CrossRef]
16. Chen, Y.-C.; Chang, Y.-H. Dielectric properties of B_2O_3-doped La $(Mg_{0.5}Sn_{0.5})$ O_3 ceramics at microwave frequencies. *J. Alloys Compd.* **2009**, *477*, 450–453. [CrossRef]
17. Huang, C.-L.; Chen, Y.-B.; Tasi, C.-F. Influence of V_2O_5 additions to 0.8(Mg0.95Zn0.05)TiO3-0.2Ca0.61Nd0.26TiO3 ceramics on sintering behavior and microwave dielectric properties. *J. Alloys Compd.* **2008**, *454*, 454–459. [CrossRef]
18. Shen, C.-H.; Huang, C.-L. New dielectric material system of $Mg_{0.95}$Co0.05TiO3$-$Zn0.975Ca0.025TiO3 at microwave frequencies. *J. Alloys Compd.* **2009**, *477*, 712–715.
19. Chen, J.-Y.; Tseng, Y.-W.; Huang, C.-L. Improved high Q value of $(1-x)Ca(Mg_{1/3}Ta_{2/3})O_3$–$xCa_{0.8}Sm_{0.4/3}TiO_3$ solid solution with zero temperature coefficient of resonant frequency. *J. Alloys Compd.* **2010**, *494*, 205–209. [CrossRef]
20. Cha, H.J.; Kang, D.H.; Cho, Y.S. Dielectric Properties and Crystal Structure of $(Mg_{1-x}Co_x)$2$(Ti_{0.95}$Sn0.05)O4 Ceramics. *Mater. Res. Bull.* **2007**, *42*, 265–273. [CrossRef]
21. Wise, P.L.; Reaney, I.M.; Lee, W.E.; Price, T.J.; Iddles, D.M.; Cannell, D.S. Optical Phonon Modes and Dielectric Behavior of $Sr_{1-3x/2}$Ce$_x$TiO$_3$ Microwave Ceramics. *J. Eur. Ceram. Soc.* **2001**, *21*, 1723. [CrossRef]
22. Huang, C.-L.; Lin, S.-H.; Liu, S.-S.; Chen, Y.-B. x $(Mg_{0.7}Zn_{0.3})_{0.95}Co_{0.05}TiO_3$-$(1-x)(La_{0.5}Na_{0.5})TiO_3$ ceramic at microwave frequency with a near zero temperature coefficient of resonant frequency. *J. Alloys Compd.* **2010**, *489*, 541–544. [CrossRef]
23. Kim, M.H.; Lim, J.B.; Kim, J.C.; Nahm, S.; Paik, J.H.; Kim, J.H.; Park, K.S. Synthesis of BaCu (B_2O_5) ceramics and their effect on the sintering temperature and microwave dielectric properties of $Ba(Zn_{1/3}Nb_{2/3})O_3$ ceramics. *J. Am. Ceram. Soc.* **2006**, *89*, 3124–3128. [CrossRef]
24. Kim, M.H.; Nahm, S.; Lee, W.-S.; Yoo, M.-J.; Kang, N.-K.; Kim, H.-T.; Lee, H.-J. Effect of B_2O_3 and CuO on the sintering temperature and microwave dielectric properties of $Ba(Zn_{1/3}Ta_{2/3})O_3$ ceramics. *Jpn. J. Appl. Phys.* **2005**, *44*, 3091. [CrossRef]
25. Kim, M.H.; Jeong, Y.H.; Nahm, S.; Kim, H.T.; Lee, H.J. Effect of B_2O_3 and CuO additives on the sintering temperature and microwave dielectric properties of $Ba(Zn_{1/3}Nb_{2/3})O_3$ ceramics. *J. Eur. Ceram. Soc.* **2006**, *26*, 2139. [CrossRef]
26. Zhang, S.; Sahin, H.; Torun, E.; Peeters, F.; Martien, D.; Dilley, N.; Newman, N. Fundamental mechanisms responsible for the temperature coefficient of resonant frequency in microwave dielectric ceramics. *J. Am. Ceram. Soc.* **2017**, *100*, 1508–1516. [CrossRef]
27. Hakki, B.W.; Coleman, P.D. A dielectric resonator method of measuring inductive capacities in the millimeter range. *IEEE Trans. Microw. Theory Tech.* **1960**, *8*, 402. [CrossRef]
28. Courtney, W.E. Analysis and evaluation of a method of measuring the complex permittivity and permeability microwave insulators. *IEEE Trans. Microw. Theory Tech.* **1970**, *18*, 476. [CrossRef]
29. Lichtenecker, K. Die dielektrizitatskonstante naturlicher und kunstlicher mischkorper. *Phys. Z.* **1926**, *27*, 115.

30. Kim, Y.-I.; Woodward, P.M. Crystal structures and dielectric properties of ordered double perovskites containing Mg^{2+} and Ta^{5+}. *J. Solid State Chem.* **2007**, *180*, 2798–2807. [CrossRef]
31. Chen, Y.-B.; Liu, S.-S. Dielectric properties of low Zr-substituted $BaTi_4O_9$ at microwave frequencies. *J. Mater. Sci. Mater. Electron.* **2019**, *30*, 5567–5572. [CrossRef]
32. Reaney, I.M.; Colla, E.L.; Setter, N. Dielectric and Structural Characteristics of Ba- and Sr-based Complex Perovskites as a Function of Tolerance Factor. *Jpn. J. Appl. Phys.* **1994**, *33*, 3984. [CrossRef]

ceramics MDPI

Communication

Oxygen-Bonding State and Oxygen-Reduction Reaction Mechanism of $Pr_{0.7}Ca_{0.3}Mn_{1-x}Co_xO_{3-d}$ (x = 0, 0.1, 0.2, 0.3)

Kanghee Jo, Seungjae Lee and Heesoo Lee *

School of Materials Science and Engineering, Pusan National University, Busan 46241, Republic of Korea; jokanghee@pusan.ac.kr (K.J.); bluejae2@pusan.ac.kr (S.L.)
* Correspondence: heesoo@pusan.ac.kr; Tel.: +82-051-510-2388

Abstract: We investigated the effects of Co doping on $Pr_{0.7}Ca_{0.3}MnO_{3-d}$ in the perspective of an oxygen-bonding state change. In all compositions, $Pr_{0.7}Ca_{0.3}Mn_{1-x}Co_xO_{3-d}$ (PCMCx, x = 0, 0.1, 0.2, 0.3) showed an orthorhombic structure, and the lattice gradually contracted with increasing Co content. The doped Co was mostly present as 2+ and 3+, which decreased the average oxidation value of the B site and created oxygen vacancies for charge compensation. However, as the Co content increased, the proportion of Co^{3+} increased, and the content of oxygen vacancies gradually decreased. In addition, the ratio of adsorbed oxygen in PCMC0.1 was the highest, and the B-O covalency was enhanced. Accordingly, the electrochemical reaction of oxygen with the cathode material in PCMC0.1 could occur most easily, showing the smallest polarization resistance among the Co-doped $Pr_{0.7}Ca_{0.3}MnO_{3-d}$. We can confirm the formation of oxygen vacancies via Co doping and the effect of B-O covalency on the oxygen-reduction reaction of $Pr_{0.7}Ca_{0.3}MnO_{3-d}$.

Keywords: co-doped $Pr_{0.7}Ca_{0.3}MnO_{3-d}$; oxygen vacancy; B-O bond covalency; distribution of relaxation time (DRT); ORR mechanism

Citation: Jo, K.; Lee, S.; Lee, H. Oxygen-Bonding State and Oxygen-Reduction Reaction Mechanism of $Pr_{0.7}Ca_{0.3}Mn_{1-x}Co_xO_{3-d}$ (x = 0, 0.1, 0.2, 0.3). *Ceramics* **2023**, *6*, 2386–2393. https://doi.org/10.3390/ceramics6040146

Academic Editors: Dawei Wang and Fayaz Hussain

Received: 17 November 2023
Revised: 10 December 2023
Accepted: 11 December 2023
Published: 15 December 2023

1. Introduction

A solid oxide fuel cell (SOFC) is an electrochemical cell that directly converts the chemical energy of fuel into electricity with an efficiency of up to 80% at high temperatures of 800~1000 °C using ceramic materials and is attracting attention as a tool for realizing carbon neutrality. High-temperature operating conditions cause high costs, deterioration of surrounding materials, and thermal shock during 'start–stop' cycles. Research to reduce the operating temperature to an intermediate temperature range (600~800 °C) is conducted to overcome the disadvantages [1–3]. To reduce the operating temperature, it is important to improve the sluggish oxygen-reduction reaction that occurs on the surface of the cathode. To improve it, it is important to control the material properties so that the reaction that mainly occurs at the air–electrode–electrolyte interface (triple-phase boundary, TPB) can occur at the electrode–air interface [4,5].

Co-based perovskite oxides such as $La_{0.8}Sr_{0.2}Co_{0.8}Fe_{0.2}O_{3-\delta}$ (LSCF) and $Ba_{0.5}Sr_{0.5}Co_{0.8}Fe_{0.2}O_{3-\delta}$ (BSCF) have been widely studied because of their ability to conduct oxygen ion and electrons simultaneously and high oxygen-reduction reaction (ORR) activity to reduce the operating temperature [6,7]. Double perovskite oxides such as $PrBaCo_2O_{5+d}$, $PrBaMn_2O_{5+d}$, and $NdBaMnO_{5+d}$, which show better activity by forming an oxygen pathway through cation ordering, have been recently studied [8–11]. However, Co-based perovskite oxides and double perovskite oxides have a large thermal expansion coefficient compared to ceria-based electrolytes, which can cause delamination of the electrode–electrolyte interface. Cobalt-free cathode materials such as $Ba_{0.5}Sr_{0.5}FeO_{3-d}$ are being studied [12]. In the case of $PrMnO_3$, Ishihara et al. showed that among various $Ln_{0.6}Sr_{0.4}MnO_3$ (Ln = La, Pr, Nd, Sm, etc.)-based perovskite oxides, $Pr_{0.6}Sr_{0.4}MnO_{3-d}$ exhibited the best electrical conductivity and, unlike the widely studied LSM, did not react with the electrolyte Y_2O_3-stabilized ZrO_2, making it suitable as a cathode material for low-

and intermediate-temperature SOFCs [13]. However, further research is needed to improve the electrode performance for a cobalt-free cathode. Small amounts of Co doping into a cobalt-free cathode have been studied [14].

We investigated the effect of Co doping on the B-O bonding structure and electrochemical properties in $Pr_{0.7}Ca_{0.3}MnO_{3-d}$. We synthesized $Pr_{0.7}Ca_{0.3}Mn_{1-x}Co_xO_{3-d}$ with different Co contents using an EDTA (ethylenediaminetetraacetic acid)-citric acid complex process, and the crystal structures were confirmed using X-ray diffraction patterns. Core level spectra were observed using X-ray photoelectron spectroscopy, and peak deconvolution was performed to confirm the oxidation and oxygen binding states of the metal ions. EIS was measured at 550~750 °C to confirm the electrochemical properties depending on the Co content, and changes in the ORR mechanism were analyzed using DRT analysis.

2. Materials and Methods

PCMCx ($Pr_{0.7}Ca_{0.3}Mn_{1-x}Co_xO_{3-d}$, x = 0, 0.1, 0.2, 0.3) powders with various compositions were synthesized using the EDTA-citrate complexing process (ECCP). $Pr(NO_3)_3 \cdot 6H_2O$ (99.9%, Sigma-Aldrich, St. Louis, MO, USA), $Ca(NO_3)_2 \cdot 4H_2O$ (98%, Sigma-Aldrich), $Co(NO_3)_3 \cdot 6H_2O$ (98%, Sigma-Aldrich), and $Mn(NO_3)_2 \cdot 4H_2O$ (98%, Sigma-Aldrich) metal precursors were dissolved in deionized water. Ethylenediaminetetraacetic acid (EDTA, 99.5%, Alfa Aesar, Haverhill, MA, USA) was added to a 1N NH_4OH (Junsei chemical Co., Tokyo, Japan) solution to obtain a NH_3-EDTA buffer solution. The NH_3-EDTA and crystallized citric acid monohydrate (99.5%, Samchun Chemical, Pyeongtaek, Korea) powders, as chelating agents, were applied to the mixed-metal precursor solution to make a sol for a total metal ion: EDTA: citric acid molar ratio of 1:1:2. The sol was heated with stirring to evaporate the solvent. Clear gels were obtained, and then these gels were pre-calcined at 200 °C. Then, the pre-calcined precursors were calcined in air at 950 °C.

Symmetric cells (PCMCx | SDC | PCMCx) were prepared to investigate the electrochemical properties of the PCMCx powders. The SDC pellets were sintered at 1400 °C for 4 h using Sm-doped ceria powder (SDC20-HP, Fuelcellmaterials, Lewis Center, OH, USA). The PCMCx powders were mixed with a binder prepared from α-terpineol and ethyl-cellulose to form PCMCx pastes, which were screen-printed onto both sides of the SDC pellets. After drying, the symmetric cells were calcined at 950 °C for 2 h in air.

To examine the crystal structures of the calcined and sintered powders, powder X-ray diffraction (XRD, PANalytical X'pert-Pro MPD PW3040/60) was performed at room temperature using a step-scan procedure (0.02°/2θ step, time per step 0.5 s) in the 2θ range of 10–90°. X-ray photoelectron spectroscopy (XPS) was carried out to measure the core level spectra of the PCMCx powders using the K-ALPAH+ System (HPXPS, Al Kα X-ray source (1486.6 eV), ThermoFisher Scientific, Waltham, MA, USA) at the Korea Basic Science Institute (KBSI), Busan, center. Impedance measurements were conducted using an IviumStat instrument (Ivium, Eindhoven, the Netherlands) over the frequency range of 10^6–0.01 Hz with an excitation voltage of 10 mV at an operating temperature of 600–700 °C under open-circuit conditions in air. The electrochemical impedance spectroscopy (EIS) results were multiplied by 0.5 to account for the two electrodes. The impedance spectra data were further fitted using EC-lab software (4.3 ver.). The distribution of relaxation time (DRT) method was applied to analyze the EIS data [15].

3. Results and Discussion

Figure 1 shows the XRD pattern of Co-doped $Pr_{0.7}Ca_{0.3}MnO_{3-d}$ and the (112) main peak. The XRD patterns were indexed according to Caignaert et al. [16]. Each composition showed an orthorhombic structure, indicating that it was synthesized without the secondary phase formation. The Goldschmidt tolerance factor (t) of PCMCx increased from 0.908 (PCM) to 0.913 (PCMC0.3), which was predicted to have a crystal structure close to the cubic structure but showed an orthorhombic structure. Compared to PCM, PCMC0.1 showed a lower angle shift of the main peak from 32.88° to 32.84°, but as the content of Co increased, the main peak gradually shifted to a higher angle, and PCMC0.3 showed a higher angle shift to 32.96°, indicating that the lattice contracted. The lattice contraction is caused via the substitution of Mn ions with larger atomic number Co ions: Mn^{3+} (58 pm LS and 64.5 pm HS) and Co^{3+} ions (54.5 pm LS and 61 pm HS) [17].

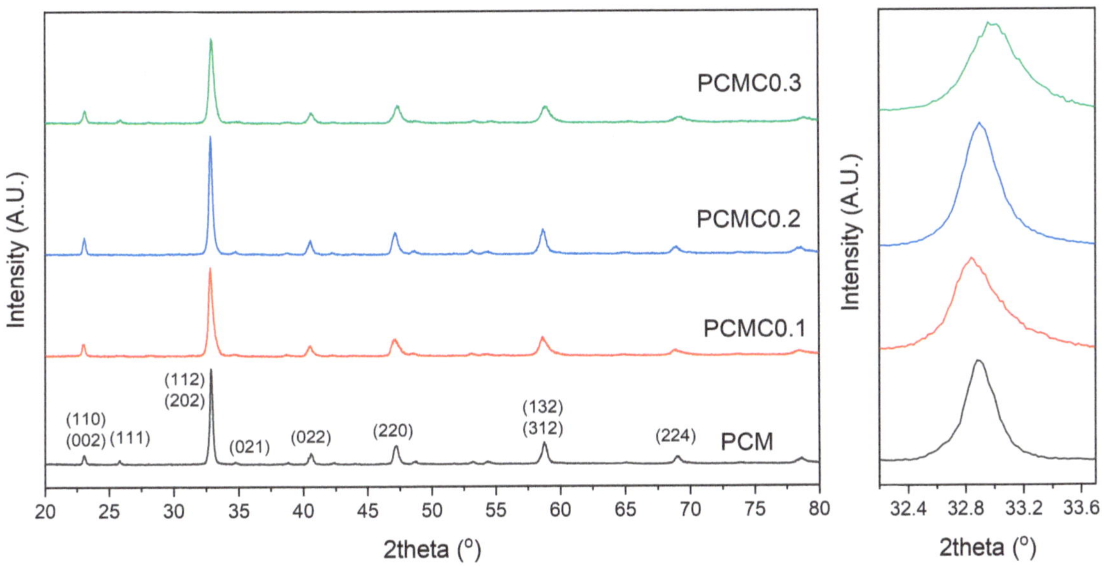

Figure 1. X-ray diffraction pattern and (112) peak of Co-doped $Pr_{0.7}Ca_{0.7}MnO_{3-d}$.

Figure 2 shows the core level spectra of B site metal ions and oxygen ions of Co-doped $Pr_{0.7}Ca_{0.3}MnO_{3-d}$. The core level spectra of the Co 2p1/2 states were observed at 790 eV to 810 eV and consisted of three peaks at 796.4 eV, 794.9 eV, 797.4 eV, and 802.5 eV that can be assigned to Co^{2+}, Co^{3+}, Co^{4+}, and satellite [18,19]. The Mn 2p2/3 spectrum consists of three peaks at 640.3, 641.7, and 643.8 eV that can be assigned to Mn^{2+}, Mn^{3+}, and Mn^{4+} ions [20]. The O 1s core level spectra consist of peaks formed at 528.86 eV, 530.95 eV, and 533.4 eV, which correspond to lattice oxygen (O1), oxygen vacancy or surface absorbed oxygen (O2), and absorbed water species (O3), respectively [21]. The fitting results of the above Mn, Co, and O core level spectra are summarized in Table 1, and Pr 3d and Ca 2p also fitted according to Mekki et al. and Wan et al., respectively, and maintained oxidation values of 3+ and 2+, respectively [22,23].

The fitting results of the O 1s spectra show that the intensity of O 1s decreased and then gradually increased with Co doping, indicating the formation of oxygen vacancies in the lattice. Co, which has a similar oxidation value to Mn, was able to form oxygen vacancies, according to Lv et al. In addition, the ratio of the O2 peak increased contrary to the trend of O1, which indicates that the B-O covalency was improved, and it is known that improved covalency indicates better ORR reactivity. O3 represents an absorbed water species [22,24].

It was found that the Co ions in PCMC0.1 had oxidation values of 2+ and 3+, but as the Co content gradually increased, the ratio of Co^{2+} decreased and the ratio of Co^{3+} increased, which is consistent with the trend of the O1 peak observed in the O 1s spectra. In addition, the overall oxidation value of the B site ions gradually increased with the increase in the Co doping amount, which is consistent with the lattice contraction with the increase in Co content confirmed using XRD results.

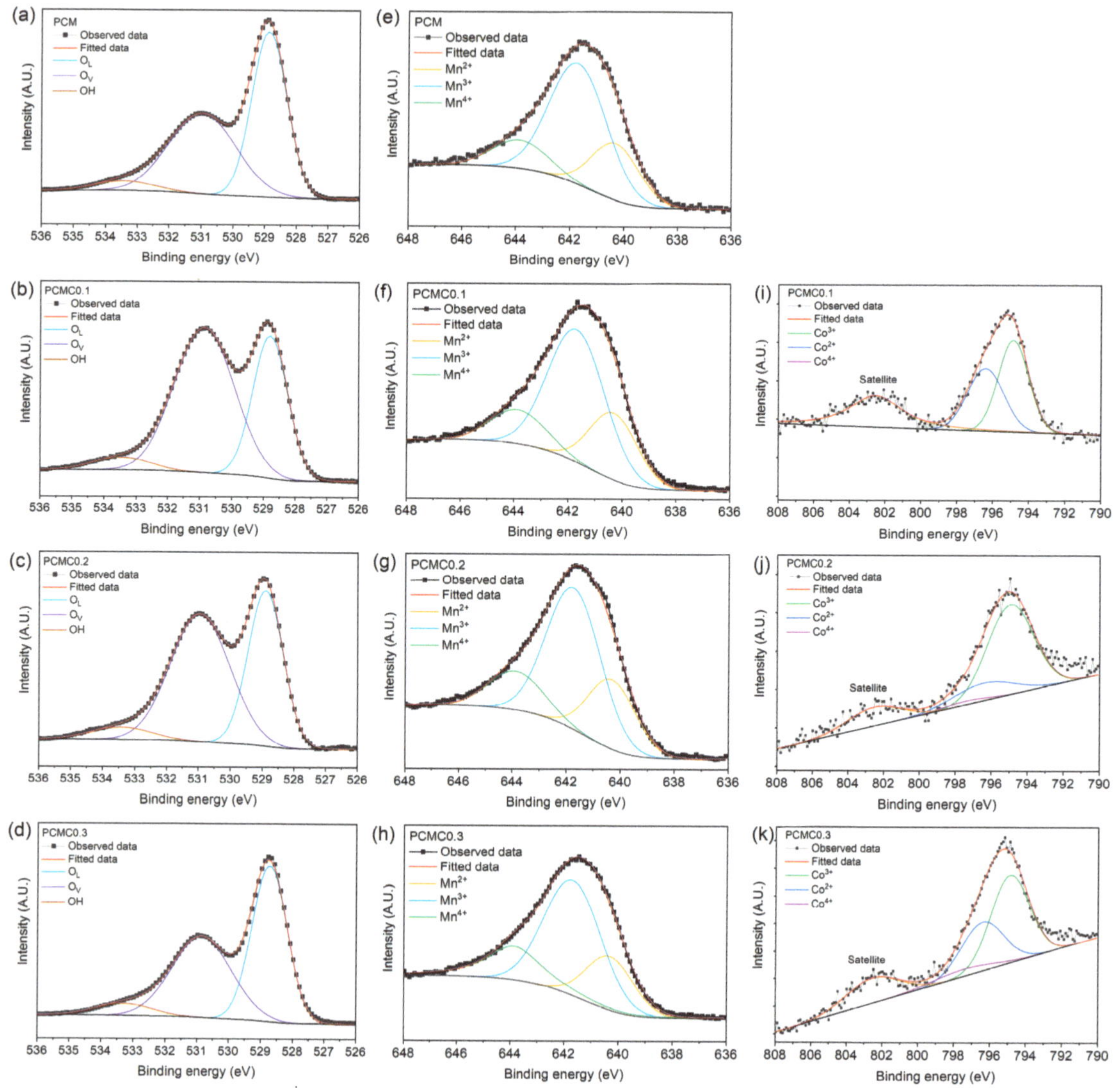

Figure 2. X-ray photoelectron spectra of PCMCx and peak deconvolution result: (**a–d**) O 1s core level spectra, (**e–h**) Mn 2p 3/2 core level spectra of PCM and PCMC0.1–0.3, and (**i–k**) Co 2p 1/2 core level spectra of PCMC0.1–0.3.

Figure 3 shows the electrochemical impedance spectra of each composition from 550 to 750 °C, and the ohmic resistance is subtracted to compare the polarization resistance. The polarization resistances are summarized in Table 2. PCMC0.1 exhibited the smallest polarization resistivity at all temperatures and was reduced by more than 60% compared to PCM. This is consistent with the formation of oxygen vacancies and improvement in B-O covalency, as confirmed using XPS results, and this trend is also confirmed by the increase in polarization resistance with increasing Co content.

Table 1. O 1s, Mn 2p 3/2, and Co 2p 1/2 XPS fitting results of PCMCx and average oxidation number.

O 1s	O1	O2		O3
PCM	49.400%	45.567%		5.033%
PCMC0.1	34.805%	60.031%		5.165%
PCMC0.2	39.679%	54.841%		5.479%
PCMC0.3	49.868%	43.843%		6.288%
Mn 2p 3/2	**Mn^{2+}**	**Mn^{3+}**	**Mn^{4+}**	**Average Oxidation Number**
PCM	24.142%	61.829%	14.029%	2.899
PCMC0.1	25.265%	59.626%	15.109%	2.898
PCMC0.2	25.295%	58.368%	16.337%	2.910
PCMC0.3	21.958%	55.977%	22.065%	3.001
Co 2p 1/2	**Co^{2+}**	**Co^{3+}**	**Co^{4+}**	**Average Oxidation Number**
PCM	-	-	-	-
PCMC0.1	44.894%	55.106%	0.000%	3.165
PCMC0.2	35.004%	57.447%	7.549%	3.275
PCMC0.3	19.789%	76.885%	3.326%	3.449

Average oxidation state of A site metal: Pr = 3+, Ca = 2+.

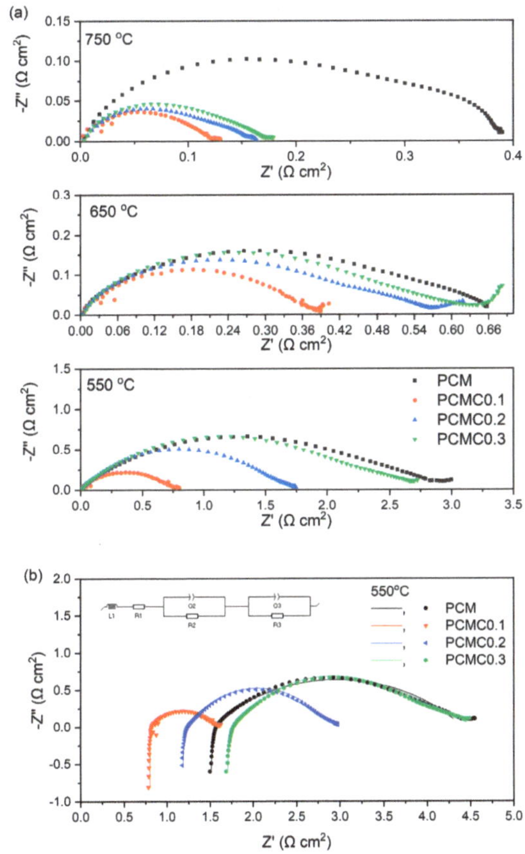

Figure 3. (**a**) Electrochemical impedance spectra of PCMCx. Rohm was subtracted to emphasize the polarization resistance and (**b**) fitting result at 550 °C.

Table 2. Ohmic resistance (R_{ohm}) and polarization resistance (R_{pol})of PCMCx via temperature.

		550 °C	650 °C	750 °C
PCM	R_{ohm} R_{pol}	1.564 Ω cm^2 2.997 Ω cm^2	0.597 Ω cm^2 0.655 Ω cm^2	0.366 Ω cm^2 0.390 Ω cm^2
PCMC0.1	R_{ohm} R_{pol}	0.823 Ω cm^2 0.787 Ω cm^2	0.539 Ω cm^2 0.38 Ω cm^2	0.320 Ω cm^2 0.127 Ω cm^2
PCMC0.2	R_{ohm} R_{pol}	1.230 Ω cm^2 1.746 Ω cm^2	0.500 Ω cm^2 0.567 Ω cm^2	0.312 Ω cm^2 0.161 Ω cm^2
PCMC0.3	R_{ohm} R_{pol}	1.766 Ω cm^2 2.670 Ω cm^2	0.814 Ω cm^2 0.682 Ω cm^2	0.461 Ω cm^2 0.179 Ω cm^2

The electrochemical kinetics of PCMCx using DRT analysis is shown in Figure 4. The electrochemical process appears as a peak in the DRT spectrum, and the impedance data of the oxygen-reduction reaction can be used to analyze the main processes involved in the electrode reaction. In Figure 4, the DRT spectrum can be divided into a high-frequency (HF) region, which is found at frequencies above 10^3 Hz, an intermediate frequency (IF) region between 10^3 and 10^0 Hz, and a low-frequency (LF) region below 10^0 Hz, with each region corresponding to a separate electrochemical reaction. The characteristic frequency of a reaction is inversely proportional to the relaxation time of the electrode reaction; therefore, the higher the characteristic frequency, the faster the relaxation rate of the electrode process. In addition, the integral area of the peak represents the resistance of the corresponding reaction [25,26].

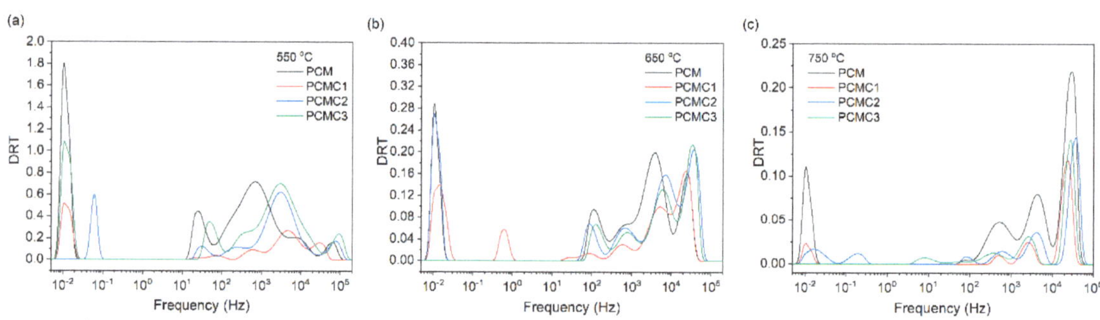

Figure 4. Distribution of relaxation time spectra of PCMCx via temperature: (**a**) 550 °C, (**b**) 650 °C, and (**c**) 750 °C.

The LF region is the region for the reaction in which oxygen molecules are diffused and adsorbed onto the electrode, and at 550 °C, PCM exhibited the largest LF resistance, which rapidly decreases with increasing temperature. The processes contributing to the HF region are likely related to one or more charge-transfer processes across the interface, such as the transfer of oxygen ions (O^{2-}) at the TPB/electrode–electrolyte interface. This region is more dominant at higher temperatures than at lower temperatures, as evidenced by the particularly high peak at 750 °C.

Electrochemical reactions corresponding to the peaks in the IF region can include reactions such as adsorption/desorption of oxygen, dissociation, and surface migration, accounting for most of the polarization resistance in EIS. This region is strongly influenced by the formation of oxygen vacancies, which is confirmed by the decrease in O1 in XPS, and the enhancement in B-O covalency, which is confirmed by the O2 peak, which corresponds to the smallest O1 ratio and the largest O2 ratio in PCMC0.1. Therefore, the polarization resistance is lower than that of PCM due to the synergetic effect of oxygen vacancy formation and B-O covalency enhancement via Co doping, and as the Co content increases again, the oxygen vacancy decreases, and the B-O covalency decreases, gradually increasing the resistance.

4. Conclusions

In this study, the effect of Co doping on the electrochemical properties of $PrCaMnO_{3-d}$ was analyzed from the perspective of the change in the bonding structure of oxygen. PCMCx showed a gradual lattice contraction with increasing Co content, which is believed to be because Co ions are smaller than Mn ions, and the B site ions are gradually oxidized and become smaller in size with increasing Co content. Among PCMCx, PCMC0.1 showed a lower average oxidation value of the B site than PCM, and the lattice oxygen decreased, indicating that oxygen vacancies were formed in the lattice. As the content of Co increased, the oxidation value of Mn and Co gradually increased, and the content of oxygen vacancies in the lattice decreased. In addition, the absorbed oxygen species in PCMC0.1 increased, resulting in an increase in B-O covalency. In EIS, PCMC0.1 showed the smallest polarization resistance, and DRT analysis showed that the decrease in resistance in the IF region, which represents the electrochemical reaction with oxygen, was the main cause.

Author Contributions: Conceptualization, H.L.; methodology, S.L.; software, K.J.; validation, K.J., S.L., and H.L.; formal analysis, S.L.; investigation, S.L.; writing—original draft preparation, K.J.; writing—review and editing, H.L.; visualization, K.J.; project administration, H.L.; funding acquisition, H.L. All authors have read and agreed to the published version of the manuscript.

Funding: This work was supported by the Korea Institute for Advancement of Technology (KIAT) grant funded by the Korea Government (MOTIE) (P0008335, HDR Program for Industrial Innovation).

Institutional Review Board Statement: Not applicable.

Informed Consent Statement: Not applicable.

Data Availability Statement: The data presented in this study are available in the article.

Conflicts of Interest: The authors declare no conflict of interest.

References

1. Singh, M.; Zappa, D.; Comini, E. Solid oxide fuel cell: Decade of progress, future perspectives and challenges. *Int. J. Hydrogen Energy* **2021**, *46*, 27643–27674. [CrossRef]
2. Jiang, L.; Wei, T.; Zeng, R.; Zhang, W.-X.; Huang, Y.-H. Thermal and electrochemical properties of $PrBa_{0.5}Sr_{0.5}Co_{2-x}Fe_xO_5+\delta$ (x = 0.5, 1.0, 1.5) cathode materials for solid-oxide fuel cells. *J. Power Sources* **2013**, *232*, 279–285. [CrossRef]
3. Zhu, C.; Liu, X.; Yi, C.; Yan, D.; Su, W. Electrochemical performance of $PrBaCo_2O_5+\delta$ layered perovskite as an intermediate-temperature solid oxide fuel cell cathode. *J. Power Sources* **2008**, *185*, 193–196. [CrossRef]
4. Tabish, A.; Patel, H.; Chundru, P.; Stam, J.; Aravind, P. An SOFC anode model using TPB-based kinetics. *Int. J. Hydrogen Energy* **2020**, *45*, 27563–27574. [CrossRef]
5. Yang, G.; Jung, W.; Ahn, S.-J.; Lee, D. Controlling the Oxygen Electrocatalysis on Perovskite and Layered Oxide Thin Films for Solid Oxide Fuel Cell Cathodes. *Appl. Sci.* **2019**, *9*, 1030. [CrossRef]
6. Agun, L.; Rahman, H.A.; Ahmad, S.; Muchtar, A. Durability and Stability of LSCF Composite Cathode for Intermediate-Low Temperature of Solid Oxide Fuel Cell (IT-LT SOFC): Short Review. *Adv. Mater. Res.* **2014**, *893*, 732–737. [CrossRef]
7. Shao, Z.; Haile, S.M. A high-performance cathode for the next generation of solid-oxide fuel cells. *Nature* **2004**, *431*, 170–173. [CrossRef]
8. Wei, B.; Schroeder, M.; Martin, M. Surface cation segregation and chromium deposition on the double-perovskite oxide $PrBaCo_2O_{5+\delta}$. *ACS Appl. Mater. Interfaces* **2018**, *10*, 8621–8629. [CrossRef]
9. Afroze, S.; Abdalla, A.; Radenahmad, N.; Nam, Q.H.; Azad, A. Synthesis, structural and thermal properties of double perovskite NdSrMn2O6 as potential anode materials for solid oxide fuel cells. In Proceedings of the 7th Brunei International Conference on Engineering and Technology 2018 (BICET 2018), Bandar Seri Begawan, Brunei, 12–14 November 2018; p. 91.
10. dos Santos-Gómez, L.; Zamudio-García, J.; Porras-Vázquez, J.M.; Losilla, E.R.; Marrero-López, D. Nanostructured $BaCo_{0.4}Fe_{0.4}Zr_{0.1}Y_{0.1}O_{3-\delta}$ Cathodes with Different Microstructural Architectures. *Nanomaterials* **2020**, *10*, 1055. [CrossRef]
11. Garcia-Garcia, F.J.; Sayagués, M.J.; Gotor, F.J. A Novel, Simple and Highly Efficient Route to Obtain $PrBaMn_2O_{5+\delta}$ Double Perovskite: Mechanochemical Synthesis. *Nanomaterials* **2021**, *11*, 380. [CrossRef]
12. Lim, T.; Jo, K.; Lee, H. Oxygen Vacancy and Valence Band Structure of $Ba_{0.5}Sr_{0.5}Fe_{1-x}Cu_xO_{3-\delta}$ (x = 0–0.15) with Enhanced ORR Activity for IT-SOFCs. *Materials* **2023**, *16*, 3231. [CrossRef] [PubMed]

13. Ishihara, T.; Kudo, T.; Matsuda, H.; Takita, Y. Doped $PrMnO_3$ Perovskite Oxide as a New Cathode of Solid Oxide Fuel Cells for Low Temperature Operation. *J. Electrochem. Soc.* **1995**, *142*, 1519–1524. [CrossRef]
14. Wang, J.; Lam, K.Y.; Saccoccio, M.; Gao, Y.; Chen, D.; Ciucci, F. Ca and In co-doped $BaFeO_{3-\delta}$ as a cobalt-free cathode material for intermediate-temperature solid oxide fuel cells. *J. Power Sources* **2016**, *324*, 224–232. [CrossRef]
15. Wan, T.H.; Saccoccio, M.; Chen, C.; Ciucci, F. Influence of the Discretization Methods on the Distribution of Relaxation Times Deconvolution: Implementing Radial Basis Functions with DRTtools. *Electrochim. Acta* **2015**, *184*, 483–499. [CrossRef]
16. Caignaert, V.; Maignan, A.; Simon, C.; Raveau, B.; Suard, E. Variation of the Jahn Teller distortion of Mn at the ferromagnetic transition in the CMR perovskite $Pr_{0.7}Sr_{0.1}MnO_3$. *Comptes Rendus De L'academie Des Sci. Ser. 2 Mec. Phys.Chim. Astron.* **1995**, *321*, 515–520.
17. Shannon, R.T.; Prewitt, C.T. Effective ionic radii in oxides and fluorides. *Acta Crystallogr. Sect. B Struct. Crystallogr. Cryst. Chem.* **1969**, *25*, 925–946. [CrossRef]
18. Flores-Lasluisa, J.; Huerta, F.; Cazorla-Amorós, D.; Morallón, E. $LaNi_{1-x}Co_xO_3$ perovskites for application in electrochemical reactions involving molecular oxygen. *Energy* **2023**, *273*, 127256. [CrossRef]
19. Mondal, R.; Mishra, N.K.; Singh, M.; Gupta, A.; Singh, P. Perovskite $La_{1-x}K_xCoO_{3-\delta}$ ($0 \leq x \leq 0.5$): A novel bifunctional OER/ORR electrocatalyst and supercapacitive charge storage electrode in a neutral Na_2SO_4 electrolyte. *Phys. Chem. Chem. Phys.* **2022**, *24*, 28584–28598. [CrossRef]
20. Huang, Z.; Zhou, W.; Ouyang, C.; Wu, J.; Zhang, F.; Huang, J.; Gao, Y.; Chu, J. High performance of Mn-Co-Ni-O spinel nanofilms sputtered from acetate precursors. *Sci. Rep.* **2015**, *5*, srep10899. [CrossRef]
21. Lv, Y.; Li, Z.; Yu, Y.; Yin, J.; Song, K.; Yang, B.; Yuan, L.; Hu, X. Copper/cobalt-doped $LaMnO_3$ perovskite oxide as a bifunctional catalyst for rechargeable Li-O_2 batteries. *J. Alloys Compd.* **2019**, *801*, 19–26. [CrossRef]
22. Mekki, A.; Ziq, K.A.; Holland, D.; McConville, C.F. Magnetic properties of praseodymium ions in Na_2O–Pr_2O_3–SiO_2 glasses. *J. Magn. Magn. Mater.* **2003**, *260*, 60–69. [CrossRef]
23. Wan, Y.; Yang, J.; Hou, H.; Xu, S.; Liu, G.; Hussain, S.; Qiao, G. Synthesis and microstructures of $La_{1-x}Ca_xCrO_3$ perovskite powders for optical properties. *J. Mater. Sci. Mater. Electron.* **2019**, *30*, 3472–3481. [CrossRef]
24. Suntivich, J.; Gasteiger, H.A.; Yabuuchi, N.; Nakanishi, H.; Goodenough, J.B.; Shao-Horn, Y. Design principles for oxygen-reduction activity on perovskite oxide catalysts for fuel cells and metal–air batteries. *Nat. Chem.* **2011**, *3*, 546–550. [CrossRef] [PubMed]
25. Yang, Q.; Tian, D.; Liu, R.; Wu, H.; Chen, Y.; Ding, Y.; Lu, X.; Lin, B. Exploiting rare-earth-abundant layered perovskite cathodes of $LnBa_{0.5}Sr_{0.5}Co_{1.5}Fe_{0.5}O_{5+\delta}$ (Ln=La and Nd) for SOFCs. *Int. J. Hydrogen Energy* **2020**, *46*, 5630–5641. [CrossRef]
26. Chen, Y.; Bu, Y.; Zhang, Y.; Yan, R.; Ding, D.; Zhao, B.; Yoo, S.; Dang, D.; Hu, R.; Yang, C.; et al. A Highly Efficient and Robust Nanofiber Cathode for Solid Oxide Fuel Cells. *Adv. Energy Mater.* **2016**, *7*, 1601890. [CrossRef]

ceramics

MDPI

Article

Effects of Replacing Co^{2+} with Zn^{2+} on the Dielectric Properties of Ba $[Zn_{1/3}(Nb_{1/2}Ta_{1/2})_{2/3}]O_3$ Ceramics with High Dielectric Constant and High Quality Factor

Yuan-Bin Chen *, Yu Fan, Shiuan-Ho Chang * and Shaobing Shen

School of Electronics and Electrical Engineering, Zhaoqing University, Zhaoqing 526061, China; fyu2004@sina.com (Y.F.); 13922623465@139.com (S.S.)

* Correspondence: n2890103@outlook.com (Y.-B.C.); wenrongc2000@yahoo.com.tw (S.-H.C.)

Abstract: In this study, we used solid-state synthesis to prepare $Ba[(Zn_xCo_{1-x})_{1/3}(Nb_{0.5}Ta_{0.5})_{2/3}]O_3$ microwave ceramics for mobile communications. Compared with $Ba[Zn_{1/3}(Nb_{0.5}Ta_{0.5})_{2/3}]O_3$, in the prepared materials, Co^{2+} substitution with Zn^{2+} improved the Q × f value and enabled densification and sintering at a lower temperature. We used X-ray diffraction (XRD) and scanning electron microscopy (SEM) to analyze the obtained microstructure. $Ba[(Zn_xCo_{1-x})_{1/3}(Nb_{0.5}Ta_{0.5})_{2/3}]O_3$ was found to have a 1:2 ordered hexagonal structure, and its Q × f value increased with the increase in sintering temperature. In this work, excellent microwave dielectric properties—$\tau_f = -0.7$ ppm/°C, $\varepsilon_r = 34.5$, and Q × f = 110,000 GHz—were obtained by sintering $Ba[(Zn_{0.3}Co_{0.7})_{1/3}(Nb_{0.5}Ta_{0.5})_{2/3}]O_3$ at 1400 °C for 5 h.

Keywords: ceramic; quality factor; dielectric; temperature coefficient; microwave

Citation: Chen, Y.-B.; Fan, Y.; Chang, S.-H.; Shen, S. Effects of Replacing Co^{2+} with Zn^{2+} on the Dielectric Properties of Ba $[Zn_{1/3}(Nb_{1/2}Ta_{1/2})_{2/3}]O_3$ Ceramics with High Dielectric Constant and High Quality Factor. *Ceramics* **2024**, *7*, 426–435. https://doi.org/10.3390/ceramics7010027

Academic Editors: Dawei Wang and Fayaz Hussain

Received: 26 January 2024
Revised: 14 March 2024
Accepted: 15 March 2024
Published: 17 March 2024

1. Introduction

Microwave dielectric ceramics have been widely used in the electronics industry because of their near-zero temperature coefficient of resonant frequency and high Q × f value. In this study, we examined the formation of substituent-free, non-stoichiometric, vacancy-containing solid solutions of BZN in the $BaO–ZnO–Nb_2O_5$ system. The perovskite structure can accommodate the formation of vacancies at the A position to a considerable extent. However, the extent of vacancy formation at the B site is limited to concentrations below 2% [1]. Relatively large concentrations of B-site vacancies change the stacking of ABO_3 layers to form a hexagonal perovskite structure, with vacancies being arranged in an orderly manner in the resulting coplanar octahedral layer. The stoichiometric changes induced by ZnO volatilization can promote the growth of ordered domains, increase the degree of cation ordering and c/a lattice distortion, and improve the dielectric loss characteristics of $Ba(Zn_{1/3}Ta_{2/3})O_3$ (BZT) [2]. A helpful effect of cation deficiency on loss has been found in other 1:2 composite perovskites, such as $BaMg_{1/3}Ta_{2/3}O_3$ (BMT) and $BaMg_{1/3}Nb_{2/3}O_3$ (BMN). A variety of different mechanisms and even contradictory results on non-stoichiometric effects on cation ordering [3] have been reported in the literature [3]. It has been pointed out in previous studies that BZT increases the degree of ordering and c/a distortion due to the partial substitution of Zn with the cationic part of the A site of Ba. The reduction of Zn and the associated reduction in the Zn–Nb ratio increase under these conditions [4]. According to the literature, the reduction of ZnO due to cation diffusion is enhanced by the formation of zinc ion vacancies; a higher degree of cation ordering has been observed in free BZT powders compared with the bulk [5]. Two BZT phases with slightly different unit cell parameters and c/a ratios were reported, and a solid explanation for the formation of zinc-deficient BZT phase was obtained by using high-resolution X-ray scattering analysis. This observation has been explained in terms of vacancy formation during the partial melting of the phases, which has been claimed to increase the degree of anti-site disorder [6,7]. There are also conflicting reports on the effects of non-stoichiometry on grain growth and densification in these systems. According to research, the volatilization of ZnO induces the formation of a second phase, and densification has a negative impact on BZT. However, according to other works, the lack of Mg^{2+} in BMN increases density, while excess Mg^{2+} hinders grain growth and densification. Regarding A-site elements, there are also conflicting data on the effect of non-stoichiometry on microstructure and densification [8]. However, no evidence of liquid-phase sintering was found in another study on BaO-deficient BMT, where the samples were prepared with $(MgCO_3)_4{\cdot}Mg(OH)_2{\cdot}5H_2O$ instead of MgO [9]. In high-Ba^{2+}-content BNN, abnormal grain growth resulted in density differences, but uniform grain size was also observed.

In previous studies, various conflicting findings in these systems have been reported, highlighting the difficulties that can be encountered during the above-mentioned processes. It can be confirmed that perovskite materials are closely related to the current trend of extending the operating frequency range of MW wireless communications in microwave-operating frequency bands, for example, 26–38 GHz in PtP Radiolinks, 20–30 GHz in VSATs, 28–40 GHz in local multi-point distribution systems (LMDSs), and even up to 60 GHz in some applications.

Low-noise modules (LNBs) are used in various microwave dielectric components, such as local multi-point distribution system (LMDS), module upconverters (BUCs), and very-small-aperture terminals (VSATs). For various microwave system requirements, in addition to the high quality of materials, it is also important to consider the price. Therefore, considering that the dielectric loss is extremely low, new low-cost dielectric materials with appropriate dielectric constants are needed [10–16].

Perovskite ceramics have good microwave dielectric properties and can be applied to dielectric resonators. In particular, the chemical ordering of B-site cations and the tilt of the oxygen octahedron are the main factors affecting the Q and τf values in these ceramics. The dielectric properties of these materials are largely influenced by their perovskite structure. According to research, $Ba(Ni_{1/3}Nb_{2/3})O_3$ and $Ba(Zn_{1/3}Nb_{2/3})O_3$ ceramics have negative temperature coefficients of resonant frequency (τ_f), while $Ba(Co_{1/3}Nb_{2/3})O_3$ [17–20] has a positive temperature coefficient of resonant frequency (τf). Therefore, a $(1-x)Ba(Co_{1/3}(Nb_{0.5}Ta_{0.5})_{2/3})O_3$–$xBa(Zn_{1/3}(Nb_{0.5}Ta_{0.5})_{2/3})O_3$ solid solution was studied, and the obtained materials showed novel τf values close to zero. $Ba(Zn_{1/3}(Nb_{0.5}Ta_{0.5})_{2/3})$, specifically, has a high dielectric constant and a high Q value, while $Ba[(Zn_xCo_{1-x})_{1/3}(Nb_{0.5}Ta_{0.5})_{2/3}]O_3$ solid solutions are expected to have excellent microwave dielectric properties. Researchers have studied $Ba[(Zn_xCo_{1-x})_{1/3}(Nb_{0.5}Ta_{0.5})_{2/3}]O_3$ microwave ceramics for a long time, but $(1-x)Ba(Co_{1/3}(Nb_{0.5}Ta_{0.5})_{2/3})O_3$–$xBa(Zn_{1/3}(Nb_{0.5}Ta_{0.5})_{2/3})O_3$ microwave ceramics have not been studied yet.

In this study, we used Zn^{2+} substitution content in $Ba(Co_{1/3}(Nb_{0.5}Ta_{0.5})_{2/3})O_3$ to form $Ba[(Zn_xCo_{1-x})_{1/3}(Nb_{0.5}Ta_{0.5})_{2/3}]O_3$, which was possible because Zn^{2+} (0.74 Å) has an ionic radius close to that of Co^{2+} (0.745 Å) [14]. This research method greatly improved $Q \times f$, allowing it to reach a very high value, and ensured an extremely reliable temperature drift coefficient for materials with stable dielectric properties. In addition, the correlations between microstructure and microwave dielectric constant/$Q \times f$ value were also studied.

2. Experimental Procedures

2.1. Synthesis of the Samples

$Ba[(Zn_xCo_{1-x})_{1/3}(Nb_{0.5}Ta_{0.5})_{2/3}]O_3$ (x = 0.1–1) samples were prepared by using solid-state mixed oxides; the starting materials were high-purity oxide powders (>99.9%), namely, BaO, ZnO, CoO, Nb_2O_5, and Ta_2O_5, all of which were purchased from Shanghai Aladdin Biochemical Technology Co., Ltd., Shanghai, China. Initially, the raw materials were meticulously weighed and mixed by using agate ball-milling in distilled water for 24 h. Subsequently, the resulting mixtures underwent a process of drying and calcination at 1100 °C for 5 h. Following this, the prepared powders were subjected to another round of drying and ball-milling, this time for 24 h, in the presence of a binder—5 wt% of 10% PVA solution. The resulting mixture was granulated by passing it through a 100-mesh sieve and then molded into pellets with dimensions of 11 mm in diameter and 5 mm in thickness using an automatic uniaxial hydraulic press at a pressure of 200 MPA. Finally, the pellets were sintered in air at temperatures ranging from 1325 °C to 1475 °C for 5 h.

2.2. Characterization of the Samples

Cu K radiation (at 30 KV and 20 mA) and a graphite monochromator were employed to collect both powder and bulk X-ray diffraction spectra in the 2θ range of 10° to 60°. The crystalline phases in the sintered ceramics were determined with XRD analysis using Cu Kα (λ = 0.15406 nm) radiation, which was generated with a Siemens D5000 diffractometer (Munich, Germany) operating at 40 kV and 40 mA. Lattice constant calculations were carried out using GSAS software by applying the Rietveld method to fit the XRD patterns [21]. To examine the microstructure and morphology of the sintered surface, a scanning electron microscope (Philips XL40FEG, Netherlands) was utilized.

Furthermore, the Archimedes method was employed to determine the bulk density of the sintered pellets. Microwave dielectric properties, including the dielectric constant and unloaded Q, were measured in the frequency range of 6–12 GHz using the post-resonant method, as recommended by Hakki and Coleman [22]. This technique involves the use of parallel conducting plates and coaxial probes operating in TE_{011} mode, where "TE" signifies transverse electric waves. In TE_{011}, the first two subscript integers indicate the waveguide mode, and the third integer indicates the order of resonance within a progressively increasing set of discrete resonant lengths. The measurement of the temperature coefficient of resonant frequency was conducted over a temperature range spanning from 20 °C to 80 °C. Additionally, for the measurement, we employed an HP8757D network analyzer and an HP8350B sweep oscillator (HP, Palo Alto, CA, USA) in combination.

3. Results and Discussion

Figure 1 shows the XRD patterns of the $Ba[(Zn_xCo_{1-x})_{1/3}(Nb_{0.5}Ta_{0.5})_{2/3}]O_3$ ceramics (with x values from 0.1 to 0.9) sintered at 1400 °C for 5 h. The 1:2 ordered structure was found in $Ba(Zn_{1/3}Nb_{2/3})O_3$ ceramics sintered at lower temperatures that presented x values from 0.1 to 0.5, but this structure disappeared in BZN ceramics sintered above 1400 °C due to the evaporation of Zn^{2+}. Therefore, the disturbance of the 1:2 ordering occurring in $Ba[(Zn_xCo_{1-x})_{1/3}(Nb_{0.5}Ta_{0.5})_{2/3}]O_3$ ceramics with an x value below 0.7 could be attributed to higher Zn^{2+} content and Zn^{2+} evaporation during the sintering process. Furthermore, the XRD patterns revealed the absence of a secondary phase, and all identified peaks successfully matched the simple perovskite unit cell. The negligible shift in the diffraction peaks with the increase in the value of x can be attributed to the nearly identical ionic diameters of Zn^{2+} and Co^{2+}. This suggests that the substitution of Co^{2+} with Zn^{2+} across the entire composition range does not significantly alter the lattice parameters.

Figure 2 shows the SEM micrographs of the $Ba[(Zn_xCo_{1-x})_{1/3}(Nb_{0.5}Ta_{0.5})_{2/3}]O_3$ ceramics sintered at 1400 °C for 5 h. With the increase in Zn^{2+} content in $Ba[(Zn_xCo_{1-x})_{1/3}(Nb_{0.5}T_{a0.5)2/3}]O_3$ ceramics, the average grain size increased and the shape became more square. At the sintering temperature of 1400 °C, with the increase in Co^{2+} content, the observed grain size was smaller than at the relatively low sintering temperature, resulting in higher porosity. Poor grain growth leads to smaller grain size, resulting in more pores and reducing bulk density.

Figure 3 shows the relative density of the $Ba[(Zn_xCo_{1-x})_{1/3}(Nb_{0.5}Ta_{0.5})_{2/3}]O_3$ ceramics as a function of their sintering temperature. It is important to observe that the relative density exhibited an initial increase with the increase in sintering temperature, reaching a peak at 1400 °C. However, a subsequent decrease in relative density occurred at sintering temperatures beyond 1400 °C, although this was not notably significant. Furthermore, it is worth noting that the relative density consistently exceeded 95.5% of the theoretical density for all specimens. The sintering temperature exhibited a decrease with the increase in Zn^{2+} content (x = 0.5), enabling the sintering of $Ba[(Zn_xCo_{1-x})_{1/3}(Nb_{0.5}Ta_{0.5})_{2/3}]O_3$ at 1400 °C.

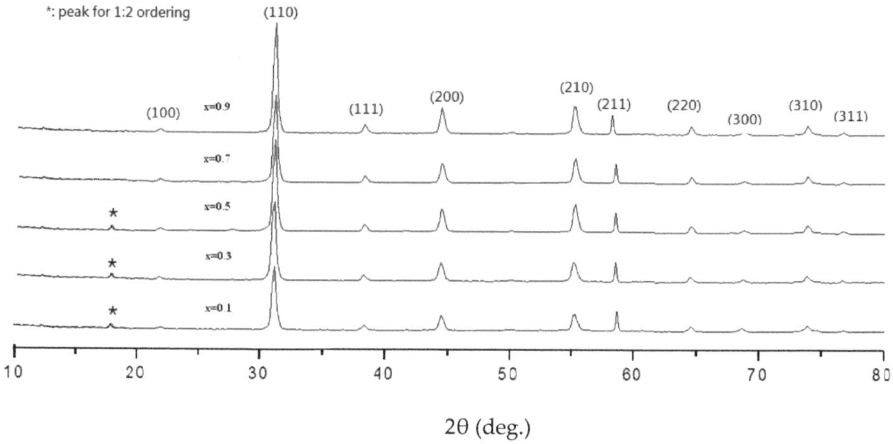

Figure 1. X-ray diffraction patterns of $Ba[(Zn_xCo_{1-x})_{1/3}(Nb_{0.5}Ta_{0.5})_{2/3}]O_3$ ceramics sintered at 1400 °C for 5 h.

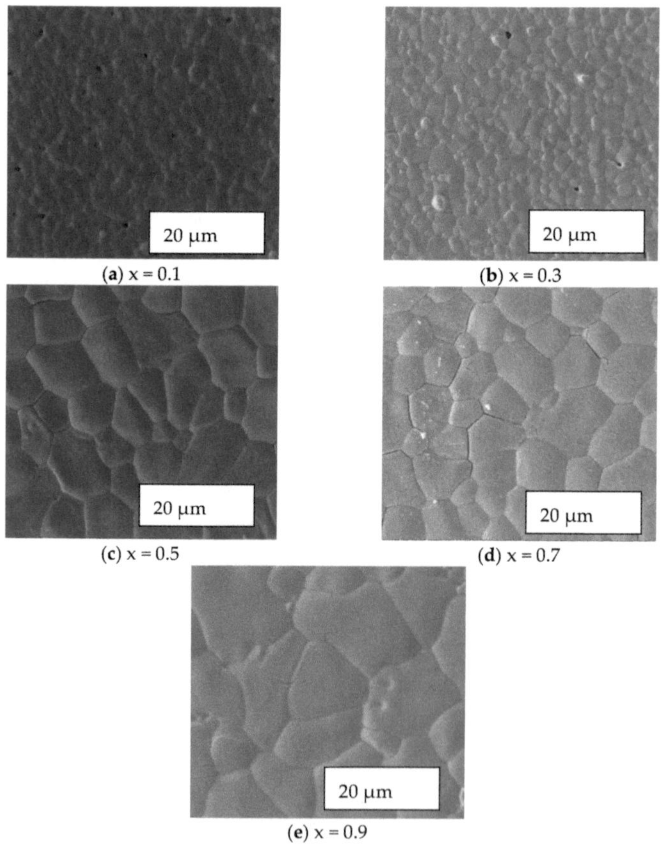

Figure 2. SEM photographs of $Ba[(Zn_xCo_{1-x})_{1/3}(Nb_{0.5}Ta_{0.5})_{2/3}]O_3$ ceramics sintered at 1400 °C for 5 h.

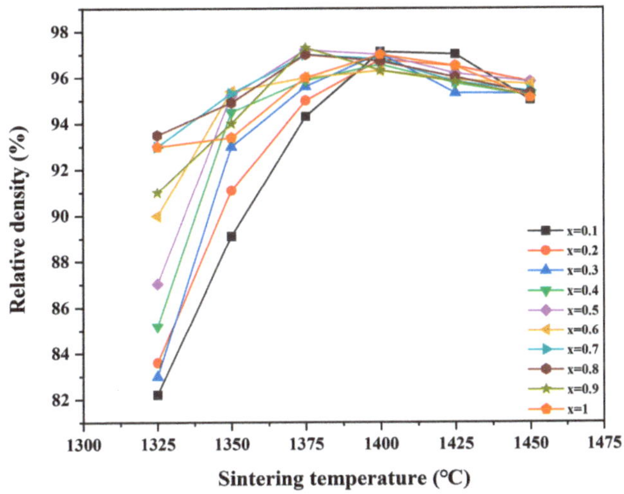

Figure 3. Relative density of $Ba[(Zn_xCo_{1-x})_{1/3}(Nb_{0.5}Ta_{0.5})_{2/3}]O_3$ ceramics sintered at different temperatures.

The dielectric property of the $Ba[(Zn_xCo_{1-x})_{1/3}(Nb_{0.5}Ta_{0.5})_{2/3}]O_3$ ceramics sintered at various temperatures was a function of their x value, as illustrated in Figure 4. The correlations observed between ε_r value and sintering temperature mirrored the trends observed in the relationship between relative density and sintering temperature. The dielectric constant exhibited a slight increase with the increase in sintering temperature. It is important to note that the dielectric constant of a microwave dielectric ceramic is influenced by ionic polarizability [23], that is, the ε_r values of $Ba[(Zn_xCo_{1-x})_{1/3}(Nb_{0.5}Ta_{0.5})_{2/3}]O_3$ increase with the Co^{2+} substitution with Zn^{2+}, as mentioned above. The ionic polarizability (α_{obs}) of the $Ba[(Zn_xCo_{1-x})_{1/3}(Nb_{0.5}Ta_{0.5})_{2/3}]O_3$ samples was estimated in order to clarify the effects of Co^{2+} substitution with Zn^{2+} on the dielectric constant by using the following equation [24]:

$$\varepsilon_r = \frac{3V_m + 8\pi\alpha_m}{3V_m - 4\pi\alpha_m} \tag{1}$$

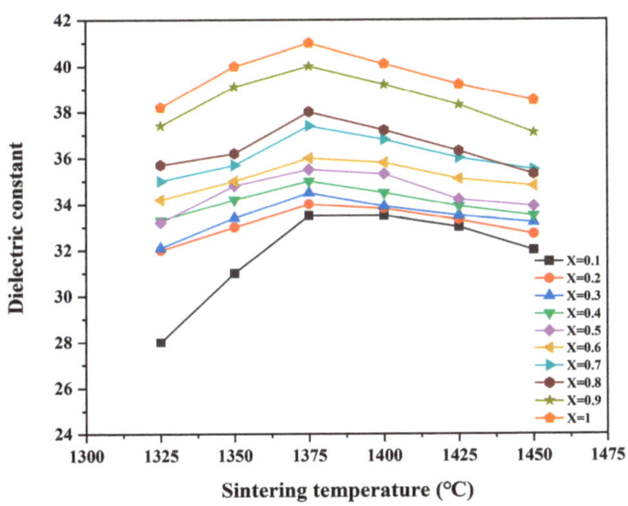

Figure 4. $\varepsilon_r$values of $Ba[(Zn_xCo_{1-x})_{1/3}(Nb_{0.5}Ta_{0.5})_{2/3}]O_3$ ceramics sintered at different temperatures.

In Equation (1), ε_r, V_m, and α_m are the relative permittivity, molar volume, and macroscopic polarizability, respectively. The macroscopic polarizability (α_m) was calculated by using the experimental relative permittivity and unit cell volume data. The data reported in Table 1 reveal an almost sigmoidal increase in theoretical polarizability with Zn^{2+} content, whereas the unit cell volume increased with x. The relative permittivity increased with α_m, and this increase was very rapid as the value of α_m approached $3V_m/4\pi$. It has previously been reported that the macroscopic polarizability of a complex system with ideal symmetry can be determined by the sum of the polarizability values of the constituent cations [25], i.e.,

$$\alpha_m = \Sigma\ \alpha\ (\text{ions}) \tag{2}$$

Table 1. Lattice parameters, cell volume, ionic polarizability, and ε_r data of sintered $Ba[(Zn_xCo_{1-x})_{1/3}(Nb_{0.5}Ta_{0.5})_{2/3}]O_3$ samples.

x=	**0.1**	**0.2**	**0.3**	**0.4**	**0.5**	**0.6**	**0.7**	**0.8**	**0.9**
a (Å)	5.7753 (2)	5.7756 (2)	5.7761 (5)	5.7784 (1)	5.78025 (3)	5.78123 (1)	5.78237 (6)	5.78242(1)	5.78415 (3)
b (Å)	5.7753 (2)	5.7756 (2)	5.7761 (5)	5.7784 (1)	5.78025 (3)	5.78123 (1)	5.78237 (6)	5.78242 (1)	5.78415 (3)
c (Å)	7.0873 (4)	7.0874 (3)	7.0873 (1)	7.0873 (4)	7.0874 (1)	7.0872 (3)	7.0873 (2)	7.0871 (2)	7.0873 (1)
V_m (Å ^{3})	204.7612	204.8312	204.9012	204.9712	205.0412	205.1112	205.1812	205.2512	205.3212
α_m (theoretical)	46.82	46.83	46.86	46.89	46.9	4.692	46.93	46.94	46.96
α_m (exp.)	44.73	44.77	44.78	44.85	44.93	44.98	45.08	45.12	45.3
ε_r (cal.)	33.5	33.8	33.9	34.5	35.3	35.8	36.8	37.2	39.2
ε_r (measured)	33.31	33.61	33.71	34.3	35.09	35.59	36.58	36.97	38.95

The theoretical polarizability (α_m(theoretical)) calculated according to Equation (2) was compared with the experimental polarizability (α_m(exp.)) determined with Equation (1), as shown in Table 1. It is worth noting that the α_m (exp.) of the $Ba[(Zn_xCo_{1-x})_{1/3}(Nb_{0.5}Ta_{0.5})_{2/3}]O_3$ end-member was larger than its α_m (theoretical) value. Shannon [25] noted that deviations from the additivity of ion polarizability can occur when the cation size changes and cation compression or vibration occurs at structural sites. Therefore, the lower αm(exp.) value of $Ba[(Zn_xCo_{1-x})_{1/3}(Nb_{0.5}Ta_{0.5})_{2/3}]O_3$ may have been due to the compression effect caused by the large difference in ion polarizability between Co^{2+} and Zn^{2+}, which is consistent with the harmonic oscillator model. Consequently, in the studied $Ba[(Zn_xCo_{1-x})_{1/3}(Nb_{0.5}Ta_{0.5})_{2/3}]O_3$ ceramics, the polarizability increased with the Zn^{2+} content, and so did the dielectric constant, according to Equation (1).

Figure 5 shows the Q × f values of the $Ba[(Zn_xCo_{1-x})_{1/3}(Nb_{0.5}Ta_{0.5})_{2/3}]O_3$ ceramics sintered at various temperatures; it can be seen that the Q × f value was a function of the x value. For the samples with x < 0.3, the Q × f value was high, above 90,000 GHz, similar to that of $Ba(Co_{1/3}(Nb_{0.5}Ta_{0.5})_{2/3})O_3$. On the contrary, for the specimens with x > 0.5, their Q × f value was below 90,000 GHz, similar to that of $Ba(Zn_{1/3}(Nb_{0.5}Ta_{0.5})_{2/3})O_3$.

Since the microstructures of the specimens with x < 0.5 and x > 0.5 were similar to those of $Ba(Co_{1/3}(Nb_{0.5}Ta_{0.5})_{2/3})O_3$ and BZN ceramics, respectively, it is considered that the Q-values of these $Ba[(Zn_xCo_{1-x})_{1/3}(Nb_{0.5}Ta_{0.5})_{2/3}]O_3$ ceramics are closely related to the microstructures of the specimens. With the increase in sintering temperature, the Q × f value increased to a maximum value but subsequently decreased. A maximum Q × f value of 120,000 GHz was obtained for a $Ba[(Zn_xCo_{1-x})_{1/3}(Nb_{0.5}Ta_{0.5})_{2/3}]O_3$ (x = 0.1) ceramic sample at 1400 °C. Microwave dielectric losses are mainly caused by lattice vibration modes, pores, second phases, impurities, and lattice defects [26]. Moreover, the relative density also increases the dielectric loss, as has been shown in other microwave dielectric materials. It is well known that the Q value of a dielectric material is influenced by intrinsic and extrinsic factors, where one factor is the interaction between polar phonon vibration and the microwave electric field in crystals while the other depends on order–disorder transition, pore density, grain size, oxygen vacancies, and impurity phases in ceramics. The intrinsic Q value sets an upper limit for pure and defect-free single crystals and can be quantitatively described by the well-known classical damped oscillator model in the microwave frequency range. In this model, when using one-phonon absorption approximation, the roughly reciprocal relationship between Q × f and permittivity (ε) can be expressed as

$$Q \times f \propto {\varepsilon_r}^{-1} \tag{3}$$

where the frequency (f) should be restricted to the vicinity of the phonon engine frequency of 10^{12} Hz at room temperature for the estimate to be valid. However, a series of experiments at room temperature demonstrated that the extrapolation of Equation (3) from microwave frequencies down to a few megahertz, that is, one to four orders of magnitude lower than optical phonon engine frequencies, can also be used to obtain a satisfactory dielectric Q order for well-processed ceramics. The experimental results, however, revealed that the dependence of Q × f on ε_r only generated Q × f α ${\varepsilon_r}^{-0.6}$, indicating that the rate of increase in the Q × f value with ε_r is very smooth compared with Equation (3). The most likely cause of this phenomenon may have exogenous origins. On the other hand, the porosity of dielectric materials has an adverse effect on the Q × f value, and different dielectric materials present different degrees of influence. For ceramics with low dielectric Q × f values of the order of 10^3 GHz, the effect of porosity on the dielectric Q value can be illustrated as [27]

$$Q = Q_o(1 - 1.5P) \tag{4}$$

where Q_o is the intrinsic quality factor measured by using microwave reflectance spectroscopy and P is the porosity. For ceramics with high Q × f values of the order of 10^5–10^6 GHz, such as polycrystalline Al_2O_3 ceramics, even a small amount of porosity can significantly reduce the dielectric Q value as [28]

$$\frac{1}{Q} = (1 - P)\frac{1}{Q_o} + A'P(\frac{P}{1 - P})^{2/3} \tag{5}$$

where Q_o is the full-density quality factor (1.565 × 10^{-5}), A' is a constant with value 9.277 × 10^{-3}, and P is the porosity. According to Equations (4) and (5), the $Ba[(Zn_xCo_{1-x})_{1/3}(Nb_{0.5}Ta_{0.5})_{2/3}]O_3$ end-component in the present study presented 8% porosity. The Q × f value was high for the specimens with $x \leq 0.5$, but it decreased when x exceeded 0.5. The Q × f value of the specimens was greatly affected by the microstructure, as shown in Figure 1. The 1:2 ordering structure only appeared in the specimens with x values from 0.1 to 0.5, while this structure does not exist in $Ba(Zn_{1/3}(Nb_{0.5}Ta_{0.5})_{2/3})O_3$ ceramics.

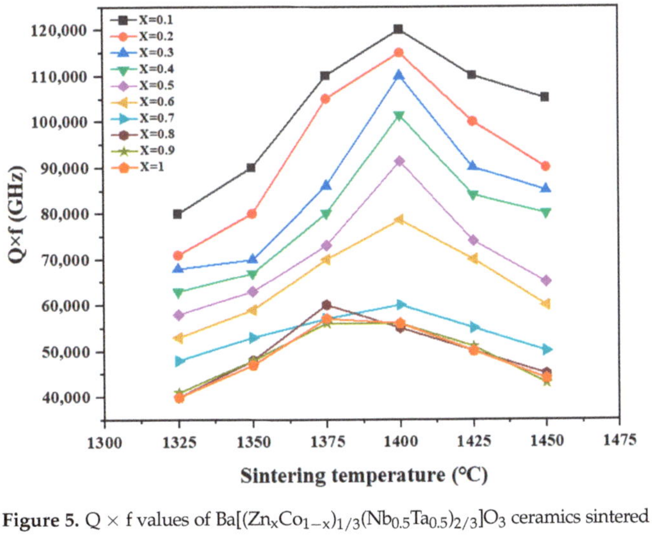

Figure 5. Q × f values of $Ba[(Zn_xCo_{1-x})_{1/3}(Nb_{0.5}Ta_{0.5})_{2/3}]O_3$ ceramics sintered at different temperatures.

Figure 6 shows that the τ_f value of $Ba[(Zn_xCo_{1-x})_{1/3}(Nb_{0.5}Ta_{0.5})_{2/3}]O_3$ ceramics sintered at various temperatures was a function of the Zn^{2+} content. The τ_f linearly increased with the x value, and it was close to zero when x = 0.3. The remarkable variations in the τ_f values of the $Ba[(Zn_xCo_{1-x})_{1/3}(Nb_{0.5}Ta_{0.5})_{2/3}]O_3$ ceramics were obtained by replacing Co^{2+} with Zn^{2+}, and these values ranged from −7.2 to 20 ppm/°C. Therefore, it is believed that for high-frequency dielectric resonator applications, the τ_f value needs to be further increased.

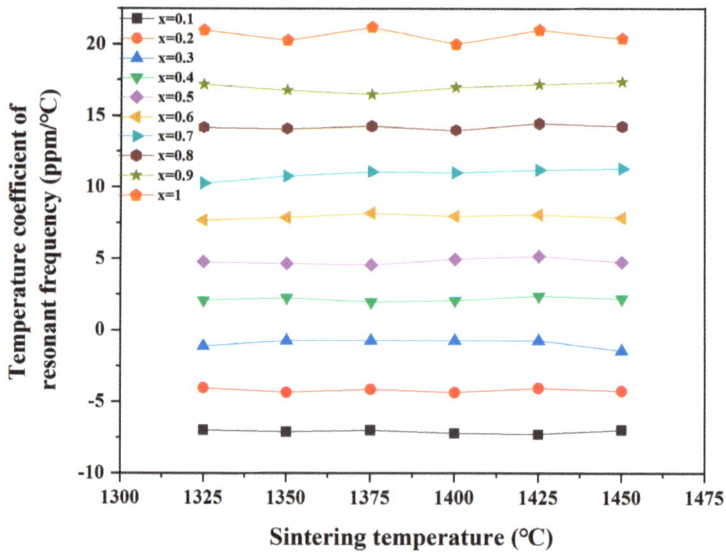

Figure 6. Temperature coefficients of resonant frequency of $Ba[(Zn_xCo_{1-x})_{1/3}(Nb_{0.5}Ta_{0.5})_{2/3}]O_3$ ceramics sintered at different temperatures.

The temperature coefficient (τ_f) of resonant frequency is defined as follows:

$$\tau_f = \alpha_l - \frac{1}{2}\tau_\varepsilon \tag{6}$$

where α_l is the linear thermal expansion coefficient and τ_ε is the temperature coefficient of permittivity. Because the α_l of a microwave dielectric material is known to be in the range of 0 to 10 ppm/°C and assuming that the α_l of a $Ba[(Zn_xCo_{1-x})_{1/3}(Nb_{0.5}Ta_{0.5})_{2/3}]O_3$ ceramic is independent of the *x* value, τ_f mainly depends on τ_ε. Moreover, by plotting the trend of τ_f values versus permittivity for the $Ba[(Zn_xCo_{1-x})_{1/3}(Nb_{0.5}Ta_{0.5})_{2/3}]O_3$ ceramics, it was found that the τ_f values increased from −7.2 to 20 ppm/°C and linearly depended on the x values increasing from 0.1 to 1. It can be seen that the τ_f value has a linear relationship with the x value.

4. Conclusions

The phase stability of $Ba(B_{1/3'}{}^{2+}\,B_{2/3'}{}'^{5+})O_3$ microwave media with complex perovskite structures was studied by using first-principles calculations. The experiment showed that many $Ba(B_{1/3'}{}^{2+}\,B_{2/3'}{}'^{5+})O_3$ ($B_{1/3'}{}^{2+}$ = Co, Mg; B''^{5+} = Nb, Ta) compounds should be completely ordered in equilibrium in the 1:2 B-site structure of the space group $P\bar{3}m1(D^3_{3d})$. Our calculations are not consistent with the experimental results on $Ba[(Zn_xCo_{1-x})_{1/3}(Nb_{0.5}Ta_{0.5})_{2/3}]O_3$ but with many experimental results on other $Ba(B_{1/3'}{}^{2+}\,B_{2/3'}{}'^{5+})O_3$ components. We found that the greater the stability of the ordered 1:2 structure relative to the disordered phase was, the higher the measured microwave quality factor was. The results show that the difference in the ionic radius and electronegativity of B' and

B″ ions exhibited no relationship with phase stability. The results show that aberrations were induced by B-site cations or ordering and were enhanced by more stable and ordered structural phases.

In the $Ba[(Zn_xCo_{1-x})_{1/3}(Nb_{0.5}Ta_{0.5})_{2/3}]O_3$ ceramics, a 1:2 ordered hexagonal structure was observed in samples with $x \leq 0.5$. τ_f increased with the increase in x, and the τ_f of the sample with x = 0.3 was close to zero. When $x \leq 0.5$, the $Q \times f$ value was higher, but it decreased when x exceeded 0.5. The $Q \times f$ value of a sample was greatly influenced by the microstructure. In this work, $Ba[(Zn_{0.3}Co_{0.7})_{1/3}(Nb_{0.5}Ta_{0.5})_{2/3}]O_3$ ceramics were sintered at 1400 °C for 5 h to obtain $\tau_f = -0.7$ ppm/°C and excellent microwave dielectric performance, with $\varepsilon_r = 34.5$ and $Q \times f = 110{,}000$ GHz. The proposed dielectric material has extremely low losses and shows promise for microwave and millimeter-wave applications.

Author Contributions: Y.-B.C.: preparation of materials; methodology; software; writing—original draft preparation; S.-H.C., Y.F. and S.S.: data curation; writing—review and editing; theoretical calculation; and validation. All authors have read and agreed to the published version of the manuscript.

Funding: This research was funded by grant number 612-22010501.

Institutional Review Board Statement: Not applicable.

Informed Consent Statement: Not applicable.

Data Availability Statement: The raw data supporting the conclusions of this article will be made available by the authors on request.

Conflicts of Interest: The authors declare that they have no known competing financial interests or personal relationships that could have appeared to influence the work reported in this study.

References

1. Millet, J.M.; Roth, R.S.; Ettlinger, L.D.; Parker, H.S. Phase Equilibria and Crystal Chemistry in the Ternary Systems $BaO-TiO_2-Nb_2$. *J. Solid State Chem.* **1987**, *67*, 259–270. [CrossRef]
2. Liu, F.; Cheng, L.-J.; Li, H.; Liu, S.-J. Ordering-induced domains in sub-micron-sized $Ba(Zn_{1/3}Ta_{2/3})O_3–BaZrO_3$ microwave ceramics. *J. Mater. Sci. Mater. Electron.* **2021**, *32*, 26126–26136. [CrossRef]
3. Siny, I.G.; Tao, R.; Katiyar, R.S.; Guo, R.; Bhalla, A.S. Raman spectroscopy of Mg-Ta order-disorder in $BaMg_{1/3}Ta_{2/3}O_3$. *J. Phys. Chem. Solids* **1998**, *59*, 181–195. [CrossRef]
4. Ioachim, A. Effect of the sintering temperature on the $Ba(Zn_{1/3}Ta_{2/3})O_3$ dielectric properties. *J. Eur. Ceram. Soc.* **2007**, *27*, 1117–1122. [CrossRef]
5. Moussa, S.M.; Ibberson, R.M.; Bieringer, M.; Fitch, A.N.; Rosseinsky, M.R. In situ Measurement of Cation Order and Domain Growth in an Electroceramic. *Chem. Mater.* **2003**, *15*, 2527–2533. [CrossRef]
6. Bensemma, N.; Trefalt, G.; Glinsek, S.; Kosec, M.; Taibi, K.; Abbaci, M. Investigation of the $BaTiO_3$-$BaMg_{1/3}Nb_{2/3}$ system: Structural, dielectric, ferroelectric and electromechanical studies. *J. Electroceram.* **2013**, *30*, 206–212. [CrossRef]
7. Kim, I.-T. Effects of non-stoichiometry and chemical inhomogeneity on the order-disorder phase formation in the complex perovskite compounds, $Ba(Ni_{1/3}Nb_{2/3})O_3$and $Ba(Zn_{1/3}Nb_{2/3})O_3$. *J. Mater. Sci.* **1995**, *30*, 514–521. [CrossRef]
8. Fan, Y.; Zhou, Z.; Liang, R.; Zhou, M.; Dong, X. The effect of A-site nonstoichiometry on the microstructure, electric properties, and phase stability of $NaNbO_3$ polycrystalline ceramics. *J. Eur. Ceram. Soc.* **2019**, *39*, 4712–4718. [CrossRef]
9. Surendran, K.P.; Sebastian, M.T.; Mohanan, P.; Moreira, R.L.; Dias, A. Effect of Nonstoichiometry on the Structure and Microwave Dielectric Properties of $Ba(Mg_{0.33}Ta_{0.67})O_3$. *Chem. Mater.* **2005**, *17*, 142–151. [CrossRef]
10. Storr, B.; Kodali, D.; Chakrabarty, K.; Baker, P.A.; Rangari, V.; Catledge, S.A. Single-Step Synthesis Process for High-Entropy Transition Metal Boride Powders Using Microwave Plasma. *Ceramics* **2021**, *4*, 257–264. [CrossRef]
11. Yang, H. The latest process and challenges of microwave dielectric ceramics based on pseudo phase diagrams. *J. Adv. Ceram.* **2021**, *10*, 885–932. [CrossRef]
12. Dernovsek, O.; Dernovsek, M.O.; Eberstein, M.; Schiller, W.A. LTCC glass-ceramic composites for microwave application. *J. Eur. Ceram. Soc.* **2001**, *21*, 1693–1697. [CrossRef]
13. Liu, L.; Chen, Y.; Feng, Z.; Wu, H.; Zhang, X. Crystal structure; infrared spectra, and microwave dielectric properties of the $EuNbO_4$ ceramic. *Ceram. Int.* **2021**, *47*, 4321–4326. [CrossRef]
14. Ji, Y.; Song, K.; Zhang, S.; Lu, Z.; Wang, G.; Li, L.; Zhou, D.; Wang, D.; Reaney, I.M. Cold sintered, temperature-stable $CaSnSiO_5$-K_2MoO_4 composite microwave ceramics and its prototype microstrip patch antenna. *J. Eur. Ceram. Soc.* **2021**, *41*, 424–429. [CrossRef]
15. Chen, C.; Peng, Z.; Xie, L.; Bi, K.; Fu, X. Microwave dielectric properties of novel $(1-x)MgTiO_3–xCa_{0.5}Sr_{0.5}TiO_3$ ceramics. *J. Mater. Sci. Mater. Electron.* **2020**, *31*, 13696–13703. [CrossRef]

16. Zaman, A.; Uddin, S. Nasir Mehboob and Asad Ali, Structural investigation and improvement of microwave dielectric properties in $Ca(Hf_xTi_{1-x})O_3$ ceramics. *Phys. Scr.* **2021**, *96*, 025701. [CrossRef]
17. Bao, J.; Du, J.; Liu, L.; Wu, H.; Zhou, Y.; Yue, Z. A new type of microwave dielectric ceramic based on K_2O–SrO–P_2O_5 composition with high quality factor and low sintering temperature. *Ceram. Int.* **2022**, *48*, 784–794. [CrossRef]
18. Nomura, S.; Kaneta, K. $Ba(Mg_{1/3}Ta_{2/3})O_3$ Ceramics with Temperature-Stable High Dielectric Constant and Low Microwave Loss. *Jpn. J. Appl. Phys.* **1982**, *21*, L624. [CrossRef]
19. Kawashima, S.; Nishida, M.; Ueda, I.; Ouchi, H. Dielectric properties of Ba(Zn,Ta)O-Ba(Zn,Nb)O ceramic. *Proc. Ferroelectr. Mater.* **1977**, *1*, 293.
20. Desu, S.B.; O'Bryan, H.M. Microwave Loss Quality of $BaZn_{1/3}Ta_{2/3}O_3$ Ceramics. *J. Am. Ceram. Soc.* **1985**, *68*, 546. [CrossRef]
21. Larson, A.C.; Von Dreele, R.B. Los Alamos National Laboratory Report LAUR-86-748. 1987. Available online: https://11bm.xray.aps.anl.gov/documents/GSASManual.pdf (accessed on 25 January 2024).
22. Hakki, B.W.; Coleman, P.D. A Dielectric Resonator Method of Measuring Inductive Capacities in the Millimeter Range. *IRE Trans. Microw. Theory Tech.* **1960**, *8*, 402. [CrossRef]
23. Belous, A.; Ovchar, O.; Durilin, D.; Krzmanc, M.M.; Valant, M.; Suvorov, D. High-Q Microwave Dielectric Materials Based on the Spinel Mg_2TiO_4. *J. Am. Ceram. Soc.* **2006**, *89*, 3441. [CrossRef]
24. Hannay, J.H. Constantino Grosse. DIBORIDE MICROPATTERNED SURFACES FOR CELL CULTURE. *Ferroelectrics* **1988**, *86*, 171–179. (United States Patent Application 20230133393). Available online: https://www.zhangqiaokeyan.com/patent-detail/06130502846548.html (accessed on 25 January 2024).
25. Shannon, S.R.D. Dielectric polarizabilities of ions in oxides and fluorides. *J. Appl. Phys.* **1993**, *73*, 348–366. [CrossRef]
26. Lichtenecker, K. Dielectric constant of natural and synthetic mixtures. *Phys. Z.* **1926**, *27*, 115.
27. Chen, Y.-B. Dielectric properties and crystal structure of $(Mg_{0.9}Zn_{0.05}Co_{0.05})_4(Nb_{(1-x)}Ta_x)_2O_9$ ceramics. *J. Alloys Compd.* **2012**, *541*, 283–287. [CrossRef]
28. Penn, S.J.; Alford, N.M.; Templeton, A.; Wang, X.; Xu, M. Michael Reece and Kevin Schrapel. *Am. Ceram. Soc.* **1997**, *80*, 1885–1888. [CrossRef]

MDPI AG
Grosspeteranlage 5
4052 Basel
Switzerland
Tel.: +41 61 683 77 34

Ceramics Editorial Office
E-mail: ceramics@mdpi.com
www.mdpi.com/journal/ceramics

www.ingramcontent.com/pod-product-compliance
Lightning Source LLC
Chambersburg PA
CBHW041101270926
51850CB00025B/3175
9783725839223